面向高等职业院校基于工作过程项目式系列教材
企业级卓越人才培养解决方案规划教材

微服务 Spring Boot 实战

天津滨海迅腾科技集团有限公司　编著

天津大学出版社
TIANJIN UNIVERSITY PRESS

图书在版编目(CIP)数据

微服务Spring Boot实战 / 天津滨海迅腾科技集团有限公司编著. — 天津 : 天津大学出版社, 2015.5（2023.3重印）
面向高等职业院校基于工作过程项目式系列教材 企业级卓越人才培养解决方案规划教材
ISBN 978-7-5618-6925-3

Ⅰ. ①微… Ⅱ. ①天… Ⅲ. ①JAVA语言－程序设计－高等学校－教材 Ⅳ. ①TP312.8

中国版本图书馆CIP数据核字(2021)第078944号

WEIFUWU SPRING BOOT SHIZHAN

出版发行	天津大学出版社
地　　址	天津市卫津路92号天津大学内(邮编:300072)
电　　话	发行部:022-27403647
网　　址	www.tjupress.com.cn
印　　刷	廊坊市海涛印刷有限公司
经　　销	全国各地新华书店
开　　本	185mm×260mm
印　　张	20.25
字　　数	513千
版　　次	2021年5月第1版
印　　次	2023年3月第2次
定　　价	69.00元

面向高等职业院校基于工作过程项目式系列教材
企业级卓越人才培养解决方案规划教材

指导专家

基于工作过程项目式教程
《微服务 Spring Boot 实战》

主　编　王新强　张明宇

副主编　李小明　李世强　张仕贵　安国艳
　　　　　郜伟民　程良勇

前　言

随着互联网的快速发展,微服务这一词汇开始进入开发人员的视野中。这是一种理念,并由理念逐渐形成一种架构,主要围绕业务领域创建应用,使这些应用可以独立地进行开发和管理。微服务架构更适用于有一定扩展复杂度,拥有很大用户增量预期的应用。随着技术的不断革新,各类企业对于微服务架构的使用需求将会增多。

对于微服务架构,与之最为搭配的便是 Spring Boot(斯普瑞布特),其大量简化了 Spring 应用搭建以及开发的过程,使开发人员方便快捷地搭建、部署和测试项目。让开发人员更加专注于逻辑代码的开发是 Spring Boot 的优势。Spring Boot 并不是 Spring 的代替品,所以在学习微服务架构知识前,必须掌握 Spring Boot 和 Spring 的基础知识,才能在应用 Spring Cloud 微服务时得心应手。

本书主要以 Spring Boot 技术作为主线,全书使用个人博客项目进行讲解,包含 Spring Boot 启动原理、注解、模型－视图－控制器(Model-View-Controller, MVC)体系、安全等。之后以 Spring Cloud 作为微服务架构进行讲解,包含 Spring Cloud 服务发现、客户端的负载均衡、容错保护机制、应用程序接口(Application Programming Interface, API)网关等知识。本书对知识点的讲解由浅入深,使每一位读者都能有所收获,也保持了整本书的知识深度。

本书主要涉及八个项目:Spring Boot 简介和项目部署、环境配置、Spring Boot 的 Web 基础、数据访问、安全管理、消息队列、Spring Cloud 的基本介绍、Spring Cloud 深入理解。各项目中,严格按照生产环境中的操作流程对知识体系进行编排,使用循序渐进的方式从 Spring Boot 项目搭建学习,过渡到 Spring Cloud 微服务架构,从项目页面的构建、数据的获取、数据展示渲染在页面上,到微服务拆分模块功能展示来对知识点进行讲解。

本书中的每个项目都设有学习目标、学习路径、任务描述、任务技能、任务实施和任务总结。全书结构条理清晰、内容详细,读者可通过任务实施将所学的理论知识充分地应用到实际操作中。

本书由王新强、张明宇共同担任主编,李小明、李世强、张仕贵、安国艳、郜伟民、程良勇担任副主编,王新强、张明宇负责整书编排。项目一和项目二由李小明、李世强编写,项目三和项目四由张仕贵、安国艳编写,项目五由郜伟民编写,项目六由程良勇编写,项目七由李小明编写,项目八由李世强编写。

本书理论内容简明、扼要,实例操作讲解细致、步骤清晰,实现了理实结合;此外,每个操作步骤后均有相对应的效果图,便于读者直观、清晰地看到操作效果,牢记书中的操作步骤。本书可使读者对 Spring Boot 和 Spring Cloud 相关知识的学习更加顺利。

天津滨海迅腾科技集团有限公司

目　录

项目一　个人博客项目搭建

学习 Spring Boot 的创建方式，掌握创建 Spring Boot 项目的流程；学习 Spring Boot 的单元测试，掌握快速测试 Spring Boot 项目的方法；了解 Spring Boot 的原理，深入理解 Spring Boot；编写完善个人博客项目应用。

- 掌握 Spring Boot 的基本概念。
- 掌握 Spring Boot 的创建方式。
- 掌握 Spring Boot 的单元测试与热部署。
- 掌握 Spring Boot 的原理。
- 掌握 Spring Boot 的 Starter 使用。

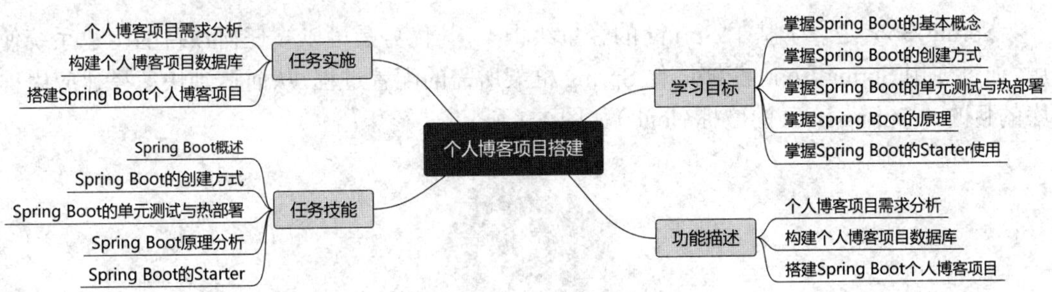

【情景导入】

随着 Spring 的出现，开发人员越来越倾向于使用轻量级框架，但是由于 Spring 框架配置比较复杂，随之诞生了 Spring Boot。Spring Boot 自正式发布起便受到了开发人员的关注。学习 Spring Boot 相关知识，首先要学习 Spring Boot 的执行流程、依赖管理和 Spring Boot 项目的搭建。

【功能描述】

- 个人博客项目需求分析。
- 构建个人博客项目数据库。
- 搭建 Spring Boot 个人博客项目。

技能点 1　Spring Boot 概述

Spring Boot 是一个基于 Spring 的轻量级框架，其优势是可以被任何项目的构建系统使用，通过使用 Spring Boot，能够简化 Spring 框架所需的配置过程，从而达到快速搭建应用程序的目的。 Spring Boot 的徽标（logo）如图 1-1 所示。

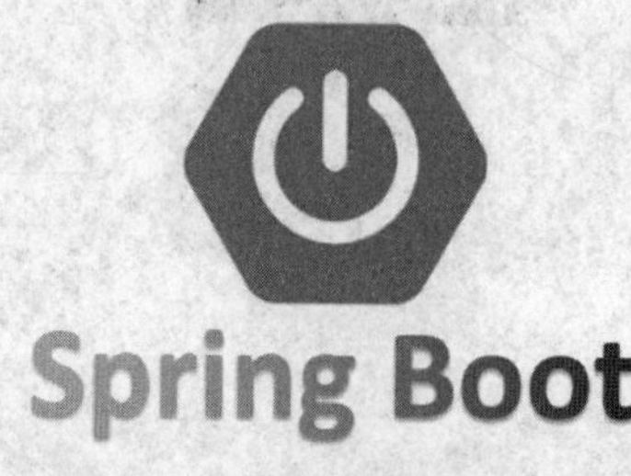

图 1-1　Spring Boot 的 logo

1. Spring Boot 的演变历史

Spring Boot 是由 Pivotal 团队研发的一款全新框架。2013 年初，该团队开始研发 Spring Boot。2014 年 4 月，Spring Boot 正式发布。Spring Boot 刚刚发布，就受到开发人员的关注，逐渐地有些企业和开发人员开始使用 Spring Boot。但直至 2016 年，Spring Boot 在国内才被使用起来，之后使用它的开发人员越来越多。Spring 官网非常重视 Spring Boot 的后续发展，将 Spring Boot 作为其公司的顶级项目进行推广，Spring Boot 的持续发展也被看好。

Spring Boot 的版本更迭速度很快，与 Java 版本密不可分，基于 Java 的发展，Spring Boot 也增加了很多组件功能，它的各版本见表 1-1。

表 1-1　Spring Boot 的版本

Spring Boot 版本	Java 版本
Spring Boot 1.2.0 RELEASE	Java 6
Spring Boot 1.2.2.RELEASE 到 Spring Boot 1.5.17.RELEASE	Java 7
Spring Boot 2.0.0.RELEASE 至今最高版本	Java 8 并且与 Java 15 兼容

Spring Boot 的核心特点包含以下几项。

①独立运行的 Spring Boot 项目。Spring Boot 可以以 jar 包的形式独立运行，运行一个 Spring Boot 项目只需通过 java-jar xx.jar 来运行。

②内嵌 Servlet 容器。Spring Boot 可选择内嵌 Tomcat、Jetty 或者 Undertow，这样开发人员无须以 war 包形式部署项目。

③提供 Starter，简化 Maven 配置。Spring Boot 会根据在类路径中的 jar 包、类，为 jar 包里的类，自动配置 Bean，这样会极大地减少开发人员要使用的配置。

④准生产的应用监控。Spring Boot 提供基于 http、ssh、telnet 的监控方法，实现运行过程中的项目监控。

⑤无代码生成和 XML 配置。Spring Boot 不是借助于代码生成来实现的，而是通过条件注解来实现的，这是 Spring 4.x 提供的新特性。

2. Spring Boot 的优点

Spring Boot 拥有开箱即用的特点，大部分 Spring Boot 应用只需要非常少量的配置代码即可完成基本的项目配置，使开发人员可以更加专注于业务逻辑。Spring Boot 的基本优势如图 1-2 所示。

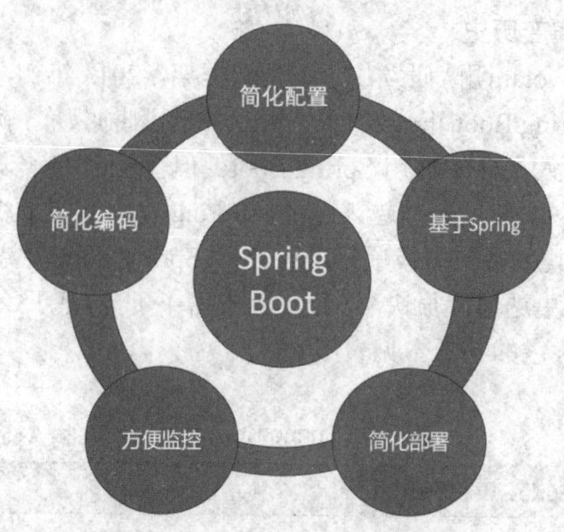

图 1-2 Spring Boot 的基本优势

（1）简化配置

虽然 Spring 使用轻量级的组件代码，但由于其烦琐的配置过程，各种配置文件会浪费开发人员的大量时间，且在出现问题时也很难找出原因。Spring Boot 更多地采用 Java Config 的配置方式，可依靠注解完成基本配置。

（2）简化编码

Spring Boot 对部分包进行了封装，只需要在 pom 文件中添加特定的依赖，即可将所需的 jar 包统一添加。

（3）简化部署

使用 Spring 时，部署项目需要在服务器上部署 Tomcat，而 Spring Boot 内置服务器，可以直接运行，降低了对运行环境的基本要求，环境变量中有 JDK 即可。

（4）整合框架

Spring Boot 可以整合很多的框架，如：Redis、MyBatis、JPA、springmvc、mongodb、swagger、mybatis-plus 等。

3. Spring Boot 与 Spring 的区别分析

Spring Boot 使用轻量级的组件代码，提供了面向切面编程（Aspect Oriented Programming，AOP）和控制反转（Inversion of Control，IOC）的技术，能够轻易地实现拦截和监控等功能。除此之外，Spring Boot 还提供了对很多第三方组件的支持，具有很高的开放性。但 Spring 创建的项目配置却是比较复杂的、重量级的。在搭建环境时，大量的 XML 文件会存在于项目中，导致项目臃肿。此外，随着项目的不断扩大，整合第三方组件所编写的代码和配置也会增加开发人员的负担，导致项目启动缓慢，所消耗的资源也会更多。

Spring Boot 对 Spring 的缺点进行了改善和优化，采用基于约定优于配置的思想，可以让开发人员不必在配置与逻辑业务之间进行思维的切换，从而可以全身心地投入逻辑业务的代码编写中，大大提高了开发效率，也在一定程度上缩短了项目周期。

技能点 2　Spring Boot 的创建方式

在开发中，常用的创建 Spring Boot 项目的方式有两种：一种是使用 Spring Initializr；另一种是创建一个 Maven 工程，导入 Spring Boot 的依赖。

1. 使用 Maven 方式构建 Spring Boot 项目

【案例】应用 Maven 构建 Spring Boot 项目。

第一步，IDEA 配置。首次打开 IDEA，进入主界面可以看到新建一个项目、打开或导入项目、版本控制和初始化设置等选项，如图 1-3 所示。

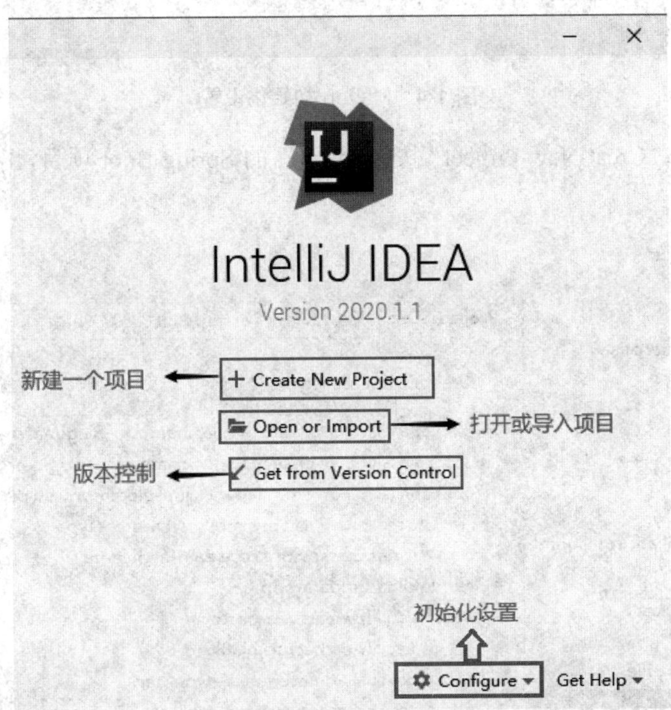

图 1-3　欢迎页

Maven 初始化设置。

点击“Configure”，选择“Settings”选项，在搜索框中搜索 Maven，在右侧对应的设置页面进行 Maven 初始化设置，如图 1-4 所示。

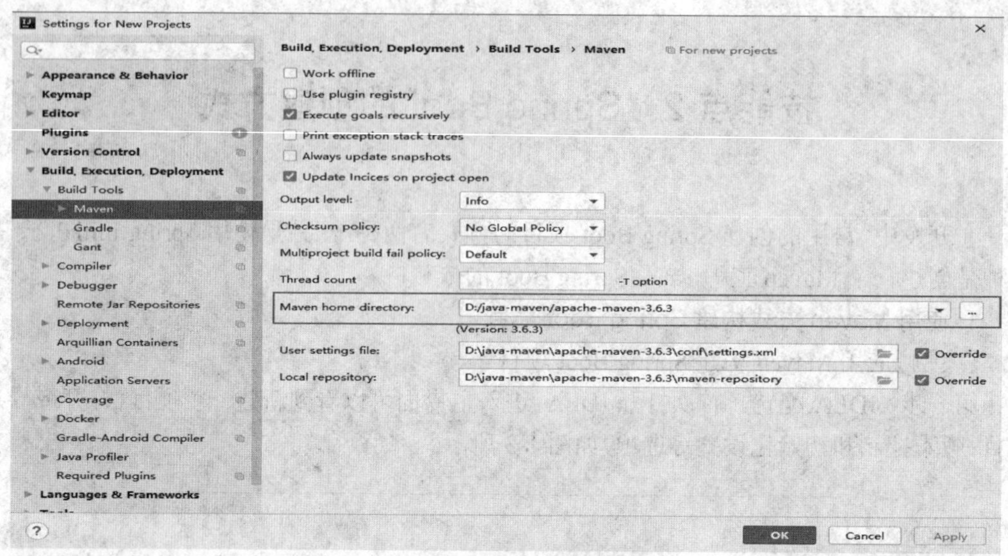

图 1-4 Maven 初始化设置

第二步，点击“Creat New Project”按钮，创建新的 Spring Boot 项目，出现如图 1-5 所示的界面。

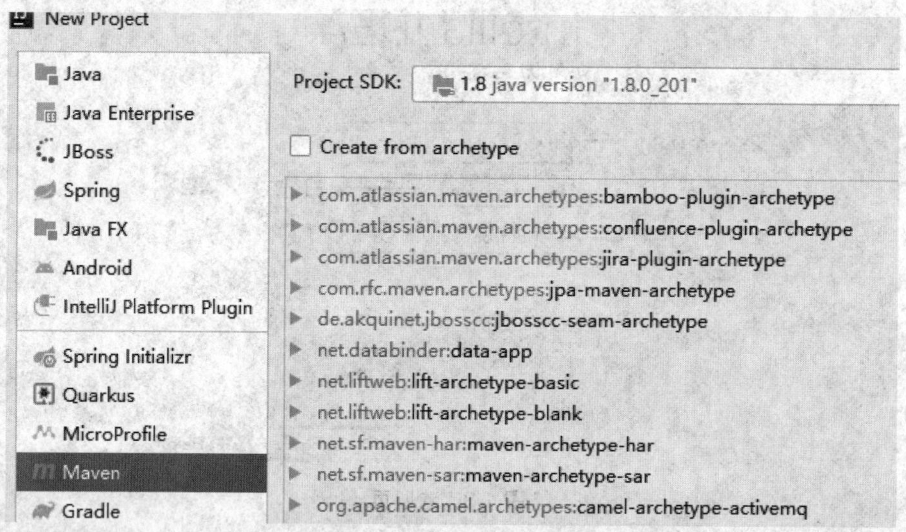

图 1-5 新建 Maven 项目

第三步，在 Project SDK 下拉列表中选择自己电脑所安装的 JDK（Java Development Kit）版本，如果是第一次使用 IDEA 需要配置 JDK 的目录，点击“New”，选择 JDK 的安装路径，如图 1-6 所示。配置好 Project SDK，然后在左面侧边栏中选择“Maven”，这样就可以使用 Maven 作为项目管理工具，如图 1-5 所示。Maven 中的“Create from archetype”是选择模板，因为我们不使用任何一种模板，所以直接点击“Next”进入下一步。

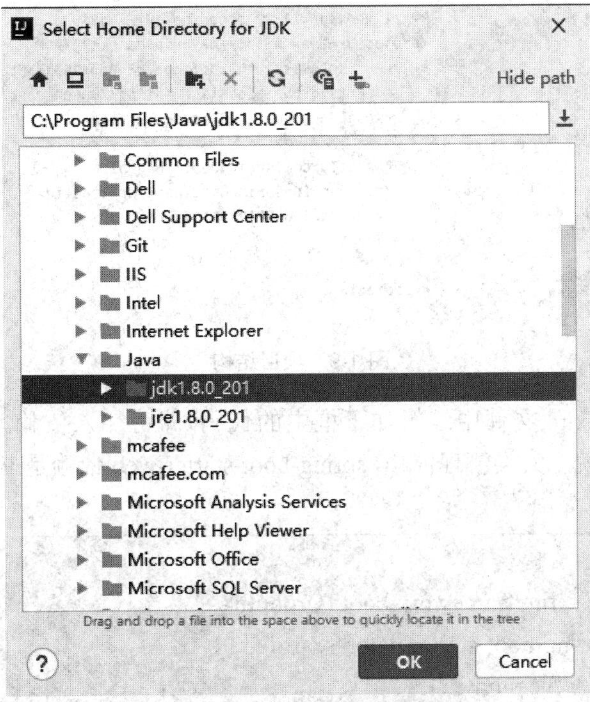

图 1-6　配置 JDK 目录

第四步，输入 GroupId、ArtifactId。GroupId 是项目组织的唯一标识符，是 Main 目录里的 Java 目录结构。ArtifactId 是项目的唯一标识符，是项目根目录的名称，如图 1-7 所示。

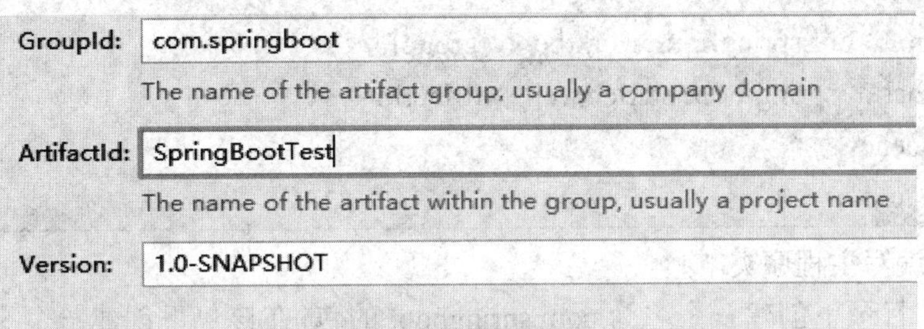

图 1-7　配置 GroupId 和 ArtifactId

第五步，在“Name”中输入与 ArtifactId 相同的项目名称，在“Location”中选择存放路径。如果不输入，系统的默认值为“untitled”，如图 1-8 所示。

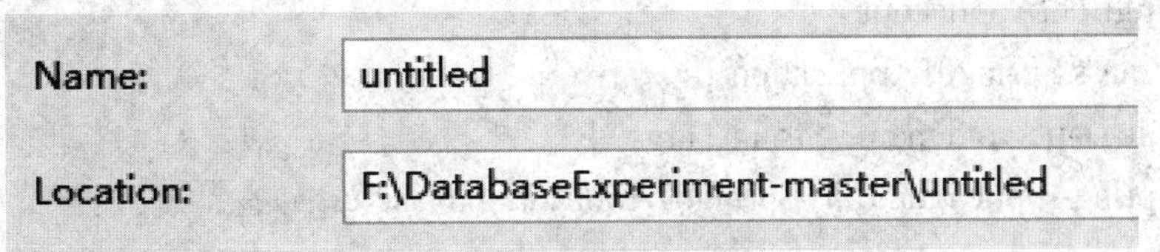

图 1-8　指定项目名称与项目位置

第六步，点击“Finish”，完成项目创建。此时，在项目中生成了“java”目录、“resources”资源目录、“test”测试目录以及一个 pom.xml 文件，如图 1-9 所示。

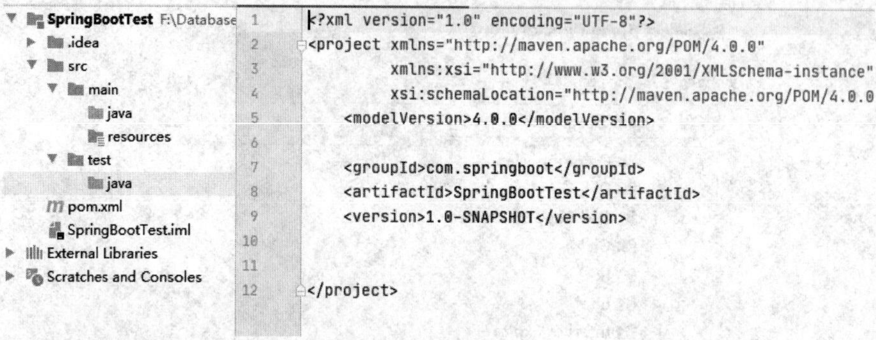

```
▼ SpringBootTest F:\Database
  ► .idea
  ▼ src
    ▼ main
        java
        resources
    ▼ test
        java
    pom.xml
    SpringBootTest.iml
► External Libraries
► Scratches and Consoles
```

```
<?xml version="1.0" encoding="UTF-8"?>
<project xmlns="http://maven.apache.org/POM/4.0.0"
         xmlns:xsi="http://www.w3.org/2001/XMLSchema-instance"
         xsi:schemaLocation="http://maven.apache.org/POM/4.0.0
    <modelVersion>4.0.0</modelVersion>

    <groupId>com.springboot</groupId>
    <artifactId>SpringBootTest</artifactId>
    <version>1.0-SNAPSHOT</version>


</project>
```

图 1-9　项目结构

第七步，在 pom.xml 文件中编辑如下框中的代码，即可注入依赖。其中，spring-boot-starter-parent 是项目统一父类管理依赖，spring-boot-starter-web 依赖是 Web 开发场景的依赖启动器。

```
<parent>
   <groupId>org.springframework.boot</groupId>
   <artifactId>spring-boot-starter-parent</artifactId>
   <version>2.4.0</version>
   <relativePath/> <！ -- lookup parent from repository -->
</parent>
<dependencies>
<dependency>
   <groupId>org.springframework.boot</groupId>
   <artifactId>spring-boot-starter-web</artifactId>
</dependency>
</dependencies>
```

第八步，编写主程序类。

在“java”目录下创建一个名为 com.springboot 的包，在该包下创建一个 Spring Boot 启动类，添加 @SpringBootApplication 注解。该注解的作用是表明该类为启动类，如示例代码 1-1 所示。

示例代码 1-1：Spring01Application 启动类

```
@SpringBootApplication
public class Spring01Application {
   public static void main（String[] args）{
      SpringApplication.run（Spring01Application.class，args）；
   }
}
```

第九步，编写 Controller 类用于 Web 访问。

在 com.springboot 的包下，创建一个 controller 包，在该包下创建一个 TestController 类，

在该类上方添加 @RestController 注解。该注解的作用相当于 @Controller+@ResponseBody 两个注解的结合，返回 JSON 数据就不需要在方法前面加 @ResponseBody 注解了。但在使用 @RestController 这个注解时，所编写的内容不经过视图解析器进行处理，直接返回字符串数据，如示例代码 1-2 所示。

示例代码 1-2：TestController 类

```
@RestController
public class TestController {
    @GetMapping("/test")
    public String aaa(){
        return "Hello!!!! ";
    }
}
```

第十步，运行启动类。

运行该项目，启动成功后打开浏览器，输入网址“http://localhost:8080/test”并访问，访问成功后如图 1-10 所示。

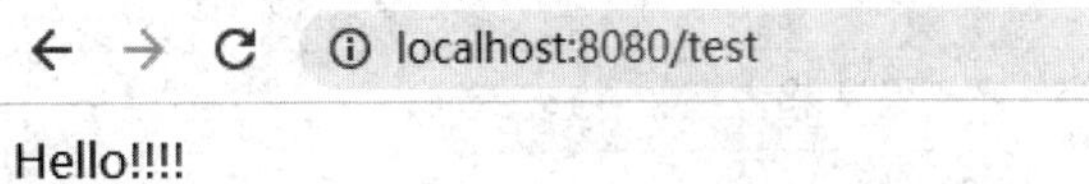

图 1-10　项目启动成功

2. 使用 Initializr 方式构建 Spring Boot 项目

【案例】使用 Initializr 创建 Spring Boot 项目。

第一步，在左面侧边栏选择“Spring Initializr”项目，“Choose starter service URL”选择“Default”为 https://start.spring.io，将会连接网络，确保网络流畅，保证查询 Spring Boot 的可用版本和组件，点击“Next”进入下一步，如图 1-11 所示。

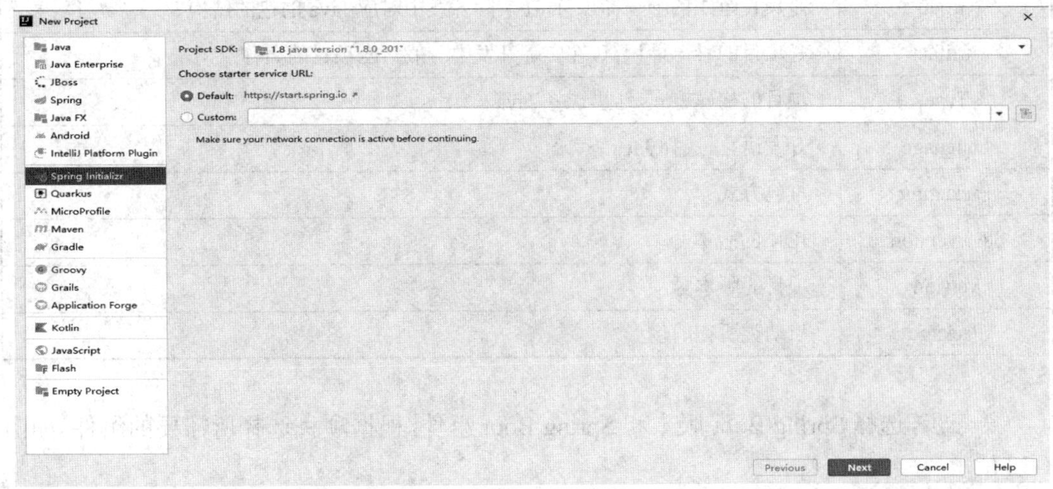

图 1-11　新建 Spring Boot 项目

第二步，选择项目类型。

将"Group"设置为 com.spring，"Artifact"设置为 demo，点击"Next"进入下一步，如图 1-12 所示。

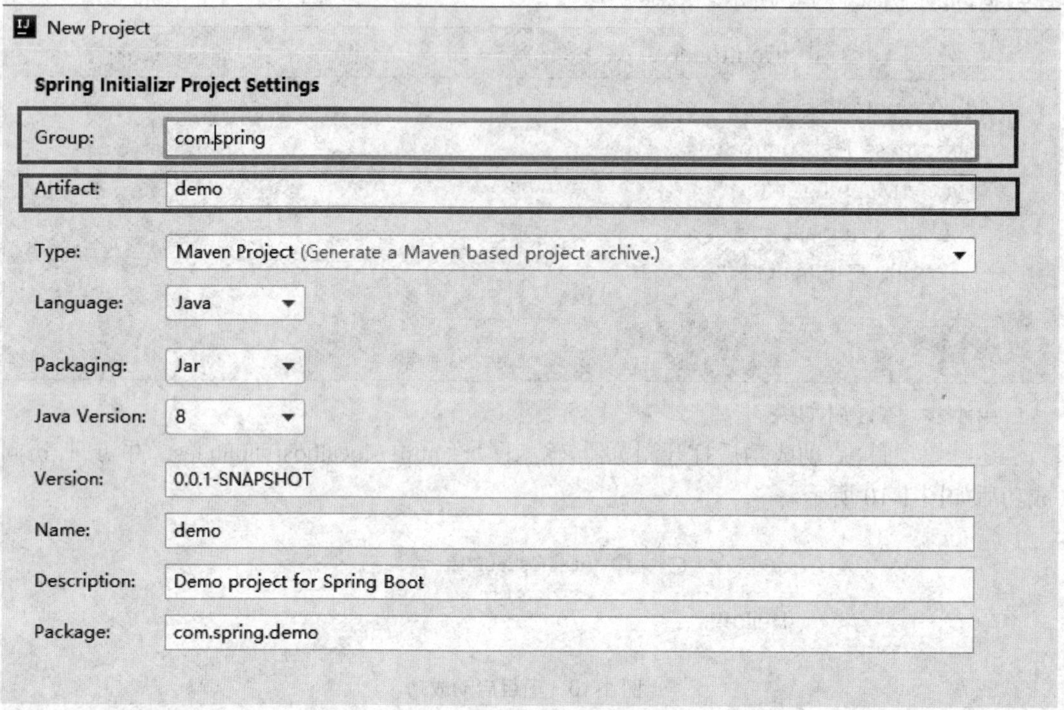

图 1-12 选择项目类型

项目类型和具体属性见表 1-2。

表 1-2 Spring Boot 项目类型和相关解释

项目类型	解释
Group	是项目组织的唯一标识符，在实际开发中对应 Java 的包的结构
Artifact	是项目的唯一标识符，在实际开发中一般对应项目的名称
Type	项目的构建方式，这里选择 Java
Language	开发语言，选择 Java
Packaging	打包方式
Java version	JDK 的版本号
Version	项目的版本号
Package	启动类所在的包

第三步，选择 Spring Boot 版本和 Spring Boot 组件，根据项目选择所需要的组件，如图 1-13 所示。

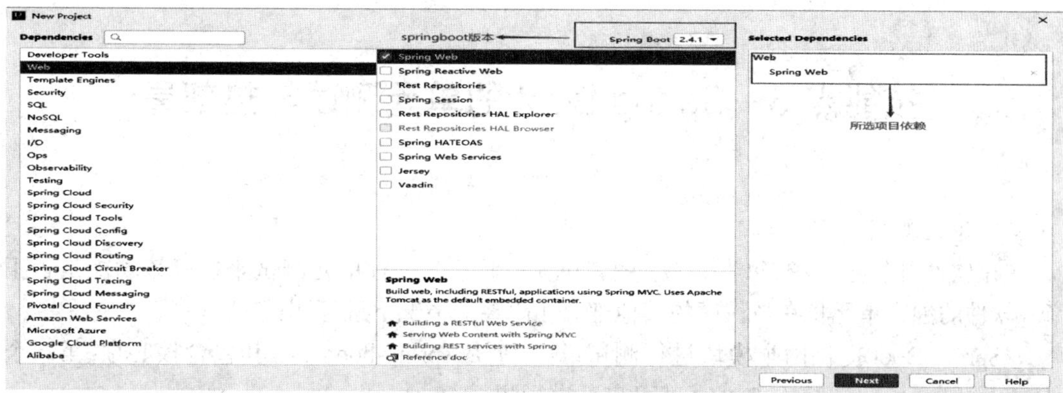

图 1-13　选择版本和组件

第四步，输入项目名和选择项目存放位置，如图 1-14 所示。点击“Finish”，创建一个 Spring Boot 项目。Spring Boot 项目无须任何代码即可运行。一个 Spring Boot 项目的结构如图 1-15 所示。

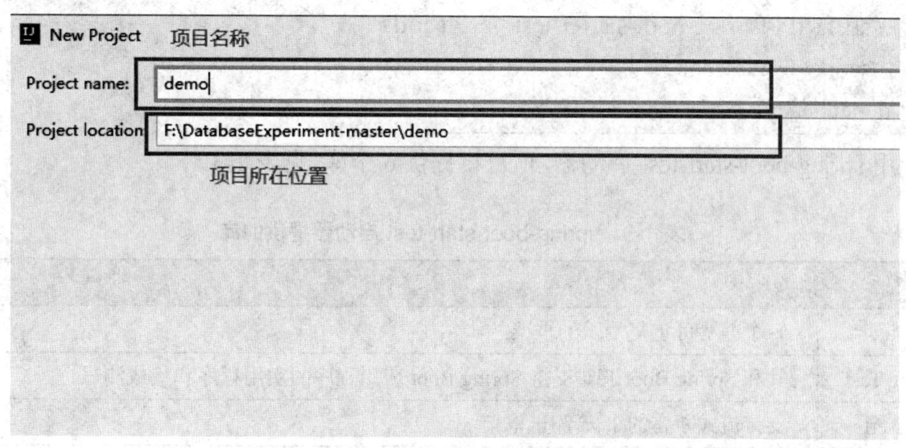

图 1-14　输入项目名称

```
demo F:\DatabaseExperiment-master\d
  .idea
  .mvn
  src
    main
      java
        com.spring.demo
          controller
            aaaaController
          DemoApplication
      resources
        static
        templates
        application.properties
    test
      java
        com.spring.demo
          DemoApplicationTests
  target
  .gitignore
  demo.iml
  HELP.md
  mvnw
  mvnw.cmd
  pom.xml
External Libraries
Scratches and Consoles
```

```xml
<?xml version="1.0" encoding="UTF-8"?>
<project xmlns="http://maven.apache.org/POM/4.0.0" xmlns:xsi="http://www.w3.org/200
        xsi:schemaLocation="http://maven.apache.org/POM/4.0.0 https://maven.apache
    <modelVersion>4.0.0</modelVersion>
    <parent>
        <groupId>org.springframework.boot</groupId>
        <artifactId>spring-boot-starter-parent</artifactId>
        <version>2.4.0</version>
        <relativePath/> <!-- lookup parent from repository -->
    </parent>
    <groupId>com.spring</groupId>
    <artifactId>demo</artifactId>
    <version>0.0.1-SNAPSHOT</version>
    <name>demo</name>
    <description>Demo project for Spring Boot</description>

    <properties>
        <project.build.sourceEncoding>UTF-8</project.build.sourceEncoding>
        <project.reporting.outputEncoding>UTF-8</project.reporting.outputEncoding>
        <druid.version>1.1.16</druid.version>
        <java.version>1.8</java.version>
    </properties>
    <dependencies>
        <dependency>
            <groupId>org.springframework.boot</groupId>
            <artifactId>spring-boot-devtools</artifactId>
            <optional>true</optional>
        </dependency>
```

图 1-15　Spring Boot 项目结构

技能点 3 Spring Boot 的单元测试与热部署

1. 单元测试

在软件开发过程中，每当完成一个功能后，通常会进行单元测试来检验该功能是否正确，软件的独立单元将在与程序的其他部分相隔离的情况下进行测试。

Spring Boot 提供两个模块进行测试：第一个是 spring-boot-test，其包含核心项；第二个是 spring-boot-test-autoconfigure，其可自动进行测试。

应用 spring-boot-start-test 启动器，项目中会导入两个测试模块以及 JUnit 和 Asssert 等库，只需要在 pom.xml 文件中写入如下所示的配置代码即可。

```
<dependency>
  <groupId>org.springframework.boot</groupId>
  <artifactId>spring-boot-starter-test</artifactId>
  <scope>test</scope>
</dependency>
```

应用 spring-boot-start-test 启动器，项目将提供 6 个库，见表 1-3。

表 1-3 spring-boot-start-test 启动器提供的库

特征	解释
JUint	一个编写可重复测试的框架
Spring Test	其和 Spring Boot Test 支持 Spring Boot 应用程序的实用程序和集成测试
Asssert	支持流式断言的 Java 测试框架
Hamcrest	一个匹配器库
Mockito	一个 Java mock 框架
JSONassert	JSON 的断言库

2. 单元测试常用注解

① @RunWith（SpringRunner.class）。将 JUnit 和 Spring 连接起来，是 JUnit4 提供的注解。

② @SpringBootTest。加载 ApplicationContext，启动 Spring，其替代了标准的 Spring-test 的 @ContextConfiguration。该注解提供了 webEnvironment 属性，该属性的值一共有 4 个，见表 1-4。

表 1-4 SpringBootTest 注解属性值

属性	解释
MOCK	提供一个模拟的 servlet 环境，开启 Mock 相关的功能。使用该注解，内嵌的服务（servlet 容器）没有真正启动

续表

属性	解释
RANDOM_PORT	提供一个真实的 servlet 环境，监听一个随机端口
DEFINED_PORT	启动一个真实的 servlet 环境，监听一个定义好的端口
NONE	加载一个非 web 的 ApplicationContext，但是并不提供任何 servlet 环境

③配置类型的注解，见表 1-5。

表 1-5　配置类型的注解

属性	解释
@TestCompontent	正式代码和测试代码在一起时使用，不运行被该注解描述的 Bean。用来指定某个 Bean
@TestConfiguration	用于覆盖已存在的 Bean 或添加额外的 Bean。不修改正式代码，使配置更加灵活
@TypeExcludeFilters	主要用于测试代码和正式代码在一起时的场景
@PropertyMapping	定义 @AutoConfigure 注解中用到的变量名称

④ Mock 类型的注解，见表 1-6。

表 1-6　Mock 类型的注解

属性	解释
@MockBean	用于 Mock 指定的 class 或被注解的属性
@MockBeans	使 @MockBean 支持在同一类型或属性上多次出现
@SpyBean	用于 Spy 指定的 class 或被注解的属性
@SpyBeans	使 @SpyBean 支持在同一类型或属性上多次出现

3. 单元测试实例

【案例】应用混合注解实现单元测试。

第一步，添加测试依赖。

在 pom.xml 文件中添加 spring-boot-start-test 测试依赖，代码如示例代码 1-3 所示。

示例代码 1-3：pom.xml 文件

```
<dependency>
  <groupId>org.springframework.boot</groupId>
  <artifactId>spring-boot-starter-test</artifactId>
  <scope>test</scope>
</dependency>
```

第二步，编写单元测试类。

使用 Initializr 创建 Spring Boot 项目，会自动生成单元测试类（class）。@SpringBootTest 注解用于加载 ApplicationContext，并标记测试类，@RunWith 注解用于加载注解 @SpringBootTest，@SpyBean 注解用于 Spy 指定的 class 或被注解的属性，TestController 类是之前案例所用的。单元测试类的代码如示例代码 1-4 所示。

示例代码 1-4：Demo2ApplicationTests 测试类

```
@RunWith（SpringRunner.class）
@SpringBootTest
class Demo2ApplicationTests {
  @SpyBean
  private TestController testController;
  @Test
  void contextLoads（）{
    System.out.println（TestController.aaa（））;
  }
}
```

第三步，选中所需的测试方法。鼠标右键点击“Run contextLoads（）”启动测试方法，控制台会显示如图 1-16 所示的信息。

```
Hello!!!

2020-12-15 14:48:02.295  INFO 16872 --- [extShutdownHook]
2020-12-15 14:48:02.296  INFO 16872 --- [extShutdownHook]
```

图 1-16 测试运行图

4. 热部署

热部署解决了在开发过程中的两个问题：第一，修改了代码之后，无须重新启动就可以看到效果，可以提高开发效率；第二，对于正在运行的程序，无须停止服务器，就可以进行功能升级，不会影响客户的使用效果。

【案例】配置 IDEA 热部署功能。

第一步，要实现热部署功能，需在 pom.xml 文件中添加依赖，如示例代码 1-5 所示。

示例代码 1-5：pom.xml 文件

```
<! --devtools 热部署 -->
<dependency>
<groupId>org.springframework.boot</groupId>
<artifactId>spring-boot-devtools</artifactId>
<optional>true</optional>
</dependency>
```

第二步，在配置文件 application.yml 中配置 devtools，如示例代码 1-6 所示。

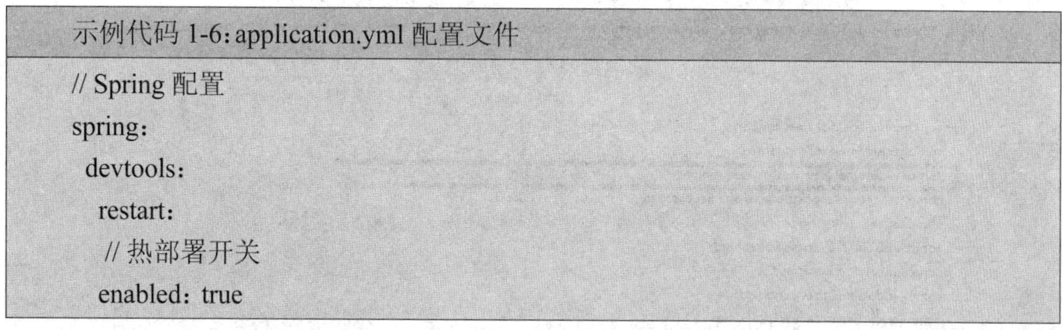

示例代码 1-6：application.yml 配置文件

```
// Spring 配置
spring:
 devtools:
  restart:
   // 热部署开关
  enabled: true
```

第三步，修改 IDEA 的设置。选择 Spring IDEA 软件下的“Settings”选项中的“Builder, Execution, Deployment ”，选择“Compiler”，勾选上“Build project automatically”选项，如图 1-17 所示。

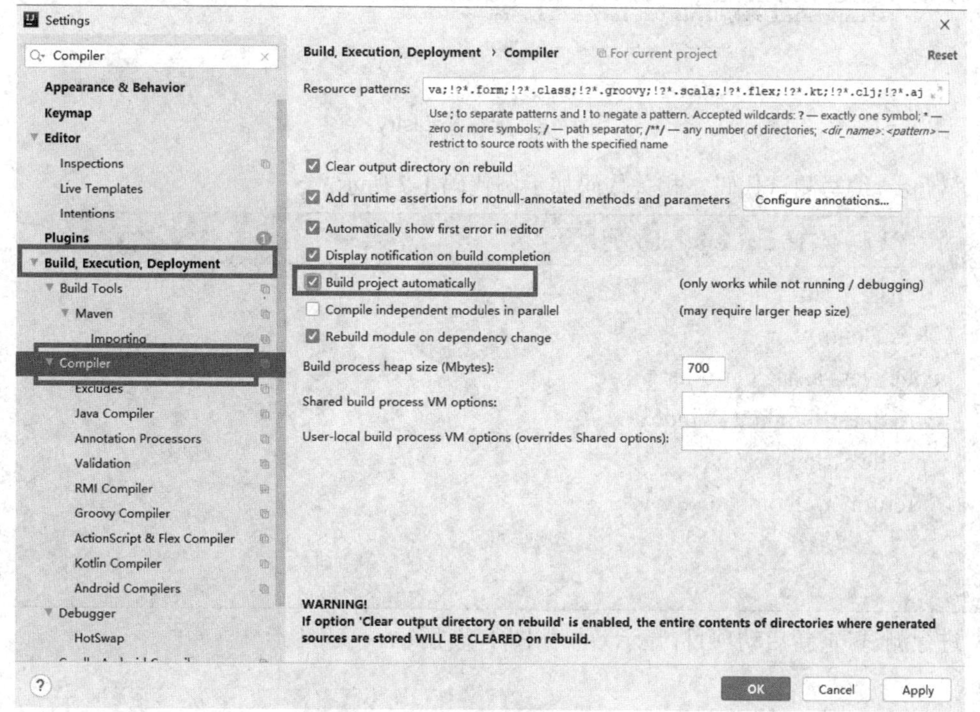

图 1-17　配置 Compiler

第四步，同时按下电脑键盘上的“Ctrl”“Shift”“Alt”“/”键，然后选择“Registry”，勾选上“compiler. automake. allow. when. app. running”选项，如图 1-18 所示。

Registry

Changing these values may cause unwanted behavior of IntelliJ IDEA. Please do not change these unless you have been asked.

Key	Value
comment.by.line.bulk.lines.trigger	100
compiler.automake.allow.parallel	☑
compiler.automake.allow.when.app.running	☑
compiler.automake.postpone.when.idle.less.than	3000
compiler.automake.trigger.delay	300
compiler.build.data.unused.threshold	30
compiler.build.report.statistics	☐
compiler.document.save.enabled	☐
compiler.document.save.trigger.delay	1500
compiler.external.javac.keep.alive.timeout	300000
compiler.process.32bit.vm.on.mac	☑
compiler.process.debug.port	-1
compiler.process.jdk	

Description:

Allow auto-make to start even if developed application is currently running. Note that automatically started make may eventually delete some classes that are required by the application.

Restore Defaults Close

图 1-18 配置 Registry

第五步,编写测试代码。相关代码如示例代码 1-7 所示。

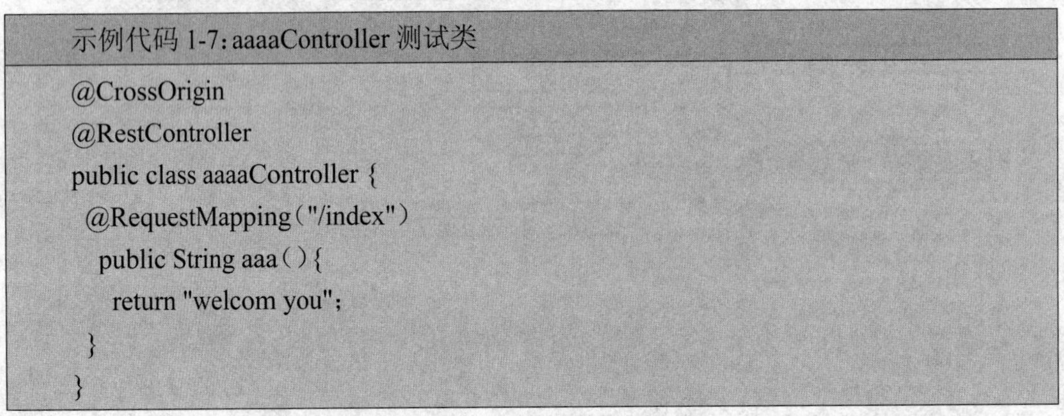

示例代码 1-7:aaaaController 测试类

```
@CrossOrigin
@RestController
public class aaaaController {
  @RequestMapping("/index")
  public String aaa(){
    return "welcom you";
  }
}
```

进行测试,重新启动项目,显示效果如图 1-19 所示。

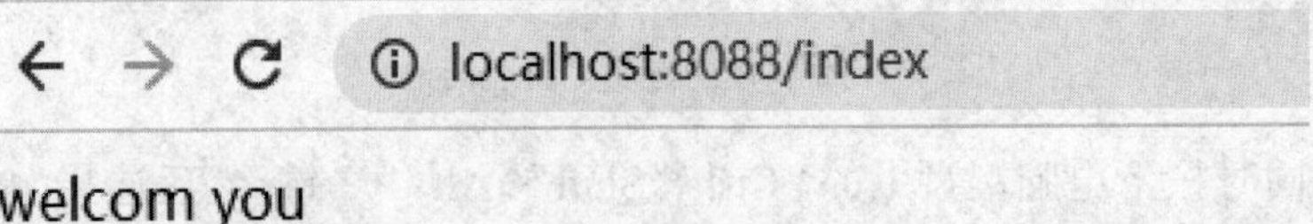

图 1-19 重新启动项目

第六步,修改信息。所用代码如示例代码 1-8 所示。

示例代码 1-8：aaaaController 类

```
@CrossOrigin
@RestController
public class aaaaController {
 @RequestMapping("/index")
  public String aaa(){
   return"Hello!!! ";
 }
}
```

无须重启项目，保存之后，重新访问，显示效果如图 1-20 所示。

← → C ⓘ localhost:8088/index

Hello!!!

图 1-20　重新访问项目

技能点 4　Spring Boot 原理分析

1. 依赖管理

Spring Boot 提供了其所支持的依赖项列表。开发人员不必在配置文件中指定依赖项的版本。Spring Boot 是自我管理的，当更新 Spring Boot 版本时，Spring Boot 会自动升级所有依赖项。

（1）Maven 依赖管理系统

Maven 项目从 spring-boot-starter-parent 继承了以下功能。

1）默认的 Java 编译器版本。

2）UTF-8 源编码。

3）从 spring-boot-dependency-pom 继承了 Dependency Section，用于管理常见依赖项的版本。

4）依赖关系，继承自 spring-boot-dependencies POM。

（2）继承父启动

在配置项目时，spring-boot-starter-parent 会自动继承，代码如下。

```
<parent>
    <groupId>org.springframework.boot</groupId>
    <artifactId>spring-boot-starter-parent</artifactId>
    <version>2.4.0</version>
    <relativePath/> <! -- lookup parent from repository -->
</parent>
```

要添加另一个启动器，只需移除“version”标记，同样也可以通过覆盖项目中的属性来添加自定义的个人依赖。

例如，如果 dependencies 中的一些引用不想使用默认的版本，可以直接加上 version 信息，把默认的覆盖掉。此外，代替默认配置的官方方法代码如下。

```
<properties>
    <project.build.sourceEncoding>UTF-8</project.build.sourceEncoding>
    <project.reporting.outputEncoding>UTF-8</project.reporting.outputEncoding>
    <druid.version>1.1.16</druid.version>
    <java.version>1.8</java.version>
</properties>
```

添加 Spring Boot Maven 插件，可以在 pom.xml 中添加 Maven，它将项目包装到可执行的 jar 文件中，代码如下。

```
<build>
    <plugins>
        <plugin>
            <groupId>org.springframework.boot</groupId>
            <artifactId>spring-boot-maven-plugin</artifactId>
        </plugin>
    </plugins>
</build>
```

可以使用“java.version”命令更改 Java 版本，代码如下。

```
<properties>
    <project.build.sourceEncoding>UTF-8</project.build.sourceEncoding>
    <project.reporting.outputEncoding>UTF-8</project.reporting.outputEncoding>
    <druid.version>1.1.16</druid.version>
    <java.version>1.8</java.version>
</properties>
```

没有父 POM 的 Spring Boot，可以使用“scope”，代码如下。

```
<dependencyManagement>
  <dependencies>
    <dependency>
      <groupId>org.springframework.boot</groupId>
      <artifactId>spring-boot-dependencies</artifactId>
      <version>2.2.2.RELEASE</version>
      <type>pom</type>
      <scope>import</scope>
    </dependency>
  </dependencies>
</dependencyManagement>
```

上面的依赖项不允许被覆盖。如果要实现覆盖，需要在 spring-boot-dependencies 条目之前的 <dependencyManagement> 标记内添加一个条目。

例如，要升级另一个 spring-data-releasetrain，请在 pom.xml 文件中添加以下依赖项，代码如下。

```
<dependencyManagement>
  <dependencies>
    <! --Override Spring Data release train from xntuto r.com -->
    <dependency>
      <groupId>org.springframework.data</groupId>
      <artifactId>spring-data-releasetrain</artifactId>
      <version>Fowler-SR2</version>
      <type>pom</type>
      <scope>import</scope>
    </dependency>
    <dependency>
      <groupId>org.springframework.boot</groupId>
      <artifactId>spring-boot-dependencies</artifactId>
      <version>2.2.2.RELEASE</version>
      <type>pom</type>
      <scope>import</scope>
    </dependency>
  </dependencies>
</dependencyManagement>
```

2. 执行流程

Spring Boot 的执行流程分为三个部分。

第一部分，SpringApplication 初始化模块，配置监听器、资源、构建器和环境变量，如图

1-21 所示。

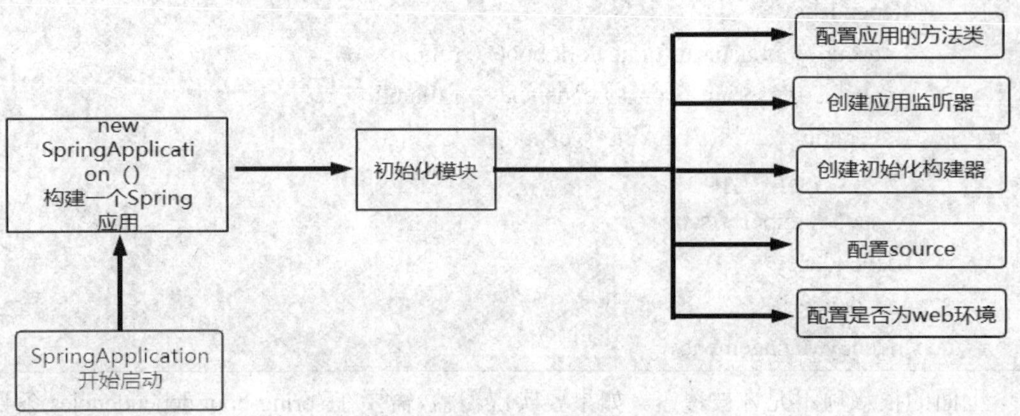

图 1-21　SpringApplication 初始化模块流程图

SpringApplication 初始化模块代码如下。

```
public SpringApplication(ResourceLoader resourceLoader, Class<? >... primarySources) {
    this.sources = new LinkedHashSet();
    this.bannerMode = Mode.CONSOLE;
    this.logStartupInfo = true;
    this.addCommandLineProperties = true;
    this.addConversionService = true;
    this.headless = true;
    this.registerShutdownHook = true;
    this.additionalProfiles = Collections.emptySet();
    this.isCustomEnvironment = false;
    this.lazyInitialization = false;
    this.applicationContextFactory = ApplicationContextFactory.DEFAULT;
    this.applicationStartup = ApplicationStartup.DEFAULT;
    this.resourceLoader = resourceLoader;
    Assert.notNull(primarySources, "PrimarySources must not be null");
    this.primarySources = new LinkedHashSet(Arrays.asList(primarySources));
  // 判断 webAppilcationType 的类型
  this.webApplicationType = WebApplicationType.deduceFromClasspath();
  this.bootstrappers = new
  ArrayList(this.getSpringFactoriesInstances(Bootstrapper.class));
  // 设置 SpringApplication 的初始化器
```

```
this.setInitializers(this.getSpringFactoriesInstances
(ApplicationContextInitializer.class));
// 设置 SpringApplication 的监听器
this.setListeners(this.getSpringFactoriesInstances(ApplicationListener.class);
// 判断主类，初始化入口类
this.mainApplicationClass = this.deduceMainApplicationClass();
}
```

①判断当前项目类型，有三种类型：NONE、SERVLET、REACTIVE。

deduceFromClasspath()方法的创建代码如下。

```
static WebApplicationType deduceFromClasspath() {
if(ClassUtils.isPresent("org.springframework.web.reactive.DispatcherHandler",
(ClassLoader)null)&& ! ClassUtils.isPresent("org.springframework.web.servlet.DispatcherServlet",(ClassLoader)null)&& ! ClassUtils.isPresent("org.glassfish.jersey.servlet.ServletContainer",(ClassLoader)null)) {
        return REACTIVE;
    } else {
        String[] var0 = SERVLET_INDICATOR_CLASSES;
        int var1 = var0.length;
        for(int var2 = 0; var2 < var1; ++var2) {
            String className = var0[var2];
            if(! ClassUtils.isPresent(className, (ClassLoader)null)) {
                return NONE;
            }
        }
        return SERVLET;
    }
}
```

②设置 SpringApplication 的初始化器和监听器。

设置初始化器的过程：从类路径下找到 META-INF/spring.factories 配置的所有 ApplicationContextInitializer，然后保存起来。

设置监听器的过程：从类路径下找到 META-INF/spring.ApplicationListener，然后保存起来，过程同设置初始化器基本一样。

③判断主类，初始化入口类，deduceMainApplicationClass()方法的创建代码如下。

```
private Class<? > deduceMainApplicationClass() {
try {
StackTraceElement[] stackTrace = (new RuntimeException()).getStackTrace();
StackTraceElement[] var2 = stackTrace;
int var3 = stackTrace.length;
for(int var4 = 0; var4 < var3; ++var4) {
StackTraceElement stackTraceElement = var2[var4];
if("main".equals(stackTraceElement.getMethodName())) {
   return Class.forName(stackTraceElement.getClassName());
 }
}
} catch(ClassNotFoundException var6)
}
 return null;
}
```

第二部分，应用的启动方案。启动涉及的模块主要有加载配置环境模块、环境模块启动流程的监听模块，如图 1-22 所示。

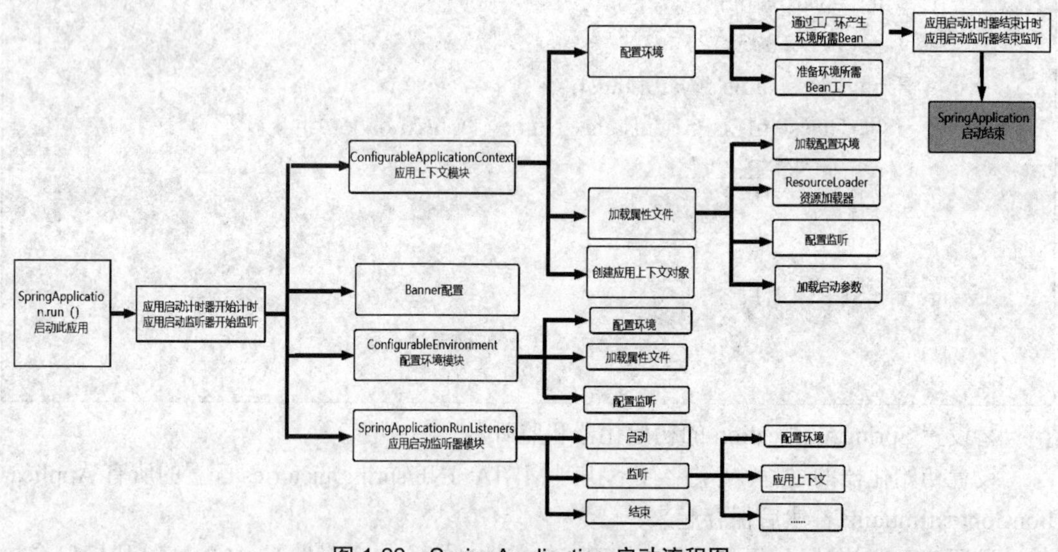

图 1-22 SpringApplication 启动流程图

项目应用的启动流程如下。

第一步，在任意创建的 Spring Boot 项目中，都会有一个自动生成的启动类，其代码如下。

```
@SpringBootApplication
public class DemoApplication {
  public static void main(String[] args) {
    SpringApplication.run(DemoApplication.class, args);
  }
}
```

第二步，SpringApplication 的 run 方法，此方法是应用主程序开始运行的第一步。首先开启一个 SpringApplicationRun-Listeners 监听器，代码如下。

```
public ConfigurableApplicationContext run(String... args) {
// 创建并启动计时监控类
        StopWatch stopWatch = new StopWatch();
        stopWatch.start();
        DefaultBootstrapContext bootstrapContext = createBootstrapContext();
        ConfigurableApplicationContext context = null;
        configureHeadlessProperty();
// 创建 SpringApplicationRunListeners 监听器对象，获取 MEAT-INF/spring.factori//es
中所有的 SpringApplicationRunListener
        SpringApplicationRunListeners listeners = getRunListeners(args);
// 启动监听，调用每一个 SpringApplicationRunListener 的 starting 方法
  }     listeners.starting(bootstrapContext, this.mainApplicationClass);
```

第三步，加载 Spring Boot 配置环境（ConfigurableEnvironment），代码如下。

```
        try {
// 将参数封装到 ApplicationArguments 对象中
            ApplicationArguments applicationArguments = new DefaultApplica-
tionArguments(args);
// 触发监听事件，调用每个 SpringApplicationRunListener 的 environmentPrepared 方法
            ConfigurableEnvironment environment = prepareEnvironment(listen-
ers, bootstrapContext, applicationArguments);
            configureIgnoreBeanInfo(environment);
// 打印 Banner
            Banner printedBanner = printBanner(environment);
            context = createApplicationContext();
  }         context.setApplicationStartup(this.applicationStartup);
```

ConfigurableEnvironment 类如图 1-23 所示。

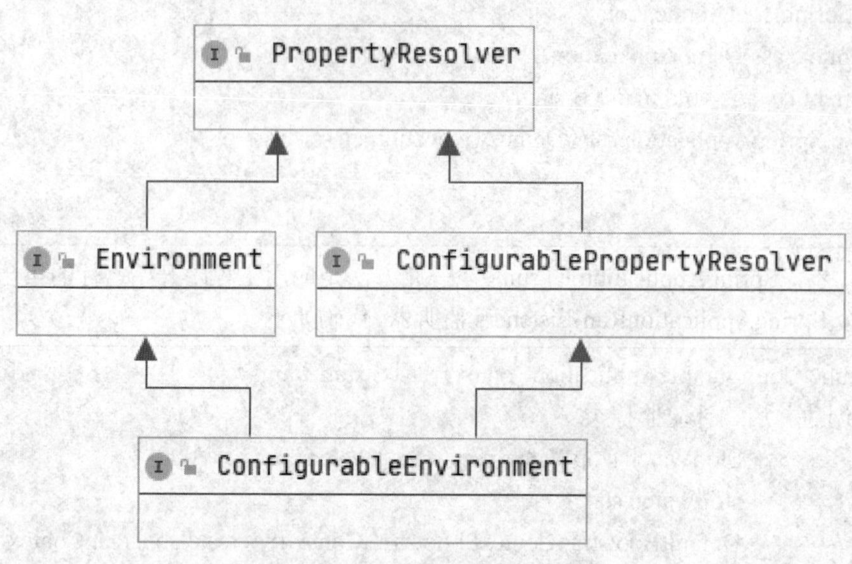

图 1-23 ConfigurableEnvironment 类

第四步，创建一个 ApplicationContext（应用上下文），通过这个上下文加载应用所需的项目运行环境、监听器、应用上下文环境、项目图标信息、项目参数，代码如下。

```
//准备上下文
    //与 Spring Cloud 相关的引导上下文
    //将 environment 保存到容器中
    //触发监听事件调用每个 SpringApplicationRunListeners 的 contextPrepared 方法
     //调用 ConfigurableListableBeanFactory 的 registerSingleton 方法向容器中注入
applicationArguments 与 printedBanner
    //触发监听事件调用每个 SpringApplicationRunListeners 的 contextLoaded 方法
            prepareContext(bootstrapContext, context, environment, listeners, applica-
tionArguments, printedBanner);
//刷新容器，完成对于项目组件的扫描、加载等
                refreshContext(context);
                afterRefresh(context, applicationArguments);
                stopWatch.stop();
                if(this.logStartupInfo){
                    new    StartupInfoLogger(this.mainApplicationClass).log-
Started(getApplicationLog(), stopWatch);
                }
```

ConfigurableApplicationContext 类主要继承了 ApplicationContext、LifeCycle 这两个类。其中，LifeCycle 是生命周期类，其定义了 start 启动、stop 结束等生命周期空值方法；ApplicationContext 是应用上下文类，其主要继承了 beanFactory（Bean 的工厂类）。

ConfigurableApplicationContext 类，如图 1-24 所示。

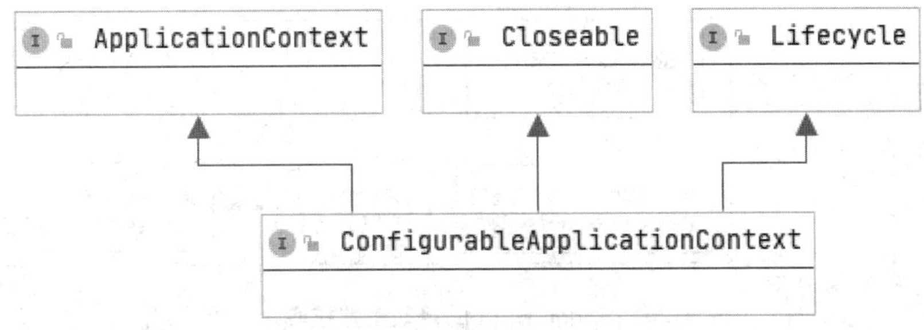

图 1-24　ConfigurableApplicationContext 类

第五步，用 listeners.started（context）方法使监听器启动配置完的应用上下文（ApplicationContext），代码如下。

```
// 触发监听事件调用每个 SpringApplicationRunListener 的 finished 方法   {
                listeners.started(context);
                callRunners(context, applicationArguments);
        }
        catch (Throwable ex) {
                handleRunFailure(context, ex, listeners);
                throw new IllegalStateException(ex);
        }
```

第六步，发送 ApplicationReadyEvent 事件，标志 SpringApplication 正在运行，即已经成功启动，可以接收服务请求，代码如下。

```
try {
// 运行应用上下文   {
                listeners.running(context);
        }
        catch (Throwable ex) {
                handleRunFailure(context, ex, null);
                throw new IllegalStateException(ex);
        }
// 返回容器
        return context;
    }
```

第三部分，自动化配置模块，该模块是 Spring Boot 的自动配置核心，如图 1-25 所示。

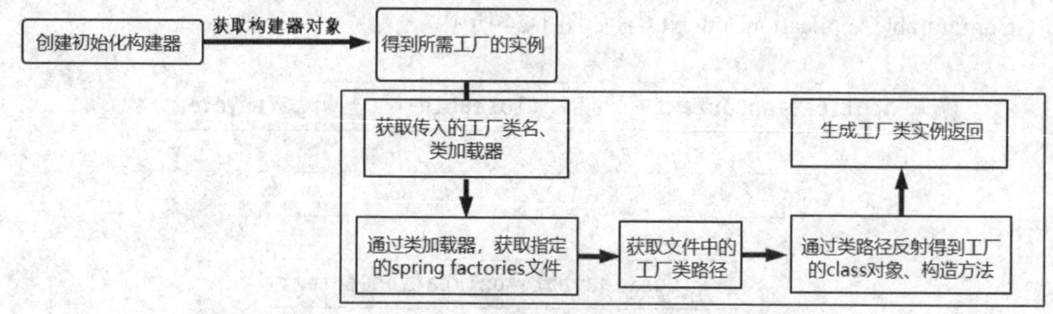

图 1-25 Spring Boot 的自动配置核心

由此分析，一个简单的 Spring Boot 主程序，通过运行一个 run 方法，就可以引发一系列复杂的内部调用和加载过程，从而初始化一个应用所需的配置、环境、资源及各种类定义等，特别是导入了一系列自动配置类，实现了强大的自动配置功能。

技能点 5 Spring Boot 的 Starter

在任何一个项目中，依赖管理都是至关重要的一部分。如果依赖管理的内容变得复杂，那么开发难度也会随之升高。虽然 Maven 已经简化了项目依赖管理，但对于众多的 artifacts 来说，这明显是不够的。

这时 Spring Boot 的 Starter 就起到了作用。它将所需的依赖全部以一致的方式进行注入并进行统一管理。开发人员在使用它时，只需在 pom.xml 文件中进行依赖注入即可，以达到快速搭建项目的目的。

1. Starter 概述

Starter 可以将需要的功能整合起来，像是一个可拔插式的插件，便于开发人员使用。例如使用 spring-boot-starter-redis 来实现 redis；使用 spring-boot-starter-jdbc 来实现 JDBC。其中，spring-boot-starter-* 起步依赖是 Spring Boot 的核心之处，它提供了 Spring 和相关技术，提供一站式服务，让开发人员不再关心 Spring 相关配置，简化了传统的依赖注入操作。当然，开发人员也可通过 application.properties 文件自定义配置。Spring Boot 常规启动都遵循类似的命名模式 spring-boot-starter-*，其中“*”是一种指定类型的应用程序，如 spring-boot-starter-web 表示应用程序依赖 SpringWeb 相关内容。另外，Spring Boot 支持第三方插件的引用，第三方启动程序通常以项目的名称开始。例如，mybatis 依赖插件引用为 mybatis-spring-boot-starter。Starter 的结构如图 1-26 所示。

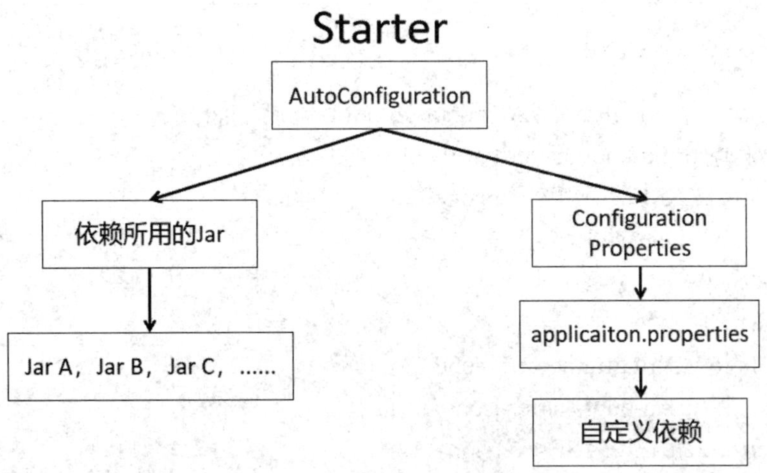

图 1-26　Starter 结构

比较常用的 Starter 见表 1-7。

表 1-7　Starter 表

名称	作用
spring-boot-starter	核心 Starter，包括自动化配置支持、日志以及 YAML
spring-boot-starter-aop	Spring AOP 和 AspectJ 相关的切面编程 Starter
spring-boot-starter-data-jpa	使用 Hibernate Spring Data JPA 的 Starter
spring-boot-starter-jdbc	使用 HikariCP 连接池 JDBC 的 Starter
spring-boot-starter-security	使用 Spring Security 的 Starter
spring-boot-starter-test	Spring Boot 测试相关的 Starter
spring-boot-starter-web	构建 restful、springMVC 的 web 应用程序的 Starter
spring-boot-starter-data-redis	支持 redis 缓存

2. Starter 的使用

基本上不同的 Starter 都会使用到两个内容：AutoConfiguration 和 ConfigurationProperties。我们使用 ConfigurationProperties 来保存配置，并且这些配置都可以有一个默认值，在没有主动覆写原始配置的情况下，默认值就会生效。除此之外，Starter 的 ConfigurationProperties 还使所有的配置属性被聚集到一个文件中，这样使开发人员就告别了 Spring 项目中的“XML 地狱”。

spring-boot-starter-web 的源代码如下。

```
<dependencies>
 <dependency>
   <groupId>org.springframework.boot</groupId>
   <artifactId>spring-boot-starter</artifactId>
   <version>2.2.2.RELEASE</version>
   <scope>compile</scope>
 </dependency>
 <dependency>
   <groupId>org.springframework.boot</groupId>
   <artifactId>spring-boot-starter-json</artifactId>
   <version>2.2.2.RELEASE</version>
   <scope>compile</scope>
 </dependency>
 <dependency>
   <groupId>org.springframework.boot</groupId>
   <artifactId>spring-boot-starter-tomcat</artifactId>
   <version>2.2.2.RELEASE</version>
   <scope>compile</scope>
 </dependency>
 <dependency>
   <groupId>org.springframework.boot</groupId>
   <artifactId>spring-boot-starter-validation</artifactId>
   <version>2.2.2.RELEASE</version>
   <scope>compile</scope>
   <exclusions>
     <exclusion>
       <artifactId>tomcat-embed-el</artifactId>
       <groupId>org.apache.tomcat.embed</groupId>
     </exclusion>
   </exclusions>
 </dependency>
 <dependency>
   <groupId>org.springframework</groupId>
   <artifactId>spring-web</artifactId>
   <version>5.2.2.RELEASE</version>
```

```
        <scope>compile</scope>
    </dependency>
    <dependency>
        <groupId>org.springframework</groupId>
        <artifactId>spring-webmvc</artifactId>
        <version>5.2.2.RELEASE</version>
        <scope>compile</scope>
    </dependency>
</dependencies>
```

分析上述代码可以了解到，spring-boot-starter-web 依赖的作用是支持全栈式 Web 开发（包括 Tomcat 和 spring-webmvc），提供 Web 开发场景所需的底层所有依赖。

在 pom.xml 文件中引入 spring-boot-starter-web 依赖时，基本就满足了日常的 Web 接口开发；而不使用 Spring Boot 时，引入 spring-web、spring-webmvc、spring-aop 等来支持项目开发，就可以实现 Web 场景开发，不需要导入其他 Web 依赖文件以及 Tomcat 服务器等。

1. 个人博客项目需求分析

个人博客系统分为前台管理和后台管理两部分。前台管理的核心功能包括文章分页展示、文章详情查看、文章评论管理；后台管理的核心功能包括系统数据展示、文章发布、文章修改、文章删除。该博客系统的后台使用 SpringMVC+SpringBoot+MyBatis 框架进行整合开发，前台使用 Spring Boot 支持的模板引擎 Thymeleaf+jQuery 完成页面信息展示，同时整合 Redis 实现缓存管理，整合 Spring Security 实现安全管理。个人博客项目的系统功能如图 1-27 所示。

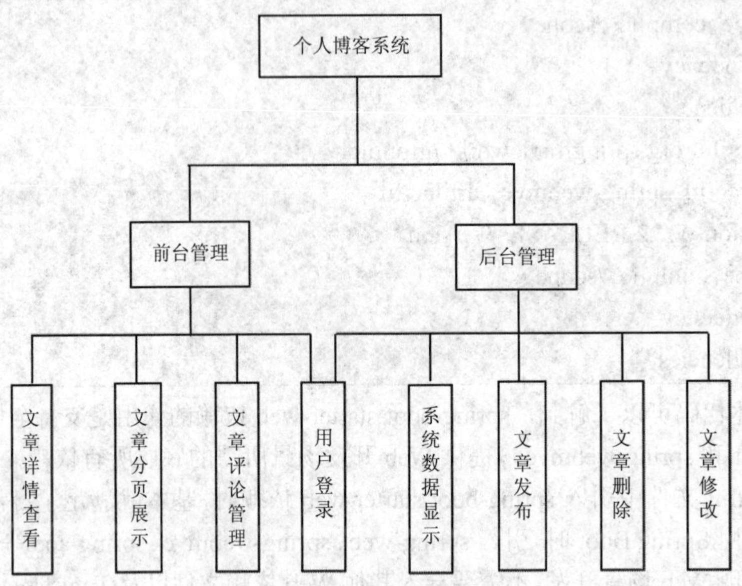

图 1-27　系统功能图

（1）用户登录

系统考虑到用户购买的真实性，规定用户必须注册会员，才能登录系统。填写相关的基本信息完成注册后进行登录。该项目的登录访问地址和预览效果如图 1-28 所示。

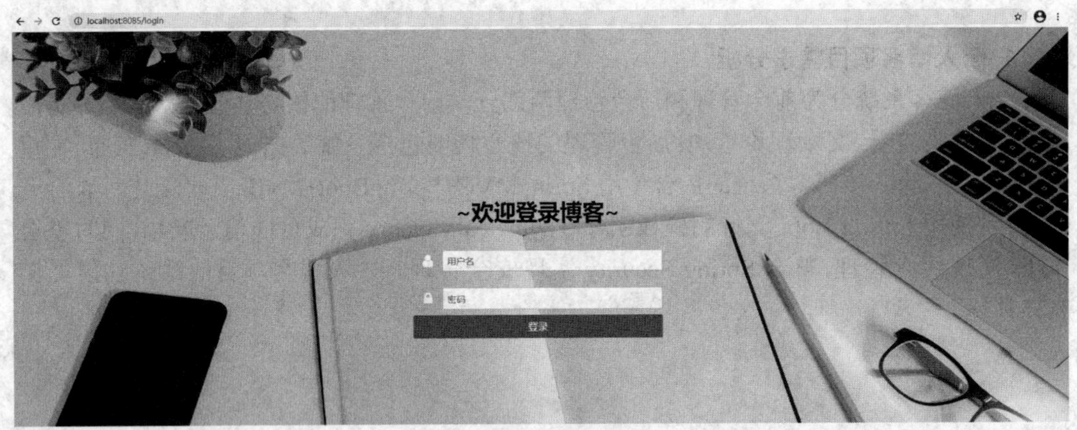

图 1-28　登录效果

（2）文章分页

在项目的首页中，分页展示了所对应的博客信息，同时页面左侧还展示了阅读的排行情况。该项目的文章分页展示访问地址和预览效果如图 1-29 所示。

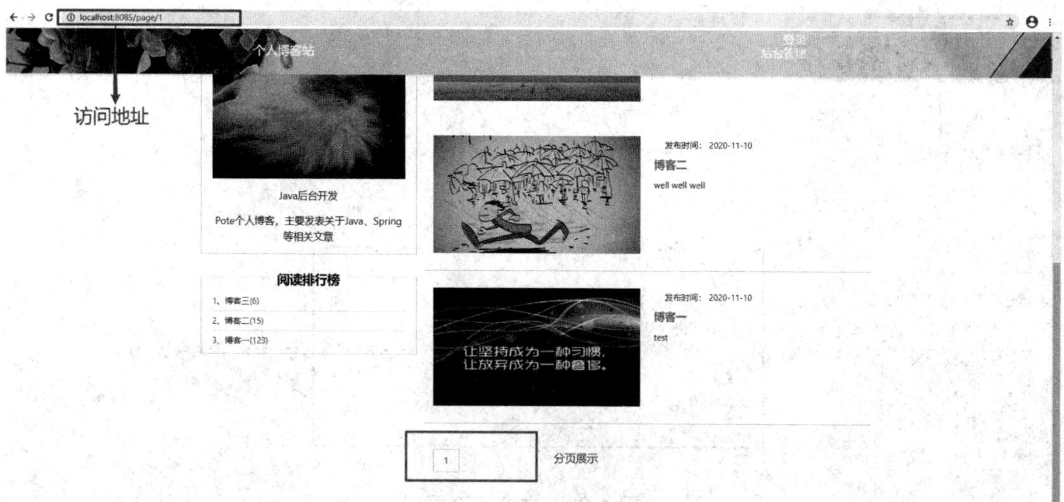

图 1-29　文章分页展示

（3）文章详情页

在该项目的文章详情页中，展示了所对应博客的详细信息，同时下方还展示了博客文章的评论信息。该项目的文章详情页访问地址和预览效果如图 1-30 所示。

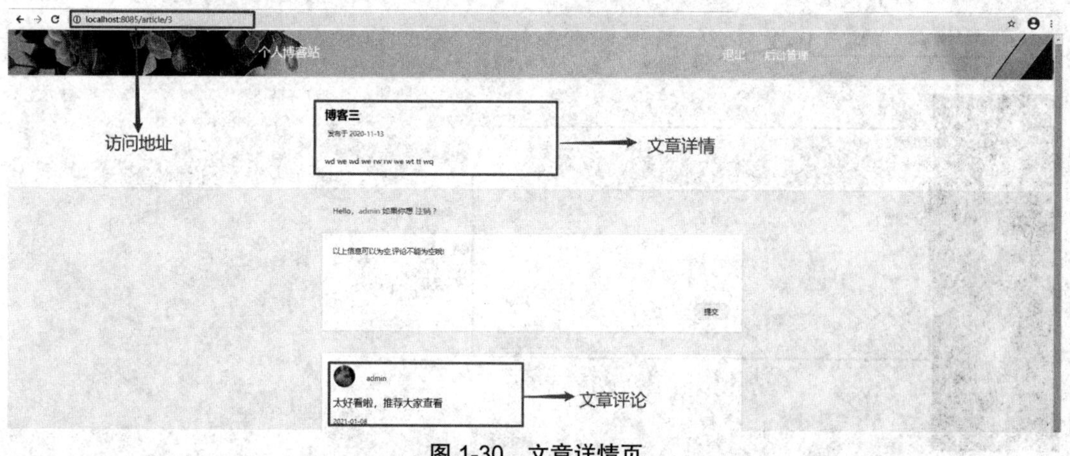

图 1-30　文章详情页

（4）文章评论页

在该项目的文章评论页中，用户只有登录后，才可评论文章并发布。文章评论页展示了文章对应的评论详情并附带分页功能。该项目的文章评论页访问地址和预览效果如图 1-31 所示。

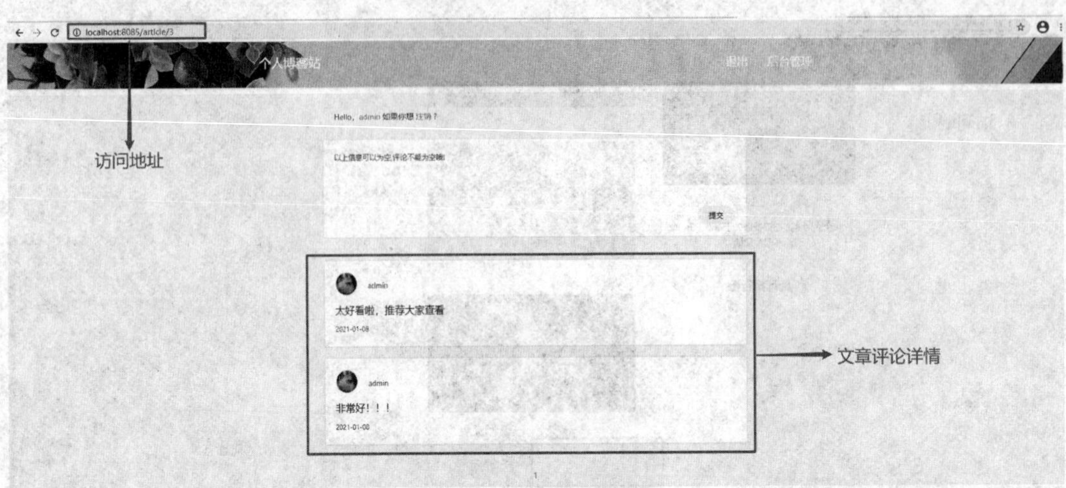

图 1-31　文章评论页

（5）系统数据页

在该项目的系统数据页中，用户只有登录后，才可进入系统管理页面。系统数据页展示了文章对应的评论详情和最新的文章信息。该项目的系统数据页访问地址和预览效果如图 1-32 所示。

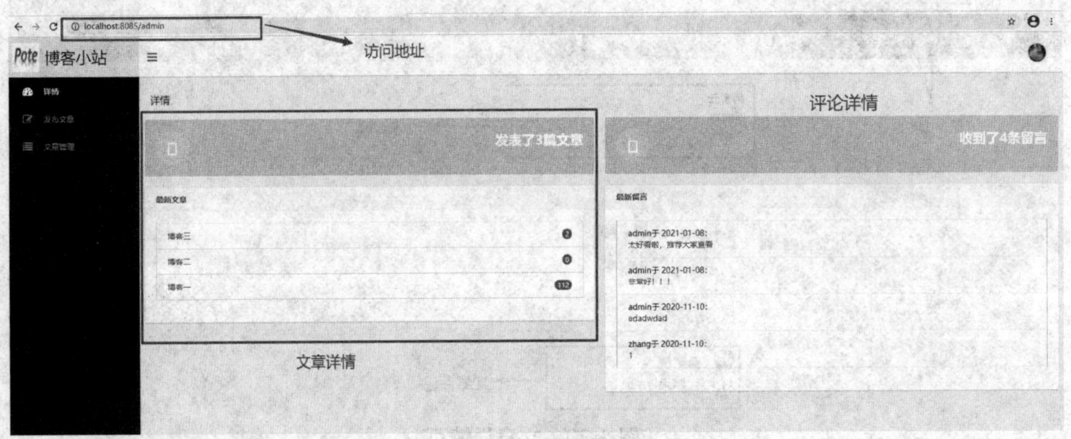

图 1-32　系统数据页

（6）文章发布页

在该项目的文章发布页中，通过不同需求的跳转，该页面不仅可以作为文章的发布页面，还可以作为文章的编辑页面。使用 Markdown 编辑器控制，用户在文章编辑过程中可以预览。该项目的文章发布页访问地址和预览效果如图 1-33 所示。

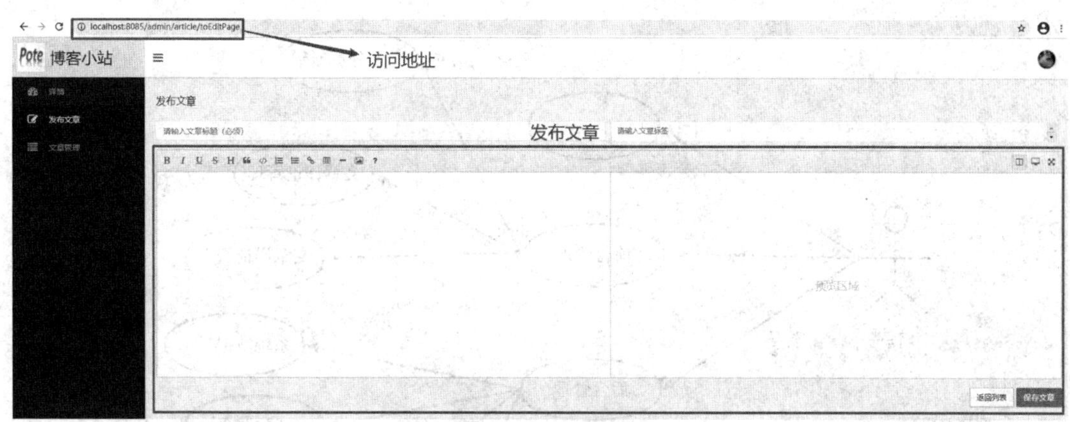

图 1-33　文章发布页

（7）文章修改和删除页

在该项目的文章修改和删除页中，提供了修改和删除功能。该页面分页展示了对应文章的信息。该项目的文章修改和删除页访问地址和预览效果如图 1-34 所示。

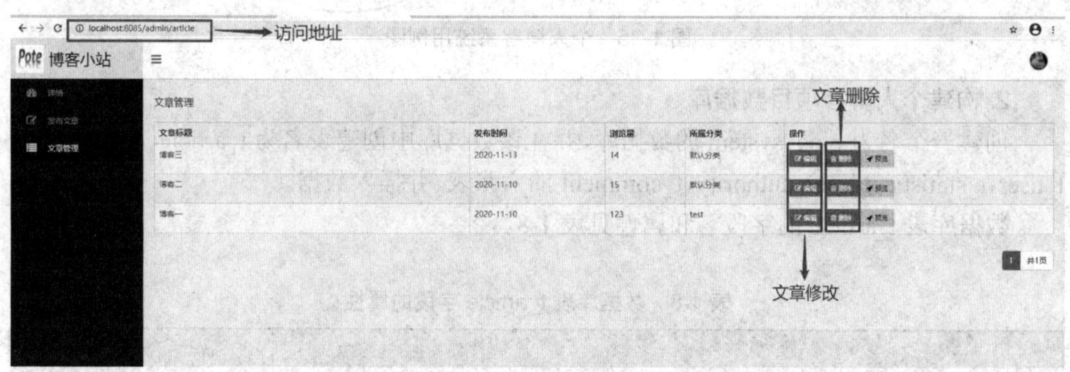

图 1-34　文章修改和删除页

根据个人博客系统的业务需求、功能需求及用户需求，绘制个人博客系统的用例图。该图描述了个人博客系统所具备的功能，如图 1-35 所示。

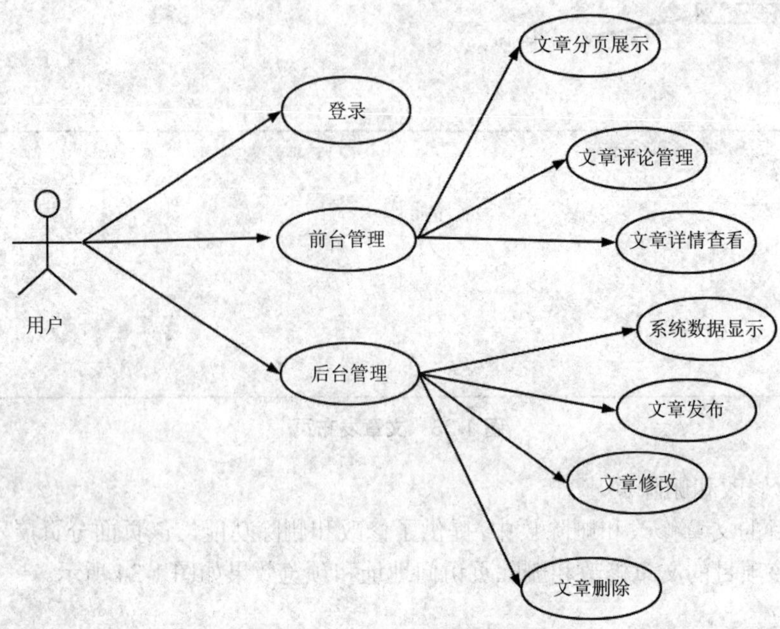

图 1-35 个人博客系统用例图

2. 构建个人博客项目数据库

创建一个名为 blog_system 的数据库，并在该数据库中创建表名为 t_article、t_authority、t_user、t_statistic、t_user_authority、t_comment 的六张表，并插入数据。

数据库表 t_article 的字段名和属性见表 1-8。

表 1-8 数据库表 t_article 字段的属性

字段名	类型	长度	说明
id	int	11	自增主键
title	varchar	255	标题
content	varchar	255	内容
created	datetime	0	创建时间
modified	datetime	0	修改时间
categories	varchar	255	类别
tags	varchar	255	标签
allow_comment	varchar	255	允许的评论
thumbnail	varchar	255	缩略图
hits	varchar	255	点击量
comments_num	int	11	评论数量

数据库表 t_authority 的字段名和属性见表 1-9。

表 1-9　数据库表 t_authority 字段的属性

字段名	类型	长度	说明
id	int	11	自增主键
authority	varchar	255	权限

数据库表 t_user 的字段名和属性见表 1-10。

表 1-10　数据库表 t_user 字段的属性

字段名	类型	长度	说明
id	int	11	自增主键
username	varchar	255	登录用户名
password	varchar	255	登录密码
email	varchar	255	用户邮箱
created	date	0	创建时间
valid	tinyint	1	是否有效

数据库表 t_statistic 的字段名和属性见表 1-11。

表 1-11　数据库表 t_statistic 字段的属性

字段名	类型	长度	说明
id	int	11	自增主键
article_id	int	11	文章 id
hits	varchar	255	
comments_num	int	11	评论数量

数据库表 t_user_authority 的字段名和属性见表 1-12。

表 1-12　数据库表 t_user_authority 字段的属性

字段名	类型	长度	说明
id	int	11	自增主键
article_id	int	11	文章 id
authority_id	int	11	作者 id

数据库表 t_comment 的字段名和属性见表 1-13。

表 1-13 数据库表 t_comment 字段的属性

字段名	类型	长度	说明
id	int	11	自增主键
article_id	int	11	文章 id
content	varchar	255	内容
created	datetime	0	创建时间
author	varchar	255	作者
ip	varchar	255	IP 地址
status	varchar	255	状态

3. 搭建 Spring Boot 个人博客项目

第一步，在 Spring Boot 的左面侧边栏选择 Spring Initializr 项目，然后在“Choose starter service URL”下选择“Default”，即 https：//start.spring.io，点击“Next”进入下一步，如图 1-36 所示。

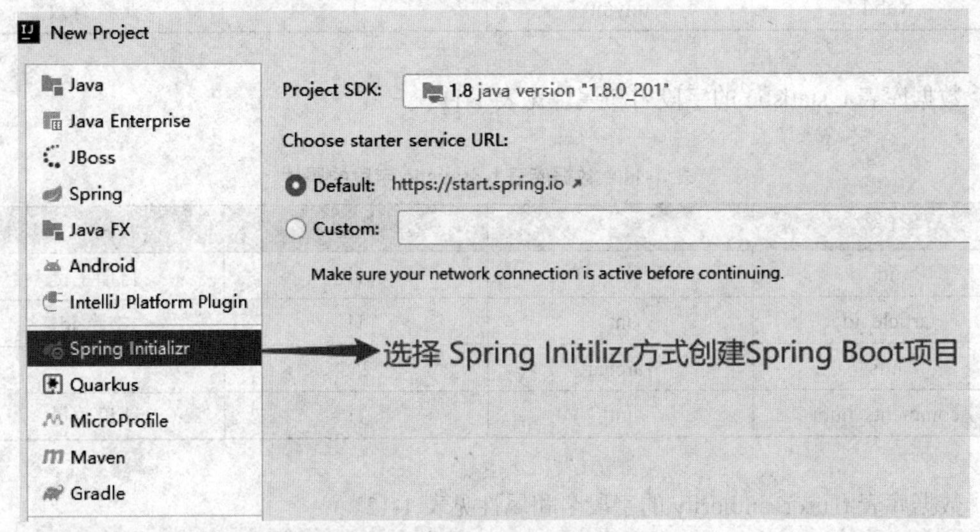

图 1-36 新建 Spring Boot 项目

第二步，选择项目类型。

将“Group”设置为 com.xtgj，“Artifact”设置为 blog_system，点击“Next”进入下一步，如图 1-37 所示。

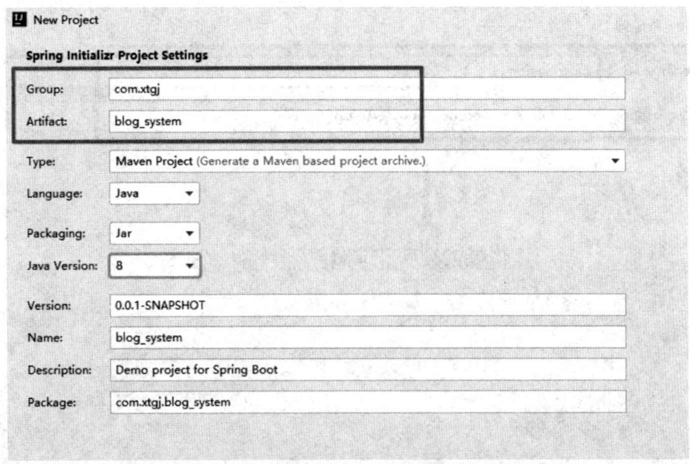

图 1-37　选择项目类型

第三步，选择 Spring Boot 版本和 Spring Boot 组件。根据项目选择所需要的组件，个人博客项目选择 Spring Web 组件，如图 1-38 所示。

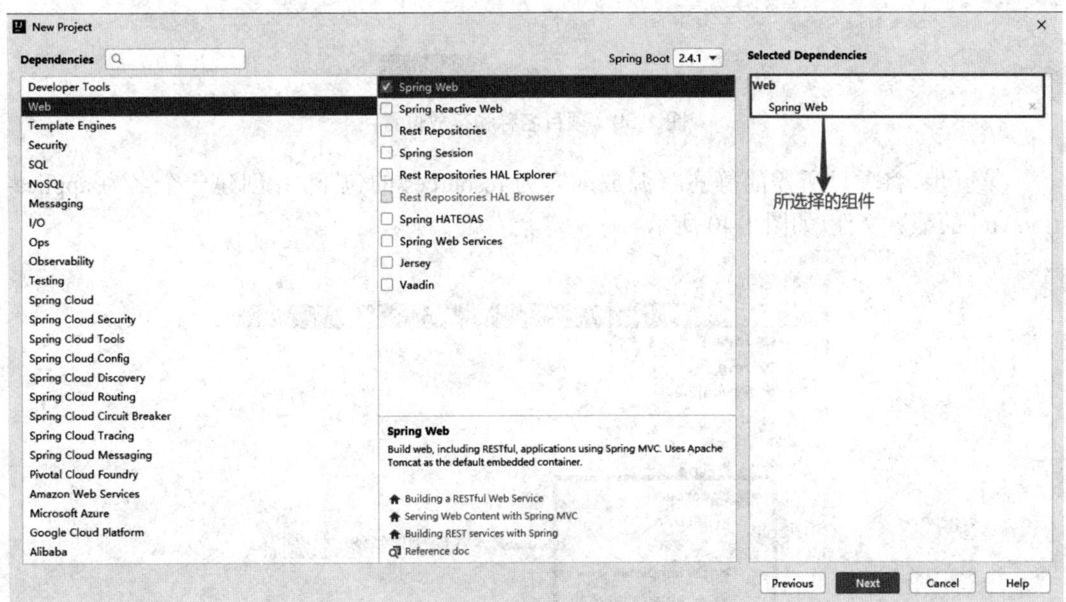

图 1-38　选择版本和组件

第四步，输入项目名“Project name”和选择项目存放位置“Project location”，如图 1-39 所示。点击“Finish”，一个 Spring Boot 项目便创建完成。

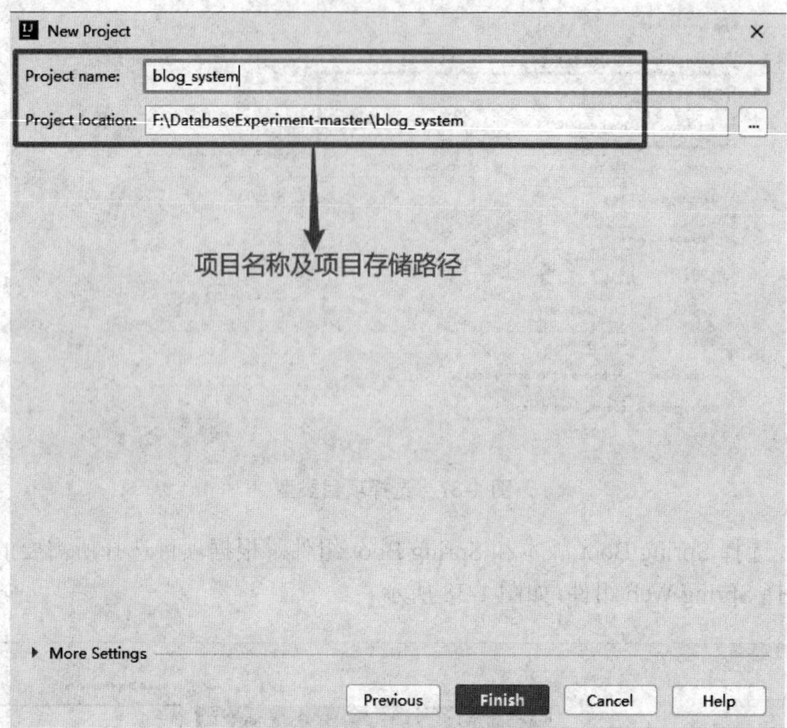

图 1-39 项目名称和存放位置

第五步，将项目所需的静态资源放入名为 resources 的包下，并创建一个名为 application.yml 的配置文件，如图 1-40 所示。

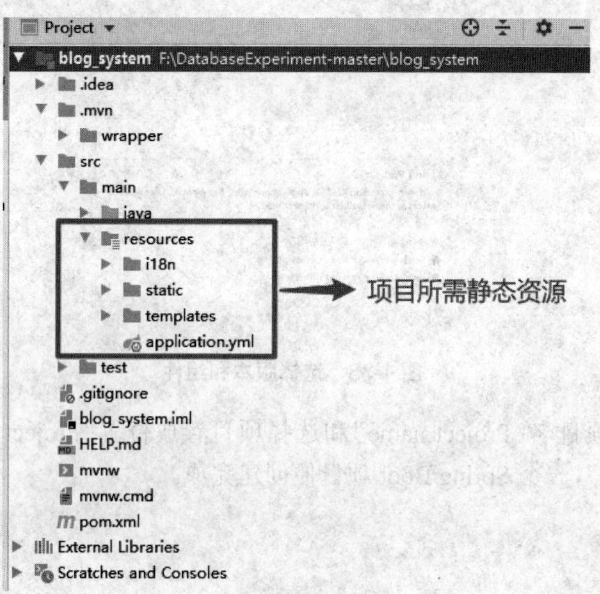

图 1-40 项目结构

第六步，在名为 template 的包下创建一个 index.html 页面，运行该页面，如图 1-41 所示。

图 1-41　运行成功示意

本任务讲解了如何搭建 Spring Boot 项目，为下一阶段的学习打下了坚固的基础。本任务使读者加深了对于 Spring Boot 的理解，掌握了基本的 Spring Boot 技术。

Configure	配置	Archetype	原型
Valid	有效	Development	开发
Execution	实施		

一、选择题

1. IDEA 界面中，“Configure”→“Project Defaults”中的 Project Structure 主要作用是（　　）。

A. 用于全局 JDK 初始化设置

B. 用于全局 Maven 初始化设置

C. 用于全局运行环境设置

D. 以上都不对

2. 下列关于 Spring Boot 项目各个包作用的说法，正确的是（　　）。

A. resources static 用于存放静态资源文件

B. resources templates 用于存放模板文件

C. application.properties 是项目的全局配置文件

D. 以上都正确

3. 下列说法正确的是（　　）。

A. Group：项目组织的唯一标识符，在实际开发中一般对应项目的名称

B. Type：项目的类型

C. Artifact：项目的唯一标识符，在实际开发中对应 Java 的包的结构

D. Packaging：打包方式

4. 下列关于 Spring Boot 自动配置原理的说法，错误的是（　　）。

A. @SpringBootApplication 只包含 @SpriingBootConfiguration、@EnableAutoConfiguration、@ComponentScan 3 个注解

B. @SpringBootConfiguration 注解表示当前类为一个配置类并可以被组件扫描器扫描

C. @EnableAutoConfiguration 注解的作用是启动自动配置，向容器中导入所有选中的自动配置类

D. @ComponentScan 注解的主要作用是扫描指定包及其子包下所有注解类文件作为 Spring 容器的组件使用

5. 下列选项中不属于 Spring Boot 的启动流程的是（　　）。

A. 应用的启动方案

B. 应用的结束方案

C. SpringApplication 初始化模块

D. 自动化配置模块

二、简答题

1. Spring Boot 的优点有哪些？

2. 简述 Spring Boot 的执行流程。

项目二　个人博客项目基本环境配置

通过学习 Spring Boot 的核心配置与注解，了解 Spring Boot 全局配置文件的使用方法，学习自定义配置，掌握配置文件属性值的注入，达到正确配置 Spring Boot 的目标。

- 了解启动类核心注解 @SpringBootApplication。
- 掌握全局配置文件。
- 掌握配置文件的注入。
- 掌握自定义配置。

【情景导入】

Spring Boot 项目在创建时会自动生成启动类和默认的配置文件，编写启动类以及配置文件可达到自动装配的目的，简化项目，使开发人员将注意力放在业务逻辑上。应用注解代替原先的配置类，使项目开发更加人性化。

【功能描述】

- 个人博客项目基本配置。

技能点 1 启动类核心注解 @SpringBootApplication

1. @SpringBootApplication 注解

在任意创建的 Spring Boot 项目中，都会有一个自动生成的启动类，其代码如下。

```
@SpringBootApplication
public class DemoApplication {
    public static void main(String[] args) {
        SpringApplication.run(DemoApplication.class, args);
    }
}
```

重点内容在于 @SpringBootApplication 注解。该注解为 Spring Boot 项目的配置注解，对于 @SpringBootApplication 注解，其代码如下。

```
@Target(ElementType.TYPE)
@Retention(RetentionPolicy.RUNTIME)
@Documented
@Inherited
@SpringBootConfiguration
@EnableAutoConfiguration
@ComponentScan(excludeFilters = { @Filter(type = FilterType.CUSTOM, classes = 
TypeExcludeFilter.class),
        @Filter(type = FilterType.CUSTOM, classes = AutoConfigurationExclude-
Filter.class) })
public @interface SpringBootApplication {
...
}
```

这些组合注解中的前四个为元注解，用于修饰当前的 @SpringBootApplication 注解，后三个为真正起作用的核心注解。它们的含义如下。

@Target 注解：标记另外的注解，用于限制此注解可以应用的 Java 元素类型。

@Retention 注解：用于修饰注解。RetentionPolicy 属性用于描述保留注释的各种策略，与元注释（@Retention）一起指定注释要保留的时长。可理解为，Rentention 搭配 Rentention-Policy 使用。其中 RetentionPolicy 有 3 个值：CLASS、RUNTIME 和 SOURCE。RUNTIME 表示需要在运行时动态地获取注解信息，该注解在运行时不仅被保存到 class 文件中，jvm 加载 class 文件之后，仍然存在。

@Documented 注解：表明这个注释是由 javadoc 记录的，在默认情况下也有类似的记录工具。如果一个类型声明被注释了文档化，它的注释成为公共 API 的一部分。

@Inherited 注解：用来修饰注解。若一个类使用了该注解，那么其子类也会继承该注解。

@SpringBootConfiguration 注解：标注当前类为配置类。该注解继承自 @Configuration，将当前类内声明的一个或者多个 @Bean 注解标记的方法实例，注入到 Spring 容器中，实例名即为方法名。其代码如下。

```
@Target(ElementType.TYPE)
@Retention(RetentionPolicy.RUNTIME)
@Documented
@Configuration
public @interface SpringBootConfiguration {
...
}
```

@EnableAutoConfiguration 注解：借助 @Import 将所有符合自动配置条件的 Bean 加载到 IOC 容器。其代码如下。

```
@Target(ElementType.TYPE)
@Retention(RetentionPolicy.RUNTIME)
@Documented
@Inherited
@AutoConfigurationPackage
@Import(AutoConfigurationImportSelector.class)
public @interface EnableAutoConfiguration {
...
}
```

@ComponentScan 注解：自动扫描并加载符合条件的组件。可通过 basePackages 等属性来指定 @ComponentScan 注解自动扫描的范围，若不指定，则默认 Spring 框架实现会从声明 @ComponentScan 注解所在类的包进行扫描。其代码如下。

```
@Retention(RetentionPolicy.RUNTIME)
@Target(ElementType.TYPE)
@Documented
@Repeatable(ComponentScans.class)
public @interface ComponentScan {
...
}
```

2. @SpringBootApplication 注解的重构

@SpringBootApplication 注解，可使用 @SpringBootConfiguration、@EnableAutoConfiguration 和 @ComponentScan 3 个注解进行重构。重构之后的入口类，与使用 @SpringBootApplication 注解的效果相同。其代码如下。

```
@SpringBootConfiguration
@EnableAutoConfiguration
@ComponentScan(excludeFilters = {@ComponentScan.Filter(type = FilterType.CUSTOM, classes =TypeExcludeFilter.class),
@ComponentScan.Filter(type=FilterType.CUSTOM, classes=AutoConfigurationExcludeFilter.class)})
public class DemoApplication {
    public static void main(String[] args) {
        SpringApplication.run(DemoApplication.class, args);
    }
}
```

技能点 2　全局配置文件

Spring Boot 项目都需要全局配置文件，目的是修改 Spring Boot 自动配置的默认值。一般使用 Properties 文件或 YAML 文件作为全局配置文件。

1. application.properties 配置文件

在创建 Spring Boot 项目时，在 resource 目录下创建名为 application.properties 的全局配置文件，如图 2-1 所示。

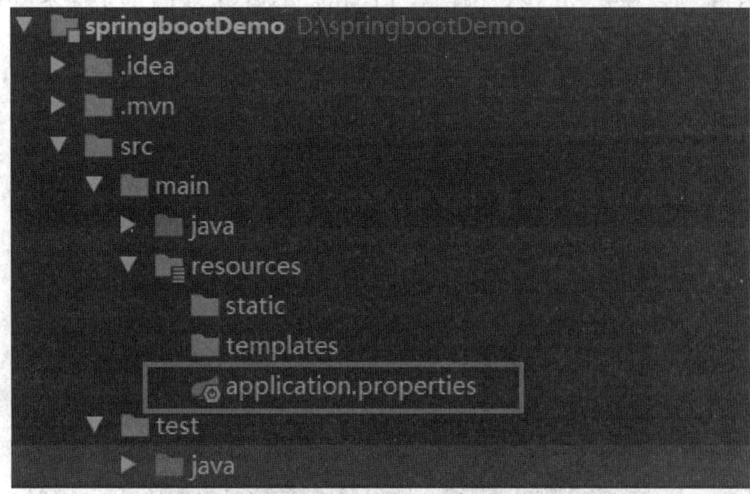

图 2-1　配置文件具体位置

application.properties 配置文件主要用于定义 Spring Boot 项目的相关属性，如环境变量、系统参数、端口地址等。application.properties 文件中的每一行都是一个键值对，键之间使用“.”进行连接，键与值之间使用“=”进行分隔，以“#”号开头的方式来进行注释。application.properties 的代码如下。

```
#properties 配置文件
# 设置服务端口
server.port=9090
```

在 application.properties 配置文件中，有一些常用的配置信息，这些配置控制着项目的端口号、应用名称等。下面就对这部分基本属性进行具体说明。

① server.port：项目的端口配置。

② spring.application.name：指定应用的名称，即在 Spring Cloud 微服务中被用来注册的服务名。

③ spring.profiles.active：用于定义多个不同的环境配置，如 dev、test 等。

④ spring.profiles.include：引入的其他配置文件名称。例如，其他配置文件的命名规则

为"applicaiton-xxx.properties",则使用 include 来引入该配置文件时,需要使用"spring.profiles.include=xxx"。

⑤ logging.config:日志配置。

⑥ server.servlet.context-path 和 server.context-path:配置项目的上下文路径。以"/"开始。在配置了上下文路径后,在访问项目时需要加入正确的路径才能正常访问接口。server.context-path 是 Spring 2.0 以下的配置方式,server.servlet.context-path 是 Spring 2.0 以上的配置方式。

2. application.properties 配置文件实例

【实例】通过 application.properties 配置文件给实体类赋值。

通过 Spring Initializr 方式创建名为 springbootDemo 的 Spring Boot 项目,在对应的包下创建名为 entity 的包。然后,在该包中创建 Elective 和 Student 两个实体类,项目结构如图 2-2 所示。Elective 实体类如示例代码 2-1 所示。

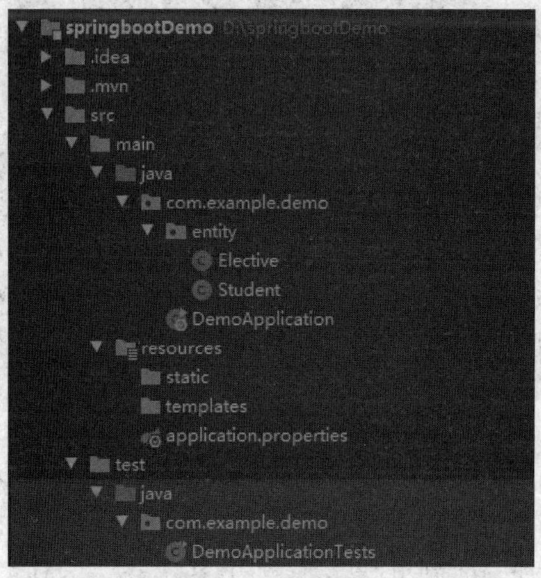

图 2-2 项目结构图

示例代码 2-1:Elective 实体类

```
public class Elective {
    private String name;        // 选修课名称
    private Date classTime;     // 课程时间

    public String getName() {
        return name;
    }
    public void setName(String name) {
        this.name = name;
```

```
    }
    public Date getClassTime() {
        return classTime;
    }
    public void setClassTime(Date classTime) {
        this.classTime = classTime;
    }
    @Override
    public String toString() {
        return "Elective{" +
                "name='" + name + '\'' +
                ", classTime=" + classTime +
                '}';
    }
}
```

Student 实体类，如示例代码 2-2 所示。

示例代码 2-2：Student 实体类

```
public class Student {
    private int id;                    //id
    private String name;               // 姓名
    private String className;          // 所属班级
    private String[] course;           // 所上课程
private Elective elective;             // 选修课程
private Map map;
// 省略 setter()和 getter()方法
// 省略 toString()方法
```

在 resources 下的 application.properties 配置文件中编写 Student 类的配置属性，如示例代码 2-3 所示。

示例代码 2-3：application.properties 配置文件

```
student.id=1                          // 设置 id
student.name=zhang                    // 设置 name 属性
student.className=ClassOne            // 设置 className 属性
student.course=Mathematics,Chinese,English        // 设置 course 属性
student.elective.name=art                   // 设置 elective 实体类中的 name 属性
student.elective.classTime=2020/11/17   // 设置 elective 实体类中的 classTime 属性
```

```
student.map.key1=value1        // 设置 map 中 key1 对应的数值
student.map.key2=value2        // 设置 map 中 key2 对应的数值
```

编辑好配置文件之后，返回 Student 类，编写 @Component 和 @ConfigurationProperties 注解。@Component 注解用于将 Student 类作为 Bean 注入 Spring 容器中；@ConfigurationProperties 注解将 prefix 属性设置为 Student，表示将配置文件中以 Student 开头的属性注入到该类中。在编写完成之后，效果如图 2-3 所示。

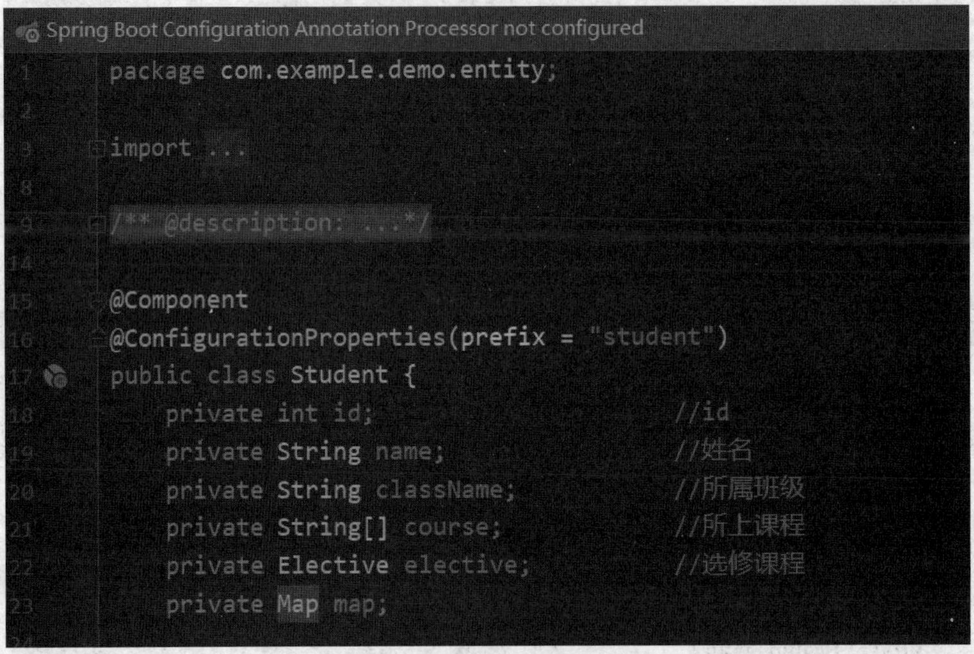

```
Spring Boot Configuration Annotation Processor not configured
package com.example.demo.entity;

import ...

/** @description: ...*/

@Component
@ConfigurationProperties(prefix = "student")
public class Student {
    private int id;                            //id
    private String name;                       //姓名
    private String className;                  //所属班级
    private String[] course;                   //所上课程
    private Elective elective;                 //选修课程
    private Map map;
```

图 2-3　IDEA 提示错误

在图中可看到 IDEA 报出提示"Spring Boot Configuration Annotation Processor not configured"，含义为未配置 Spring Boot 注释处理器。原因是在使用 @ConfigurationProperties 注解进行配置文件的属性值注入时，所定义的 Student 是用户自行定义的，Spring Boot 无法自动识别。在实际开发中，为了防止这种情况的发生，可以在 pom.xml 文件中添加配置注释处理器的依赖"spring-boot-configuration-processor"，添加之后在编写 application.properties 配置文件时，就会出现代码的书写提示，提高了编写效率。所添加的依赖如下。

```
<dependency>
        <groupId>org.springframework.boot</groupId>
        <artifactId>spring-boot-configuration-processor</artifactId>
        <optional>true</optional>
</dependency>
```

为了查看 application.properties 配置文件对应实体类是否成功，在项目中对应的 DemoApplicationTests 类中使用 @Autowired 注解将 Student 作为 Bean 注入 Spring 容器，并在 contextLoads 方法中进行输出，如示例代码 2-4 所示。

示例代码 2-4：DemoApplicationTests 测试类

```
@SpringBootTest
class DemoApplicationTests {
    @Autowired
    private Student student;
    @Test
    void contextLoads() {
        System.out.println(student);
    }
}
```

运行该测试类，在控制台上输出的信息如图 2-4 所示。这样，便完成了通过 application.properties 配置文件对实体类 Student 进行赋值的实例。

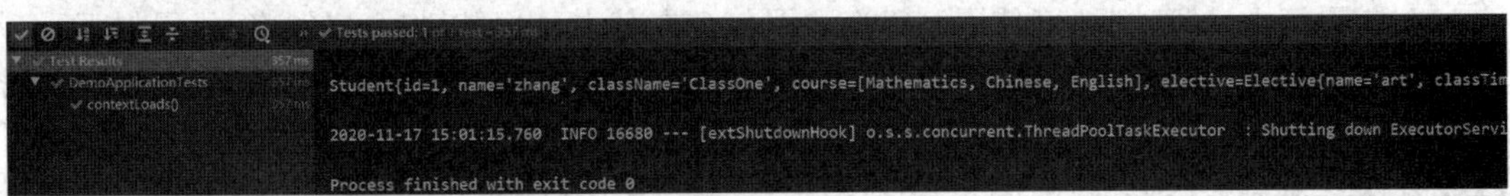

图 2-4　控制台显示实例信息

3. application.yaml 配置文件

YAML 是 Spring Boot 所支持的一种 JSON 超集文件格式，YAML 配置文件以数据为核心，是一种更为直观且容易被计算机识别的数据序列化格式。application.yaml 配置文件的工作原理和 application.properties 配置文件是一样的，但 YAML 格式的配置文件更为简洁。YAML 配置文件的扩展名可以使用 .yml 或者 .yaml。

YAML 配置文件的编写语法如下。

①“key：(空格)value”格式用于配置属性，其使用缩进控制层级关系，其中空格不可省略，且对大小写是敏感的。

②当在 YAML 配置文件中配置的属性值为普通数据类型时，可以直接配置对应的属性值；对于字符串类型的属性值，不需要额外添加引号，示例代码如下。其中，配置 server 的 port 和 path 属性，port 和 path 处于同一级。

```
server:
  port: 8080
  path: /index
```

③当 YAML 配置文件中配置的属性值为数组或单列集合类型时，主要有两种书写方式：缩进式写法和行内式写法。其中，缩进式写法也有两种。

第一种缩进式写法为“-(空格)属性值”的形式，示例代码如下。

```
student:
  course:
    - Mathematics
    - Chinese
    - English
```

第二种缩进式写法为多个属性值之间加英文逗号分隔，最后一个属性值不添加逗号，示例代码如下。

```
student:
  course:
    Mathematics,
    Chinese,
    English
```

在使用行内式写法时，可使用“[]”将数据包含在其中，也可以将“[]”省略，代码如下。

```
student:
  course: [Mathematics, Chinese, English]
```

④当 YAML 配置文件中配置的属性值为 Map 集合或对象类型时，YAML 文件的格式同样有两种书写方式：缩进式写法和行内式写法。

缩进式写法的示例代码如下。

```
student:
  map:
    key1: value1
    key2: value1
```

在使用行内式写法时，属性值要用大括号“{}”包含，其示例代码如下。

```
student:
  map: {key1: value1, key2: value1}
```

4. application.yaml 配置文件实例

【案例】在上个案例的基础上，使用 application.yaml 配置文件给实体类赋值。

在 resources 中创建 application.yaml 文件，编写 Student 类的配置属性，如示例代码 2-5 所示。

示例代码 2-5：application.yaml 配置文件

```
student:                    // 设置 student 对象
  name: zhang               // 设置 name 属性
  class-name: ClassOne      // 设置 class-name 属性
  id: 1                     // 设置 id 属性
  course: [Chinese, Mathematics, English]        // 设置 course 属性
```

```
    elective：
        name：art            // 设置 elective 对象中 name 属性
        classTime：2020/11/17  // 设置 elective 对象中 classTime 属性
  map：{key1：value1，key2：value2}  // 设置 map 属性 key1 和 key2
```

运行测试类(properties 文件的优先级高，需将 application.properties 配置文件进行注释)，在控制台上输出的信息如图 2-5 所示。

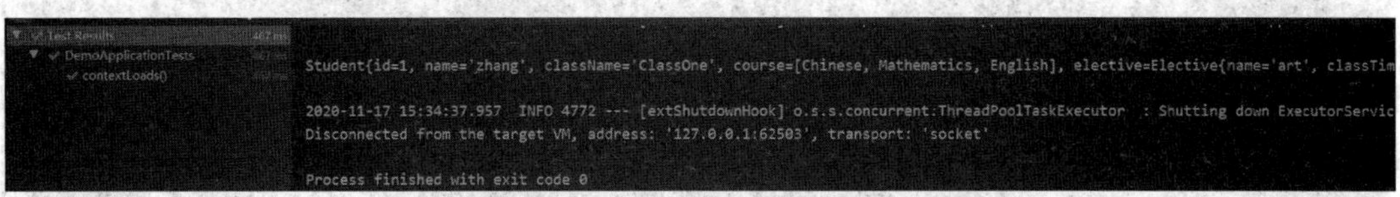

图 2-5　应用 YAML 文件赋值实例

5. 多环境配置

在实际开发中，不同的开发阶段会需要不同的生产环境配置，如开发环境、部署环境和测试环境等。当需要切换生产环境时，手动修改就成为一项庞大的工程，一旦某些参数配置错误，会带来很严重的损失。为了解决这个问题，Profile 应运而生。Profile 可以让 Spring 为不同的环境提供不同配置的功能，可以通过激活、指定参数等方式快速切换环境 。

使用 Profile 进行多环境配置时， Spring 官方文档中给出 Profile 的语法规则为 application-{profile}.properties(.yaml/.yml)。其中，{profile} 对应具体的环境标识，如 test、dev 等。相关的环境配置文件名如下。

```
application-dev.properties     // 开发环境配置
application-test.properties    // 测试环境配置
application-prod.properties    // 生产环境配置
```

当开发人员需要使用对应的环境配置文件时，可在 application.properties 配置文件中使用 spring.profiles.active={profile} 的语法格式来进行激活操作，其代码如下。

```
spring.profiles.active=dev     // 激活开发环境配置
```

【案例】使用 Profile 文件进行多环境配置。

第一步，在 resources 目录下，创建 application-dev.properties 和 application-test.properties 配置文件，并配置服务端口号信息，分别如图 2-6 和图 2-7 所示。

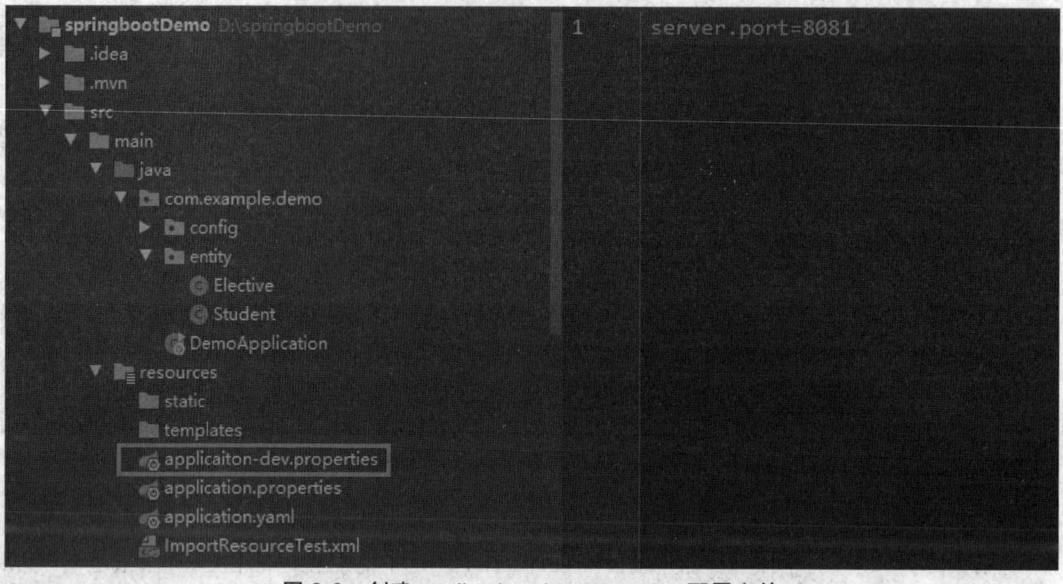

图 2-6 创建 application-dev.properties 配置文件

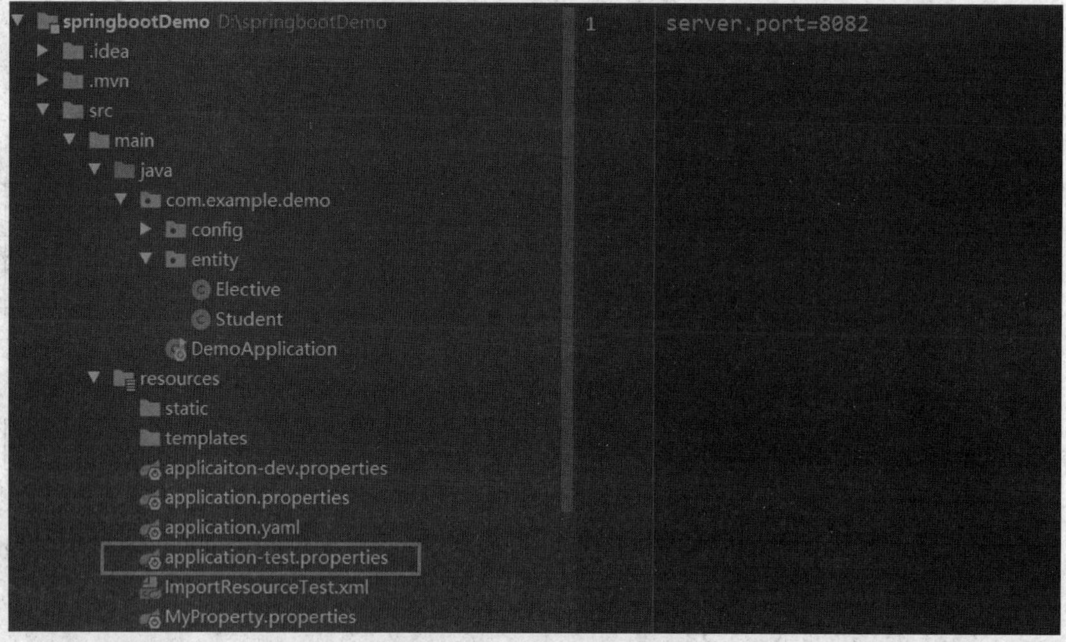

图 2-7 创建 application-test.properties 配置文件

第二步，在 application.properties 中设置 spring.profiles.active=dev（图 2-8），激活对应的 dev 环境配置，并运行测试类。

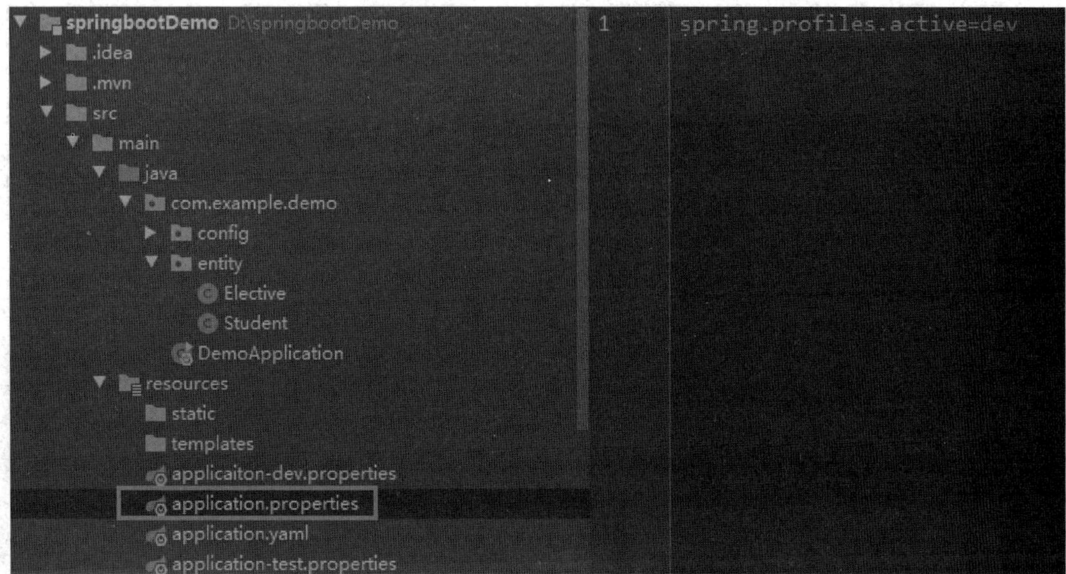

图 2-8　设置 application.properties 配置文件

第三步，运行。运行之后的效果如图 2-9 所示，可见在控制台上显示激活对应的配置文件为 dev。

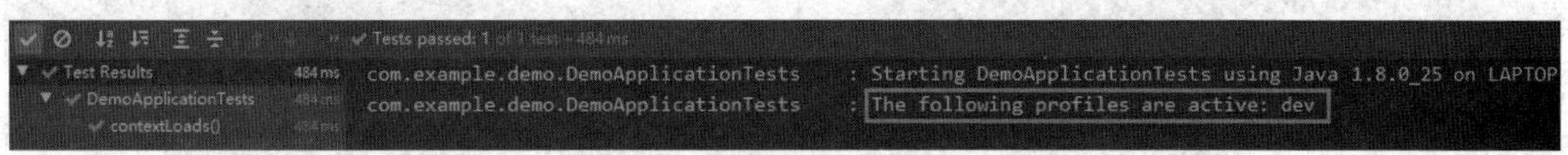

图 2-9　激活 dev 配置文件

技能点 3　配置文件的注入

在使用 Spring Boot 的全局配置文件配置属性时，如果所配置的属性是 Spring Boot 默认提供的属性，如 server.port，那么 Spring Boot 会从内部自动扫描并读取其属性值；如果所配置的是用户自定义的属性，则必须在程序中注入这些属性。

1. 使用 @ConfigurationProperties 注入属性

Spring Boot 提供 @ConfigurationProperties 注解，用于将配置文件中的自定义属性值批量注入 Bean 对象。在实体类中，需要通过编写 Setter 方法，才可以使用对应的注解进行批量注入，如示例代码 2-6 所示。

示例代码 2-6：Student 实体类

```
@Component
@ConfigurationProperties(prefix = "student")
public class Student {
```

```
    private int id;                  //id
    private String name;             // 姓名
    public void setId(int id) {          // 对应的 setter 方法
        this.id = id;
    }
    public void setName(String name) {
        this.name = name;
    }
}
```

在上述代码中，使用 @Component 和 @ConfigurationProperties(prefix =“student”)注解可将配置文件上的属性映射到 Student 类中。

2. 使用 @Value 注入属性

@Value 注解是 Spring 框架提供的，用来读取配置文件中的属性值并逐个注入 Bean 对象的属性中。在 Spring Boot 中，也可以使用该注解读取和注入配置文件的属性值，如示例代码 2-7 所示。

示例代码 2-7：Student 实体类

```
@Component
public class Student {
    @Value("${student.name}")
private String name;             // 姓名
}
```

3. @ConfigurationProperties 和 @Value 的区别

有两种注解可以实现配置文件的注入，分别是 @ConfigurationProperties 和 @Value。这两者之间有一些区别。

（1）底层框架

@ConfigurationProperties 注解属于 Spring Boot 框架，@Value 注解则属于 Spring 框架。由于 Spring Boot 默认支持 Spring 框架，所以可以在 Spring Boot 框架中使用 @Value 注解的相关功能。

（2）功能

@ConfigurationProperties 注解将配置文件中的属性批量注入 Bean 对象，@Value 为单独注入。

（3）松散语法

松散语法指在 Properties 文件中设置属性值时，可变更对应的属性名，如使用“-”、“_”等分隔名称。例如，在 Student 实体类中，设置所属班级为 private String className，其中 className 在属性配置时可以使用如下的配置方式。

```
student.className=ClassOne       // 使用标准方式
student.class-name=ClassOne       // 使用 "-" 分隔单词
student.class_name=ClassOne       // 使用 "_" 分隔单词
STUDENT.CLASS_NAME=ClassOne      // 使用大写和 "_" 分隔单词
```

这种方式只能在应用了 @ConfigurationProperties 注解的情况下使用，@Value 注解不可使用。

（4）SpEL

@Value 注解支持 SpEL 表达式，对应的语法为 @Value（“#{ 表达式 }”）。应用这种方法可以直接注入 Bean 的属性值，而 @ConfigurationProperties 注解不支持此功能。@Value 注解的示例代码如下。

```
@Value（"#{2*7}"）
private int id;            //id
@Value（"#{' 张 Pote'}"）
private String name;       // 姓名
```

（5）JSR303 数据校验

JSR303 数据校验是 @ConfigurationProperties 注解所支持的，主要用来校验 Bean 中的属性是否符合相关的数值规定。如若属性值不符合相关的校验规定，程序就会自动报错。以下是几种基本的校验规则，完整校验规则请另外查阅相关资料。

①空校验。

@Null：校验对象是否为空值（null）。

@NotNull：校验对象是否不为 null。

@NotBlank：校验约束字符串是不是 null 以及被 Trim 去空的长度是否大于 0，只对应字符串，并且会去掉前后的空格。

@NotEmpty：校验对应的约束元素是否为 null 或者是 empty。

② Booelan：校验。

@AssertTrue：校验 Boolean 对象是否为 true。

@AssertFalse：校验 Boolean 对象是否为 false。

③长度校验。

@Size（min=，max=）：校验对象长度是否在给定的范围之内。

@Length（min=，max=）：校验带注释的字符串是否在最小和最大之间。

④日期校验。

@Past：校验 Date 以及 Calendar 对象是否在当前时间之前。

@Pattern：校验 String 对象是否符合正则表达式的规则。

⑤数值校验。

@Min：校验 Number 和 String 对象是否大于等于指定的值。

@Max：校验 Number 和 String 对象是否小于等于指定的值。

@Email：校验是否为邮件地址，如果为 null，则不进行验证，表示通过验证。

（6）Setter 方法

使用 @ConfigurationProperties 注解进行配置文件的属性值读取注入时，必须设置 Setter 方法才能匹配注入对应的 Bean 属性上。@Value 注解进行配置文件属性值注入时，无须设置 Setter 方法，只需通过表达式读取对应的信息，自动注入下方的 Bean 属性。

（7）复杂类型封装

@ConfigurationProperties 注解支持任意数据类型的属性注入，包括复杂类型；@Value 只能注入基本数据类型。

以上对两种注解进行了详细的区别说明，两者的区别总结见表 2-1。

表 2-1　@ConfigurationProperties 和 @Value 的区别

项目	@ConfigurationProperties	@Value
框架	Spring Boot	Spring
功能	批量注入	单个注入
松散语法	支持	不支持
SpEL	不支持	支持
JSR303 数据校验	支持	不支持
Setter 方法	必须	无须
复杂类型封装	支持	不支持

技能点 4　自定义配置

在创建 Spring Boot 项目时自动进行装配，简化了很多手动配置，开发人员可通过修改全局配置文件来适应具体的生产环境。但对于一些特殊的情况，仍需要自定义配置文件，但 Spring Boot 无法自动识别这些自定义的配置文件，这时就需要开发人员手动进行加载。

1. @Configuration 注解

@Configuration 注解标注在类上，用于配置 Spring 容器应用上下文，将该类作为 Spring 的 XML 文件中的“beans”。在该注解的类中，使用标注 @Bean 注解的方法，其返回的类型都会被注册为 Bean。使用 @Bean 注解时，方法名就是组件名。

【案例】使用 @Configuration 注解编写自定义配置类。

第一步，在 com.example.demo 的包下，创建名为 config 的包，该包中创建两个类 myConfig 和 myConfigTest，如示例代码 2-8、2-9 所示，项目结构如图 2-10 所示。

示例代码 2-8：MyConfigTest 类

```
public class MyConfigTest {

    public void start()
    {
        System.out.println(" 开始初始化 ....");
    }

    public void running()
    {
        System.out.println(" 开始运行 ....");
    }

    public void ending(){
        System.out.println(" 结束 .....");
    }
}
```

MyConfig 是 @Configurtion 注解的配置类，该类会被 Spring 容器自动识别，使用 @Bean 注解的是 MyConfigTest()方法，其返回值对象会被添加到 Spring 容器中。

示例代码 2-9：MyConfig 配置类

```
@Configuration  // 使用该注解定义该类为一个配置类
public class MyConfig {
    // 使用 @Bean 标注在方法上，注册该 Bean 对象到 Spring 容器，id 为默认的方法名
    @Bean
public MyConfigTest myConfigTest(){
        System.out.println(" 执行 myConfigTest 初始化 ");
        return new MyConfigTest();
    }
}
```

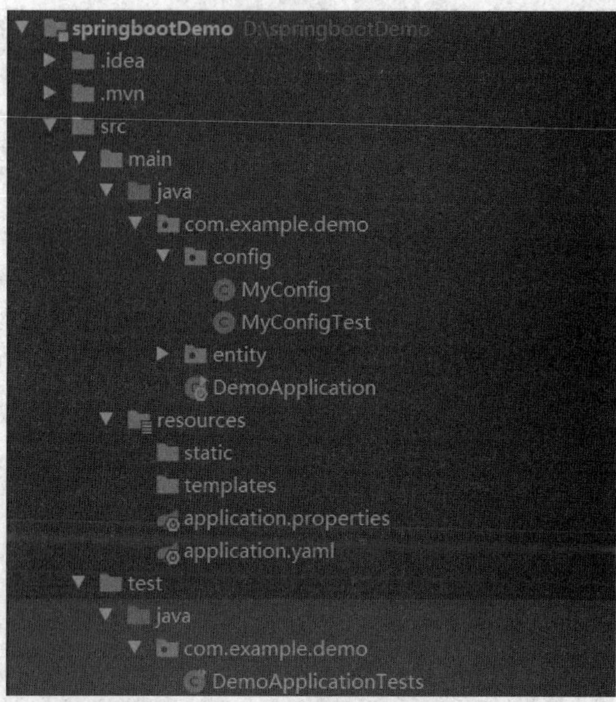

图 2-10　项目结构图

第二步，编写 DemoApplicationTests 测试类，在完成了配置文件的定义后，需要对声明的配置类进行测试，如示例代码 2-10 所示。

示例代码 2-10：DemoApplicationTests 测试类

```
@SpringBootTest
class DemoApplicationTests {
  @Autowired
  private ApplicationContext applicationContext;
  @Test
void contextLoads() {
//@Configuration 注解的 Spring 容器加载方式
    ApplicationContext context = new AnnotationConfigApplicationContext(MyConfig.
class);
     // 获取 Bean，调用对应的方法
          MyConfigTest myConfigTest = (MyConfigTest) context.getBean("myCon-
figTest");
     myConfigTest.start();
     myConfigTest.running();
     myConfigTest.ending();
  }
}
```

第三步，查看控制台。可以看到输出“执行 myConfigTest 初始化”，表示 @Configurtion 注解应用成功，Spring 容器已经成功启动了。之后出现的“开始初始化 ...”等提示词，表示 @Configuration 已启动容器以及已成功应用 @Bean 注解注册 Bean，如图 2-11 所示。

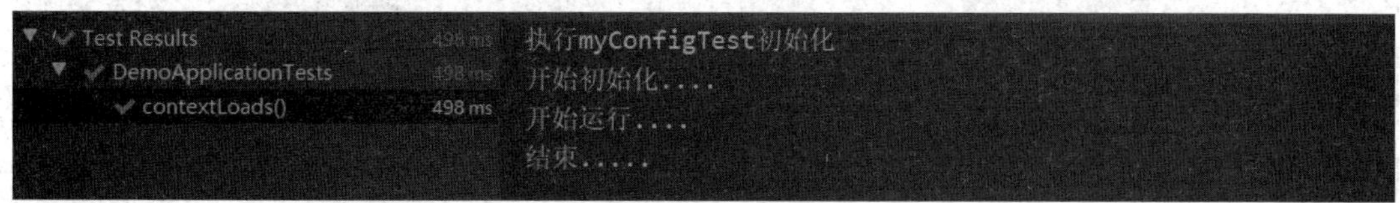

图 2-11　测试效果

2. @PropertySource 注解

加载自定义配置文件可以使用 @PropertySource 注解和 @Configurtion 注解来实现。@PropertySource 主要用于指定自定义配置文件的位置和名称，@Configurtion 可以用来指定类为自定义配置类。

【案例】应用 @PropertySource 注解加载自定义配置文件。

第一步，在 resource 目录下创建 MyProperty.properties 文件，在配置文件中编写 Student 属性信息，如示例代码 2-11 所示。

示例代码 2-11：MyProperty.properties 配置文件

```
student.id=10  // 设置 id 属性
student.name=wang   // 设置 name 属性
student.className=ClassTwo   // 设置 className 属性
student.course=Mathematics，Chinese，English  // 设置 course 属性
student.elective.name=music   // 设置 elective 实体类的 name 属性
student.elective.classTime=2020/11/19   // 设置 elective 实体类的 classTime 属性
student.map.key1=value1  // 设置 key1 属性
student.map.key2=value2  // 设置 key2 属性
```

第二步，继续沿用 Student 类，并修改其注解配置。添加 @PropertySource 注解和 @Configurtion 注解，将对应的类路径（classpath）设置为 MyProperty.properties，如示例代码 2-12 所示。

示例代码 2-12：Student 实体类

```
@Component
// 指明配置源文件位置
@PropertySource(value = {"classpath:MyProperty.properties"})
// 将配置文件中开头名为 student 的属性值注入该配置类中
@ConfigurationProperties(prefix = "student")
public class Student {
  private int id;             //id
  private String name;            // 姓名
```

```
    private String className;           // 所属班级
    private String[] course;            // 所上课程
    private Elective elective;          // 选修课程
    private Map map;
 // 省略 setter()和 getter()方法
 // 省略 toString()方法
 }
```

第三步，在对应的测试类中引入 Student 对象，如示例代码 2-13 所示。

示例代码 2-13：DemoApplicationTests 测试类

```
@SpringBootTest
class DemoApplicationTests {
    @Autowired
    private Student student;
    @Test
    void contextLoads() {
        System.out.println(student);
    }
}
```

第四步，执行该测试类，在控制台上显示的效果，如图 2-12 所示。

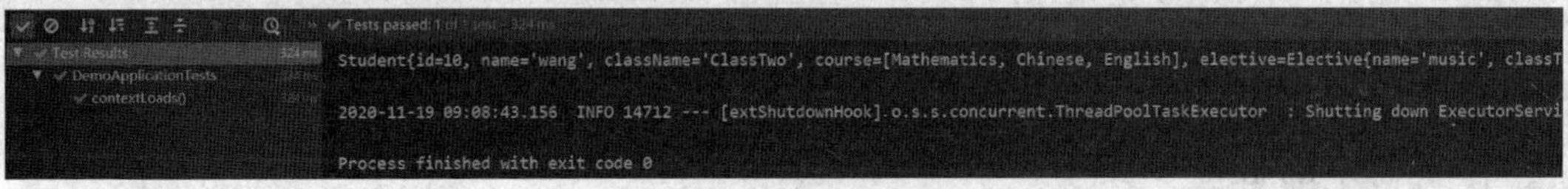

```
Tests passed: 1
Test Results
  DemoApplicationTests
    contextLoads()
Student{id=10, name='wang', className='ClassTwo', course=[Mathematics, Chinese, English], elective=Elective{name='music', classT
2020-11-19 09:08:43.156  INFO 14712 --- [extShutdownHook] o.s.s.concurrent.ThreadPoolTaskExecutor  : Shutting down ExecutorServi
Process finished with exit code 0
```

图 2-12　控制台效果

结果表明，使用 @PropertySource 注解和 @Configurtion 注解后，可将配置文件中所设置的属性值显示在控制台上。

3. @ImportResource 注解

对于传统的 XML 配置文件，在 Spring Boot 项目中同样可以使用 @ImportResource 注解进行手动加载。

@ImportResource 注解通常放置在启动类上，可在注解中编写 locations =“classpath: ...”来标记 XML 配置文件的路径和名称。

【案例】使用 @ImportResource 注解加载 XML 配置文件。

第一步，在 resources 目录下，创建 ImportResourceTest.xml 文件，编写对应的约束语句以及 <bean> 标签。对应 Student 实体类，注入属性值数据，如示例代码 2-14 所示。

示例代码 2-14：Student 实体类

```
public class Student {
  private int id;                //id
  private String name;           // 姓名
  private String className;        // 所属班级
  private String[] course;         // 所上课程
private Elective elective;         // 选修课程
private Map map;
// 省略 setter()和 getter()方法
// 省略 toString()方法
}
```

ImportResourceTest.xml 文件，如示例代码 2-15 所示。

示例代码 2-15：ImportResourceTest.xml 文件

```
<? xml version="1.0" encoding="UTF-8"? >
<beans xmlns="http://www.springframework.org/schema/beans"
    xmlns:xsi="http://www.w3.org/2001/XMLSchema-instance"
      xsi: schemaLocation="http: //www.springframework.org/schema/beans http: //www.
springframework.org/schema/beans/spring-beans.xsd">
  <bean id="student" class="com.example.demo.entity.Student">
    <property name="name" value="zhang"/>
    <property name="className" value="ClassThree"/>
  </bean>
</beans>
```

第二步，编写 XML 配置文件之后，修改启动类 DemoApplication 注解，添加 @ImportResource（locations = "classpath: ImportResourceTest.xml"），如示例代码 2-16 所示。

示例代码 2-16：DemoApplication 启动类

```
@ImportResource(locations = "classpath: ImportResourceTest.xml")
@SpringBootApplication
public class DemoApplication {
  public static void main(String[] args) {
    SpringApplication.run(DemoApplication.class, args);
  }
}
```

第三步，编写测试类 DemoApplicaitonTests，如示例代码 2-17 所示。

示例代码 2-17：DemoApplicaitonTests 测试类

```
@SpringBootTest
class DemoApplicationTests {
    @Autowired
    private Student student;
    @Test
    void contextLoads() {
        System.out.println(student);
    }
}
```

第四步，运行该测试类，在控制台上显示的效果，如图 2-13 所示。

```
Test Results
  DemoApplicationTests
    contextLoads()
Student{id=0, name='zhang', className='ClassThree', course=null, elective=null, map=null}

2020-11-19 09:25:26.156  INFO 2760 --- [extShutdownHook] o.s.s.concurrent.ThreadPoolTaskExecutor  : Shutting
Disconnected from the target VM, address: '127.0.0.1:52097', transport: 'socket'
```

图 2-13 控制台效果

通过控制台可以看到，在 Student 中注入了 name 和 className 的属性值，说明成功应用 @ImportResource 注解将 XML 配置文件注入到了 Spring 容器中。

在项目一的基础上，完善个人博客项目，配置对应的项目环境。在 resources 目录下，除了 application.yml 默认的配置文件之外，继续创建新的配置文件 application-jdbc.properties，对应的项目结构如图 2-14 所示。

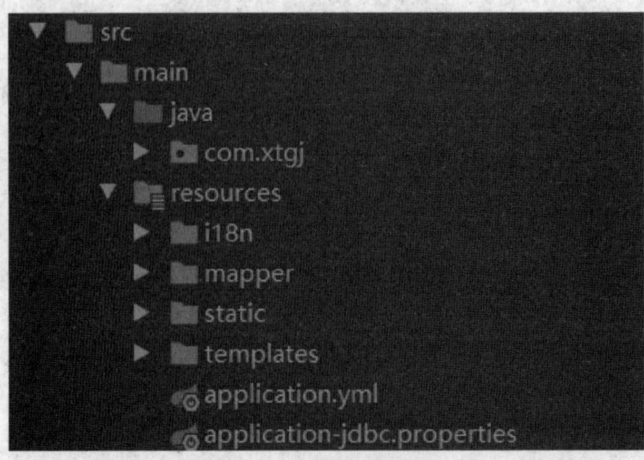

图 2-14 项目结构图

第一步，编写 application.yml 配置文件，如示例代码 2-18 所示。

示例代码 2-18：application.yml 配置文件

```
server:
  port: 8085
spring:
  profiles:
    // 配置 jdbc、redis 和 mail 配置文件
    active: jdbc,redis,mail
  // 关闭 thymeleaf 页面缓存
  thymeleaf:
    cache: false
  // 配置国际化资源文件
  messages:
    basename: i18n.logo
  security:
    user:
      password: 1234
      name: admin
      roles: admin

// MyBatis 配置
mybatis:
  configuration:
    // 开启驼峰命名匹配映射
    map-underscore-to-camel-case: true
  // 配置 MyBatis 的 XML 映射文件路径
  mapper-locations: classpath:mapper/*.xml
  // 配置 XML 映射文件中指定的实体类别名路径
  type-aliases-package: com.xtgj.model.domain
//pagehelper 分页设置
pagehelper:
  helper-dialect: mysql
  reasonable: true
  support-methods-arguments: true
  params: count=countSql
```

```
// 浏览器 Cookie 相关设置
COOKIE：
  // 设置 Cookie 默认时长为 30 分钟
  VALIDITY：1800
```

第二步，编写 application-jdbc.properties 配置文件，用于配置数据库连接池以保证项目的效率，配置连接数据库的参数，如示例代码 2-19 所示。

示例代码 2-19：application-jdbc.properties 配置文件

```
// 添加并配置第三方数据库连接池 druid
spring.datasource.type = com.alibaba.druid.pool.DruidDataSource
spring.datasource.initialSize=20
spring.datasource.minIdle=10
spring.datasource.maxActive=100

// 数据源连接配置
spring.datasource.url = jdbc：mysql：//localhost：3306/blog_system？serverTimezone=UTC&useSSL=false
spring.datasource.username = root
spring.datasource.password = root
//driver-class-name 可以省略
//spring.datasource.driver-class-name = com.mysql.jdbc.Driver
```

本任务主要对个人博客项目进行配置，设置了端口号、数据库连接等内容。本任务使读者加深了对于配置文件编写方法的理解，也掌握了设置 Spring Boot 项目的方法。

Target	目标	Retention	保留
Inherited	继承	Initializr	初始化器
Configuration	配置		

一、填空题

1. Spring Boot 项目的全局配置文件默认位于 ________ 目录下。

2. 对于 Spring Boot 项目，项目构建之后使用 ________ 注解对启动类进行配置，即可启动项目。

3. 对于 Spring Boot 项目，若使用 application.yaml 配置文件对项目进行配置，其内容应使用 ________ 格式对属性进行配置。

4. 使用 __________ 注解支持 SPEL 表达式。

5. 在 JSR303 数据校验中，应使用 _________ 注解，对对象是否为 null 进行校验。

二、选择题

1. 下列关于 YAML 配置文件的说法，正确的是（　　）

A. YAML 配置文件的写法配置单列集合属性，其中属性值的大括号“[]”可以省略

B. YAML 配置文件的写法配置双列集合属性，其中属性值的中括号“{}”可以省略

C. YAML 配置文件的内容是“key: value”形式的键值对，并使用缩进式写法

D. 以上说法都不正确

2. 下列关于 @ConfigurationProperties 注解的说法，正确的是（　　）

A. 使用 @ConfigurationProperties 注解对 Bean 进行注入时，必须设置 Setter 方法

B. @ConfigurationProperties 注解只能作用于类上

C. @ConfigurationProperties 注解必须与 @Component 注解结合使用

D. 要想使 @ConfigurationProperties 注解生效，必须使用 @EnableConfigurationProperties 注解

3. 下列关于 @ConfigurationProperties 注解和 @Value 注解的说法，正确的是（　　）

A. @ConfigurationProperties 和 @Value 均是 Spring 框架自带的注解

B. 进行属性注入时，@ConfigurationProperties 和 @Value 均需要 Setter 方法

C. @ConfigurationProperties 将配置文件中的属性批量注入 Bean 对象，@Value 为单独注入

D. @ConfigurationProperties 注解支持 SPEL 表达式

三、简答题

简要叙述多环境配置的方式。

项目三 个人博客项目主页部分

通过学习基本的应用架构，了解 Spring Boot 中 Web 的基本概念，学习 Thymeleaf 模板的基本概念和特点，熟悉 Spring Boot 控制器，能够运用所学的相关知识，编写个人博客项目主页部分的内容。

- 掌握 Spring Boot 的 Web 开发支持。
- 掌握 Thymeleaf 模板。
- 掌握 Spring Boot 控制器。
- 了解 Spring Boot 模型概念。
- 掌握 Spring Boot 常用组件。
- 掌握 Spring Boot 文件上传与下载。

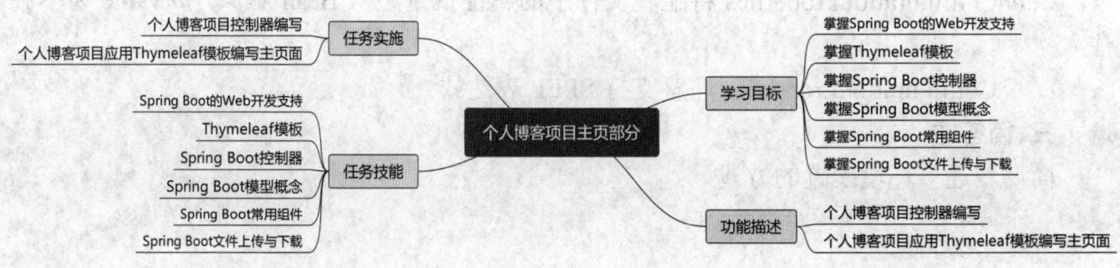

【情境导入】

对于 Web 项目,如何让用户动态地浏览到信息是其开发的重点之一。在网络发展迅速的今天,各种组件框架层出不穷,其中 Spring Boot 支持 Web 项目并整合了 MVC 框架、动态模板 Thymeleaf 和 Servlet 组件等,实现动态显示信息的功能,增强用户的体验。

【功能描述】

- 个人博客项目控制器编写。
- 应用 Thymeleaf 模板编写个人博客项目的主页面。

技能点 1　Spring Boot 的 Web 开发支持

Spring Boot 遵循规约大于配置的思想,因此大大简化了配置的复杂程度。主要表现在消除了 web.xml 和依赖注入配置的整合,经过 Spring Boot 的封装,开发人员就不需要去关注底层的实现,只需专注于业务代码的实现。但对于一些特殊需求来说,自行配置 web.xml 也是被支持的。

Spring Boot 使用 Starter 来提供依赖管理,有利于进行快速的 Web 开发。在对应的 pom.xml 文件中,编辑如下代码,即可注入依赖。

```
<dependency>
    <groupId>org.springframework.boot</groupId>
    <artifactId>spring-boot-starter-web</artifactId>
</dependency>
```

如今 Spring Boot 提供了一套完整的 Web 开发流程,其中使用 Spring 开发一个 Web 工程有以下两条路线。

①前后端分离。

在这种方式下，前端开发和后端开发完全分离，只需要协商好接口就行，前端负责开发页面并调用后端接口展示数据，后端只负责提供接口即可。

②使用 Spring Boot 自带的模板。

Spring Boot 支持多种主流模板，如 FreeMarker、Thymeleaf、Groovy 等。

1. Spirng Boot 整合支持 MVC

Spring Boot 为了实现并简化 Web 的开发过程，整合了 Spring MVC，在开发中只需引入对应的 Web 开发框架即可。

应用 Spring Boot 的自动配置特性，在开发 Spring MVC Web 时会十分便捷。Spring Boot 其提供了以下的自动配置，来完成对于 MVC 的支持。

①内置 ContentNegotiatingViewResolver 和 BeanNameViewResolver 两种视图解析器。

②支持静态资源。

③自动注册了 Converter 转换器和 Formatter 格式化器。

④获取所有的 HttpMessageConverter 并转换 Http 请求和响应。

⑤自行注册了消息代码解析器。

⑥定义错误代码生成规则。

在任何时候，如果开发人员对 Spring Boot 默认提供的 Spring MVC 组件设定不满意，都可以进行修改、扩展或替换。操作方法列举如下。

①在 IOC 容器中注册新的同类型的 Bean 定义。

②编写基于 WebMvcConfigurer 接口的 Bean 定义。

③编写基于 WebMvcConfigurationSupport 的 Bean 定义。

④使用 @EnableWebMvc 的 @Configuration 配置类完全接管所有 Spring MVC 的相关配置，自己完全重新配置。

【案例】编写 MyWebMvcConfig 配置类，实现 WebMvcConfigurer 接口以及使用 @EnableWebMvc 注解全面接管完成 Spring MVC 自动配置。

第一步，创建 MyWebMvcConfig 配置类，来扩展 Spring MVC，实现 WebMvcConfigurer 接口。使用该接口的 addViewControllers()语句，实现将发送过来的请求全部跳转至 test 页面，如示例代码 3-1 所示。

示例代码 3-1：MyWebMvcConfig 类

```
@Configuration
public class MyWebMvcConfig implements WebMvcConfigurer{
  @Override
  public void addViewControllers(ViewControllerRegistry registry) {
        registry.addViewController("/").setViewName("test");
// 浏览器发送请求来到 test 页面
  }
}
```

第二步，若想全面接管 Spring MVC 的自动配置，则在对应的自定义配置 MyWebMvc-

Config 类上使用 @EnableWebMvc 注解。此时 Spring Boot 中关于 Spring MVC 的所有配置全部失效，只会使用有开发人员自定义注解的自定义配置，如示例代码 3-2 所示。

示例代码 3-2：MyWebMvcConfig 类

```
@EnableWebMvc
@Configuration
public class MyWebMvcConfig implements WebMvcConfigurer{
    @Override
    public void addViewControllers(ViewControllerRegistry registry){
        registry.addViewController("/").setViewName("test"); // 浏览器发送请求来到test 页面
    }
}
```

Model 模型层的简介如下。

对于 Spring Boot 的 MVC 模块，其中的模型（Model）表示企业数据和业务逻辑，它是应用程序的主体部分。业务流程的处理过程对其他层来说是无关紧要的，模型接收视图请求数据，并返回最终的处理结果。

数据模型也是业务模型中非常重要的部分之一，数据模型主要指实体对象的数据持续化，如一张订单保存到数据库，从数据库获取订单，将这个模型单独列出，所有相关数据的操作只限制在该模型层中。

Model 模型主要包括实体类（Entity）、业务层（Service）和数据访问层（Dao），在需要访问数据库时，会经由 Controller 调用对应的业务层，再通过 Dao 访问数据库，获得对应的数据，然后返回给 Controller 至前端 view 页面。

2. Spring Boot 支持的视图技术

运用前端模板引擎，使前端的开发人员无须关注后端的业务逻辑编写实现，只专注于前端显示数据和呈现效果即可，这样的设计就解决了前后端交织在一起的情况。Spring Boot 基于这种开发目的，对很多模板引擎提供了支持，可以通过注入依赖关系在项目中使用这些模板引擎。具体介绍如下。

FreeMarker：一款模板引擎，是一个 Java 类库，为开发产品的组件，即一种基于模板和要改变的数据，并用来生成输出文本（HTML 网页、电子邮件、配置文件、源代码等）的通用工具。其 logo 如图 3-1 所示。

图 3-1　FreeMarker 的 logo

Thymeleaf：一款用于渲染 XML/XHTML/HTML5 内容的模板引擎，是用来开发 Web 和独立环境项目的服务器端的 Java 模板引擎，类似 JSP、Velocity、FreeMaker 等。它可以轻易地与 Spring MVC 等 Web 框架进行集成并作为 Web 应用的模板引擎。其 logo 如图 3-2 所示。

图 3-2 Thymeleaf 的 logo

Groovy：一种基于 JVM（Java 虚拟机）的敏捷开发语言，它结合了 Python、Ruby 和 Smalltalk 的许多强大的特性。Groovy 代码能够与 Java 代码很好地结合，也能用于扩展现有代码。由于其运行在 JVM 上的特性，Groovy 也可以使用其他非 Java 语言编写的库。其 logo 如图 3-3 所示。

图 3-3 Groovy 的 logo

Mustache：一款 Java Script 模板，可以应用于 HTML、配置文件、源代码等任何地方。它没有固定的逻辑模板语法。模板形式为包裹住内容，格式中写入的是模板的内容，被替换的内容字段使用双花括号包裹起来“{{}}”。其 logo 如图 3-4 所示。

Mustache

图 3-4 Mustache 的 logo

JSP：JSP（JavaServer Pages）是以 Java 语言作为脚本语言主导所创建的一种动态网页技术标准。JSP 部署于网络服务器上，以响应客户端发送的请求，并根据请求内容动态地生成文档或者网页，然后返回给请求者。其 logo 如图 3-5 所示。

图 3-5 JSP 的 logo

对于 JSP 技术，Spring Boot 官方虽然支持但不建议使用，所以在应用 JSP 时会有一些限制。以下为官方文档中给出的一些限制条件。

① 对于 Jetty 和 Tomcat，如果使用 war 包装，它应该可以工作。与 java -jar 一起启动时，可执行的 war 将起作用，并且还可部署到任何标准容器中。使用可执行 jar 时，不支持 JSP。

② Undertow 不支持 JSP。

③ 创建定制的 error.jsp 页面不会覆盖默认视图以进行错误处理，应改用自定义错误页面。

技能点 2　Thymeleaf 模板

1. Thymeleaf 模板的基本语法

Thymeleaf 提供了一种可被浏览器正确显示，且格式优雅的模板创建方式，因此也可以用作静态建模。开发人员可以使用它创建经过验证的 XML 与 HTML 模板。使用 Thymeleaf 创建的 HTML 页面可以直接在浏览器上打开并展示其静态资源，这有利于前后端分离的开发方式。

使用 Thymeleaf 模板首先需要添加对应的依赖，代码如下。

```
<dependency>
  <groupId>org.springframework.boot</groupId>
  <artifactId>spring-boot-starter-thymeleaf</artifactId>
</dependency>
```

Thymeleaf 会对 HTML 中的标签进行严格的验证，如果缺少部分标签就会报错，可以通过以下依赖去除这一验证，添加的依赖内容如下。

```
<dependency>
    <groupId>net.sourceforge.nekohtml</groupId>
    <artifactId>nekohtml</artifactId>
    <version>1.9.22</version>
</dependency>
```

在全局配置文件中，可以对 Thymeleaf 模板的参数进行一些配置，代码如下。

```
spring.thymeleaf.cache=true          // 启动模板缓存
spring.thymeleaf.encoding=UTF-8      // 将模板确定为 UTF-8
spring.thymeleaf.mode=HTML5          // 确定模板模式为 HTML5
spring.thymeleaf.prefix=classpath:/resources/templates/ // 指定模板页面存放路径
spring.thymeleaf.suffix=.html        // 指定对应模板页面名称后缀
```

2. Thymeleaf 模板的常用标签

在 HTML 页面上使用 xmlns 属性引入 Thymeleaf 标签。xmlns 属性定义一个或多个可供选择的命名空间。该属性可以放置在文档内任何元素的开始标签中。表现形式类似于 URL，定义了一个命名空间，之后浏览器会将此命名空间用于该属性所在元素内的所有内容。

使用 xmlns:th="http://www.thymeleaf.org" 引入 Thymeleaf 模板，代码如下。

```
<html lang="en" xmlns:th="http://www.thymeleaf.org">
```

Thymeleaf 模板标签的形式为“th:”，具体标签内容见表 3-1。

表 3-1 Thymeleaf 模板标签

标签	功能	示例
th:id	替换 id	<irput th:id="'xxx'+ ${collect.id}"/>
th:text	文本替换	<p th:text="${collect.description]">description</p>
th:utext	支持 html 的文本替换	<p th:utext="${htmlcontent}">conten</p>
th:object	替换对象	<div th:object="$session.user}"/>
th:value	属性赋值	<input th:value="${user.name}"/>
th:with	变量赋值运算	<div th:with="sum=${person.number}%2==0"></div>
th:style	设置样式	th:style="'display:'"
th:onclick	点击事件	th:onclick="'getSum()'"
th:each	属性赋值，如果没有显式地设置状态变量，则 Thymeleaf 将始终创建一个默认的迭代变量，该状态迭代变量名称为：迭代变量 +“Stat”	<tr th:each="user，userStat:${users}">
th:if	判断条件	<a th:if="${userId==collect.userId}">
th:unless	和 th:if 判断相反	<a th:href="@{/login]" th:unless=${session.user!=null]>Login</a>
th:href	链接地址	<a th:href="@{/login]" th:unless=${session.user!=null]>Login</a>
th:switch	多路选择配合 th:case 使用	<div th:switch="${user.role}">
th:case	th:switch 的一个分支	<p th:case="'admin'">admin</p>
th:include	布局标签，替换内容到引入的文件	<head th:include="layout::htmlhead"th:with="title='xxx'"></head>/>
th:replace	布局标签，替换整个标签到引入的文件	<div th:replace="fragments/header::title"></ div>
th:selected	selected 选择框选中	th:selected="(${xxx.id} == ${configObj.id})"
th:src	图片类地址引入	<img th:src="@{/img/logo.png}"/>

续表

标签	功能	示例
th:inline	定义 js 脚本可以使用变量	<script type="text/javascript" th:inline="javascript">
th:action	表单提交的地址	<form th: action="@{/subscribe}">
th:remove	删除某个属性	<tr th: remove="all"> 删除包含标签和所有的孩子

表 3-1 列出了部分 Thymeleaf 的属性标签，对于其中示例的标准表达式语法，有很多的语法表示，见表 3-2。

表 3-2　Thymeleaf 的语法表示

说明	表达式语法
变量表达式	${....}
消息表达式	#{....}
URL 表达式	@{....}
选择变量表达式	*{....}
片段表达式	~{....}

表达式的具体用法和含义如下。

（1）变量表达式 "${...}"

变量表达式主要用于获取上下文中的变量值，其示例代码如下。

```
<p th:text="${sum}"> 总和 </p>
```

上述代码中使用了 Thymeleaf 模板的变量表达式 ${...} 用来动态获取 p 标签中的内容，如果当前程序没有启动或者当前上下文中不存在 sum 变量，该 p 标签中会显示标签默认值“总和”；如果当前上下文中存在 sum 变量并且程序已经启动，当前 p 标签中的默认文本内容将会被 sum 变量的值所替换，从而达到动态替换模板引擎页面数据的效果。

（2）消息表达式 "#{…}"

消息表达式 #{…} 主要用于 Thymeleaf 模板页面国际化内容的动态替换和展示，使用消息表达式 #{…} 进行国际化设置时，还需要提供一些国际化配置文件。

（3）URL 表达式 "@{…}"

URL 链接表达式 @{…} 一般用于页面跳转或者资源引入，在 Web 开发中占据着非常重要的地位，并且使用也非常频繁。

（4）选择变量表达式 "*{...}"

选择变量表达式与变量表达式功能相似但是有一个重要的区别，*{...} 评估所选对象的表达式而不是整个上下文。若没有选定的对象，变量表达式和选择变量表达式的语法就会完全相同。

（5）片段表达式 "~{...}"

片段表达式用于标记一个片段模板，并根据需要，移动或者传递给其他模板，最常见用法为使用 th: fragment 属性来定义被包含的模板片段，以供其他模板使用，再使用 th: insert 或者 th:replace 属性插入一个片段。其有三个不同的格式。

①∽ {templatename：: selector}。其中 templatename 表示模板名称，即 Spring Boot 项目中 resources 目录下 templates 文件夹中的 HTML 文件的名称，它根据 Spring Boot 对 Thymeleaf 的规则进行映射， selector 既可以是 th: fragment 定义的片段名称，也可以是选择器，如标签的 id 值、CSS 选择器或者 XPath 等。

②∽ {templatename}，包含名为 templatename 的整个模板。

③∽ {：: selector} 或∽ {this：: selector}，包含在同一模板中的指定选择器的片段。

具体使用方法如下。

代码定义了一个名为 myThymeleaf 的片段，然后可以使用 th: insert 或 th: replace 属性（Thymeleaf 3.0 不再推荐使用 th: include）包含进需要的页面中，示例代码如下。

```
<div th: fragment="myThymeleaf">
    Hello，myThymeleaf！
</div>
<body>
    ... <div th: insert="~{header：: myThymeleaf}"></div>    ...
</body>
```

common/index 为模板名称，此时 Thymeleaf 会查询 resources 目录下的 index 模板页面，title 即为片段的名称。

3. Thymeleaf 模板应用实例

【案例】应用 Thymeleaf 模板，将学生信息显示在浏览器上，当显示静态页面时，无须运行整体项目，效果如图 3-6 所示。

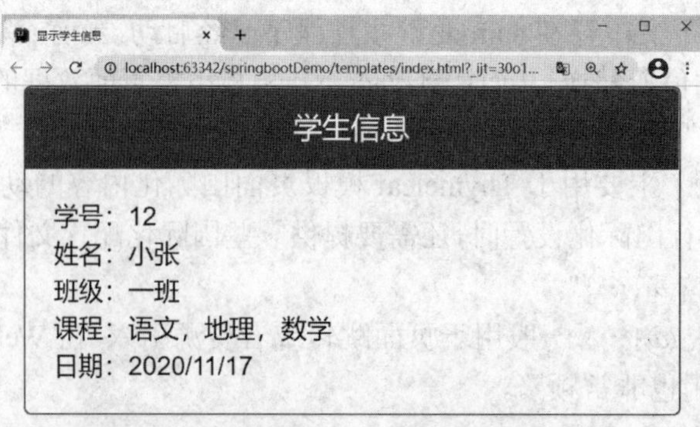

图 3-6 静态应用 Thymeleaf 模板显示信息

第一步，编写 index.html 页面，如示例代码 3-3 所示。

示例代码 3-3：index.html 页面

```
<! DOCTYPE html>
<html lang="en" xmlns:th="http://www.thymeleaf.org">
<head>
    <title> 显示学生信息 </title>
    <meta charset="UTF-8"/>
    <meta name="viewport" content="width=device-width，initial-scale=1"/>
        <link th:href="@{/css/bootstrap.css}" href="../static/css/bootstrap.css" rel="-
stylesheet"/>
    <link th:href="@{/css/bootstrap-theme.css}" href="../static/css/bootstrap-theme.css"
        rel="stylesheet"/>
    <script src="https://s3.pstatp.com/cdn/expire-1-M/jquery/3.1.1/jquery.min.js"></script>
    <script th:src="@{/js/bootstrap.js}" src="../static/js/bootstrap.js"></script>
</head>
<body>
<div class="container">
    <div class="row">
        <div class="col-md-4">
            <div class="panel panel-primary">
                <div class="panel-heading text-center">
                    <span class="panel-title"> 学生信息 </span>
                </div>
                <div class="panel-body">
                    学号:<span th:text="#{student.id}">12</span><br/>
                    姓名:<span th:text="#{student.name}"> 小张 </span><br/>
                    班级:<span th:text="#{student.className}"> 一班 </span><br/>
                    课程:<span th:text="#{student.course}"> 语文，地理，数学 </span><br/>
                    日期:<span th:text="#{student.elective.classTime}">2020/11/17</span><br/>
                </div>
            </div>
        </div>
    </div>
</div>
</body>
</html>
```

第二步，应用 Thymeleaf 模板，使用 th: text="#{...}" 将 MyProperty.properties 文件中 student 数据信息显示在页面上。

MyProperty.properties 配置文件如示例代码 3-4 所示。

示例代码 3-4：MyProperty.properties 配置文件

```
student.id=10
student.name=wang
student.className=ClassTwo
student.course=Mathematics，Chinese，English
student.elective.name=music
student.elective.classTime=2020/11/19
student.map.key1=value1
student.map.key2=value2
```

第三步，由于需要动态获取对应的信息，需要应用 Web 开发的有关知识，编写最简单的 Controller 控制器并渲染名为 index 页面，如示例代码 3-5 所示。具体的 Controller 控制器知识会在后续的技能点中进行讲解。

示例代码 3-5：ControllerTest 类

```
@Controller
public class ControllerTest {
    @RequestMapping（"/index"）
    public String IndexView（Model model）{
        return "index"；
    }
}
```

第四步，在浏览器中输入对应的地址，显示 index.html 页面。由于 index.html 页面读取 Properties 文件中的属性，Spring Boot 会检查当前使用者的语言，会产生乱码现象，效果如图 3-7 所示。

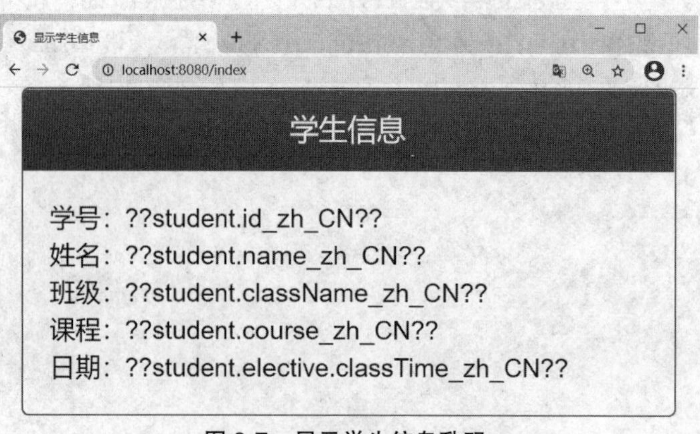

图 3-7 显示学生信息乱码

解决此类问题需要进行相关的配置，此处有两种解决方法。

第一种，编写解决，即使用自定义配置类的方法重写 ResourceBundleMessageSource 中的方法，加载 Properties 文件，如示例代码 3-6 所示。

示例代码 3-6：ControllerTestConfig 配置类

```
@Configuration
public class ControllerTestConfig {
  @Bean
public ResourceBundleMessageSource messageSource() {
  // 实例化对象，对于支持国际化的应用程序，能够为不同的语言环境解析文本消息
        ResourceBundleMessageSource messageSource = new ResourceBundleMessage-
Source();
    messageSource.setUseCodeAsDefaultMessage(true);
    messageSource.setFallbackToSystemLocale(false);
    messageSource.setBasename("MyProperty");// 将 basename 指向 MyProperty 文件
    messageSource.setDefaultEncoding("UTF-8");
    messageSource.setCacheSeconds(2);
    return messageSource;
  }
}
```

第二种，使用 Spring Boot 的自动装配特性，即在 application.properties 文件中应用 spring.messages.basename 关键字对配置文件进行装配，也可以进行字符编码、缓存等内容的设置，如示例代码 3-7 所示。

示例代码 3-7：application.properties 配置文件

```
spring.messages.basename=MyProperty  # 如有多个文件可以使用逗号进行分隔
spring.messages.encoding=UTF-8
```

上述的方法任选其一，完成后，再次访问 index.html 页面，即可正确显示配置文件信息。正确显示后，配置国际化学生信息，将学号、姓名、班级、课程和日期的内容分别改为“10”“wang”“ClassTwo”“Mathematics，Chinese，English”和“2020/11/19”，再次访问 index.html 页面。效果如图 3-8 所示。

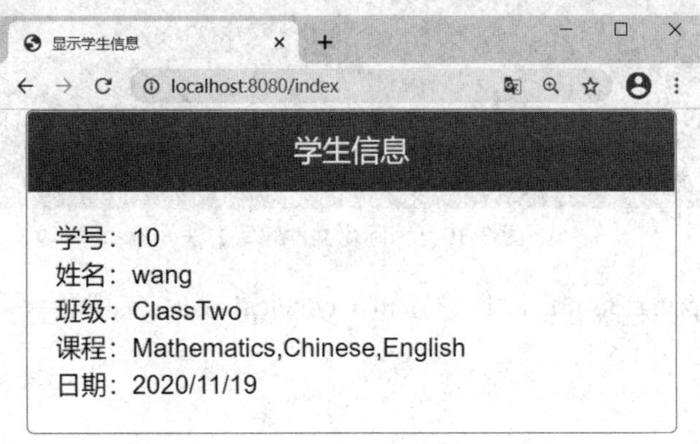

图 3-8　配置国际化学生信息

4. Spring Boot 国际化配置

国际化（简称为 i18n）是设计和制造容易适应不同区域要求的产品的一种方式。它要求从产品中抽离所有地域语言、国家 / 地区和文化相关的元素。换言之，应用程序的功能和代码设计无须做大的改变就能够适应不同的语言和地区的需要，其代码简化了不同本地版本的生产。

在对应的项目 resources 目录下创建名为 i18n 的文件夹，在其中创建 Properties 文件：login.properties、login_en_US 和 login_zh_CN。Spring Boot 默认识别的语言配置文件路径为 resources 下的 login.properties，其国际化语言文件名称必须按照“文件前缀名称 _ 语言代码 _ 国家代码 .properties”的形式进行命名。创建完成后，如图 3-9 所示。

图 3-9 国际化目录

打开 login.properties 文件，选择“Resource Bundle”，切换至编写国际化的文件模式，均以 key-value 的形式进行编写，点击左上角“+”可添加 key 值，相对应的不同语言 value 效果在页面右侧进行编写，效果如图 3-10 所示。

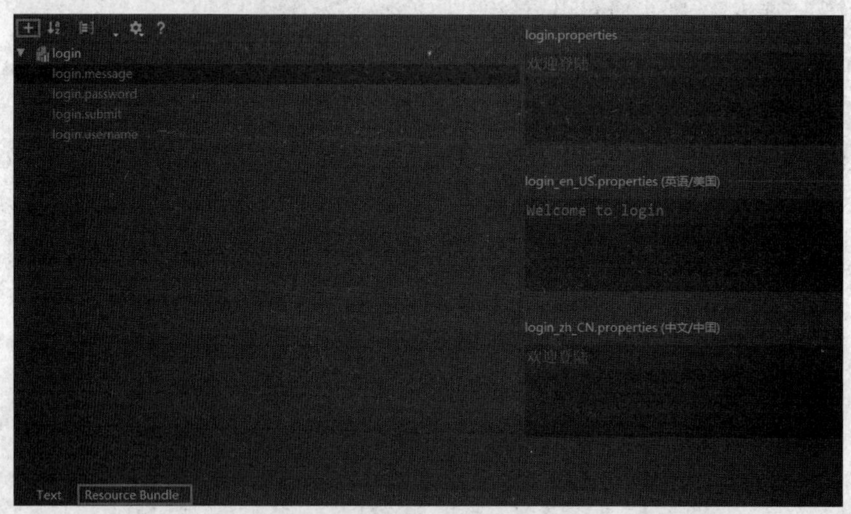

图 3-10 国际化文件编写方法

【案例】应用 Spring Boot 国际化设置和 Thymeleaf 模板，实现登录页面的编写，效果如图 3-11 所示。

图 3-11　使用国际化应用登录界面

第一步,编写前端 login.html 页面,使用 th：href="@{/login（l='zh_CN'）}" 来切换不同的语言，/login 表示对应的 Controller 控制器，Thymeleaf 模板中使用小括号来表示“?”即在 login 后使用“l='zh_CN'”或者“l='en_US'”来进行语言选择,如示例代码 3-8 所示。

示例代码 3-8：login.html 页面

```
<! DOCTYPE html>
<html lang="en" xmlns：th="http：//www.thymeleaf.org">
<head>
  <meta http-equiv="Content-Type" content="text/html；" charset="utf-8">
  <title> 登录 </title>
      <link  th：href="@{/css/bootstrap.css}"  href="../static/css/bootstrap.css"  rel="-
stylesheet"/>
  <link th：href="@{/css/bootstrap-theme.css}" href="../static/css/bootstrap-theme.css"
     rel="stylesheet"/>
  <script src="https：//s3.pstatp.com/cdn/expire-1-M/jquery/3.1.1/jquery.min.js"></script>
  <script th：src="@{/js/bootstrap.js}" src="../static/js/bootstrap.js"></script>
</head>
<body>
<div class="container">
  <div class="row">
    <div class="col-md-4">
```

```
            <div class="panel panel-primary">
                <div class="panel-heading text-center">
                    <h3 class="panel-title" th:text="#{login.message}"></h3>
                </div>
                <div class="panel-body">
                        <input type="text" th:placeholder="#{login.username}" name="username"
/><br/>
                        <input type="text" th:placeholder="#{login.password}" name="password"
/><br/>
                            <input type="submit" class="btn btn-success col-sm-offset-3" th:val-
ue="#{login.submit}"/><br/>
                    <a class="btn btn-sm" th:href="@{/login(l='zh_CN')}"> 中文 </a>
                    <a class="btn btn-sm" th:href="@{/login(l='en_US')}">English</a>
                </div>
            </div>
        </div>
    </div>
</div>
</body>
</html>
```

第二步，使用 th：href="@{...}" 的方法选择语言，需要在后台编写对应的 resolveLocale 即进行国际化支持，根据传递的字符串信息，对 message.properties 文件进行匹配，如示例代码 3-9 所示。

示例代码 3-9：MyLocaleResolver 类

```
public class MyLocaleResolver implements LocaleResolver {
  @Override
  public Locale resolveLocale(HttpServletRequest httpServletRequest) {
    String l = httpServletRequest.getParameter("l"); // 获取传递的 'en_US' 字符串
    Locale locale = Locale.getDefault();
    if(l! =null) {
      String[] split = l.split("_");
      locale = new Locale(split[0], split[1]);
    }
    return locale;
  }

  @Override
```

```
        public void setLocale(HttpServletRequest httpServletRequest, HttpServletResponse
httpServletResponse, Locale locale) {
        }
    }
```

第三步，在 application.properties 文件中，对国际化进行配置。设置 spring.messages.basename 属性，其中 i18n 表示国际化文件所对应的项目类路径 resources 的位置，如示例代码 3-10 所示。

示例代码 3-10：application.properties 配置文件

```
spring.messages.basename=i18n.login   // 如有多个文件可以使用逗号进行分隔
spring.messages.encoding=UTF-8
```

第四步，完成这些文件的编写之后，运行该项目，在地址栏中输入 localhost：8080/login，进入登录界面，点击下方的“中文”后，观察地址栏的变化改为 http：//localhost：8080/login？l=zh_CN，效果如图 3-12 所示。

图 3-12　应用语言“中文”显示登录界面

第五步，再次点击登录界面下方的“English”，观察地址栏的变化，变为 http：//localhost：8080/login？ l=en_US，效果如图 3-13 所示。至此，完成了国际化配置的案例。

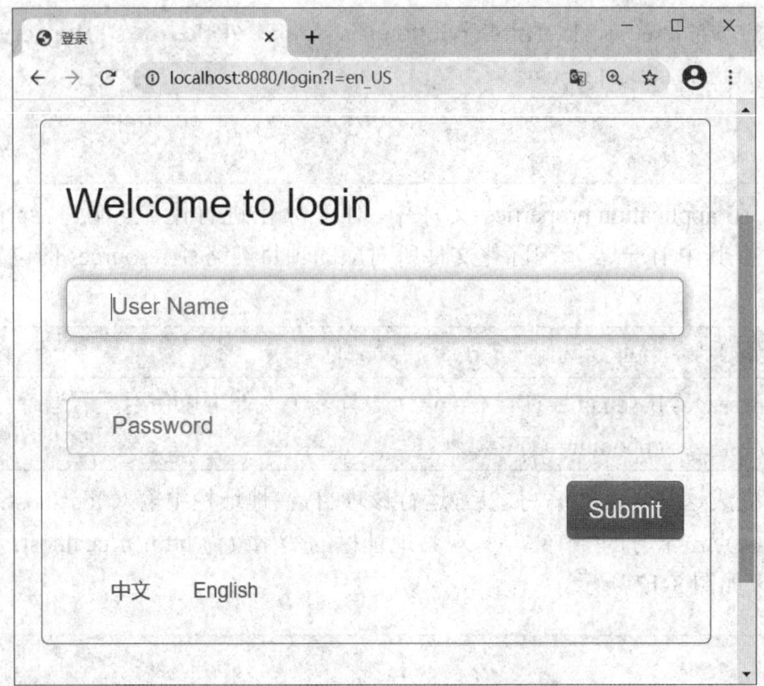

图 3-13 应用语言“English”显示登录界面

技能点 3 Spring Boot 控制器

Spring Boot 项目引入了 spring-boot-starter-web 模块，默认集成了 Spring MVC，同样负责功能处理部分。与传统 Spring 项目相同，Controller 控制器用于收集请求与请求参数，并绑定到命令对象中，再交由业务对象，业务对象处理完毕后，返回模型数据，之后返回视图层进行页面渲染。

1. 初识 Controller

在 Spring Boot 的 MVC 模块中，控制器 Controller 负责处理由 DispatcherServlet 分发的请求，它把用户请求的数据经过业务处理层处理之后封装成一个 Model，然后再把该 Model 返回给对应的 View 进行展示。

在 Spring Boot 的 MVC 模块中提供了一个非常简便的定义 Controller 的方法，开发人员无须继承特定的类或实现特定的接口，只需使用 @Controller 标记一个类是 Controller，然后使用 @RequestMapping 和 @RequestParam 等一些注解用以定义 URL 请求和 Controller 方法之间的映射。这样定义的 Controller 就能被外界访问到。此外，Controller 不会直接依赖于 HttpServletRequest 和 HttpServletResponse 等 HttpServlet 对象，可以通过 Controller 的方法参数灵活地获取到。

定义一个 Controller 控制器名为 MyController，在类上标记 @Controller 注解，然后使用 @RequestMapping（"/showView"）标记在 Controller 方法上，表示当请求 /showView 的时候

访问的是 MyController 的 showView 方法。该方法返回了一个包括 Model 和 View 的 ModelAndView 对象，示例代码如下。

```
@Controller
public class MyController {
    @RequestMapping("/showView")
    public ModelAndView showView() {
    ModelAndView modelAndView = new ModelAndView();
    modelAndView.setViewName("viewName");
     modelAndView.addObject(" 需要放到 model 中的属性名称 ", " 对应的属性值，它
是一个对象 ");
    return modelAndView;
    }
}
```

2. Controller 常用注解

在学习常用的注解之前，首先要了解 Spring Boot 的 MVC 模块中控制器接收参数的方式，并根据不同的参数选择适合的注解，才能完成功能需求。部分参数列举如下。

普通参数：只要保证前端参数名称和传入控制器的参数名称一致即可，适合参数较少的情况。

pojo 类型：如果前端传的是一个 pojo 对象，只要保证参数名称和 pojo 属性一致即可。

注解方式：当前端参数名和控制器名称不一致时可以使用注解方式，如前端是 param_id，而控制器中是 paramId 的情况。

URL 方式：当前端以 restful 格式传递参数时，后端使用这种方式接收。

JSON 方式：前后端如果用 JSON 方式交互时，可以使用这种方式，这是常用的方式。

列表方式：如果前端传的是一个列表，列表元素可以是基本类型也可以是引用类型，可使用这种方式接收。

（1）@RequestMapping 注解

@RequestMapping 是一个处理请求的注解，一般用在类或方法上。该注解作用在类上时，表示该类中所有响应请求的方法都以该地址作为父路径；作用在方法上时，表示在类的父路径下追加方法上注解中的地址来访问该方法。@RequestMapping 注解有六个属性，见表 3-3。

表 3-3　@RequestMapping 注解的属性

属性	类型	作用
value	String[]	指定实际请求路径。将请求和方法相对应。如果 @RequestMapping 至少有一个属性时，必须写上 value 属性的名称
method	RequestMethod[]	映射指定请求的方法类型，如 GET、POST 等。当请求类型为空时，表示可以处理各种类型的请求
consumes	String[]	指定处理请求的提交内容类型

续表

属性	类型	作用
produces	String[]	指定返回的内容类型，仅当 request 请求头中包含该类型时，才会调用该方法处理请求
params	String[]	指定 request 中包含某些参数时，才调用该方法处理
headers	String[]	指定 request 中包含某些 headers 时，才调用该方法处理

使用 @RequestMapping 注解方法（method）时，将 @RequestMapping 的 value 属性的值映射到 method 上，访问"http://localhost:8080/ 项目名 /test"地址时，就会在 test()中进行处理。其中，method 属性指定访问的方式，RequestMethod.POST 表示该方法只处理 POST 请求，示例代码如下。

```
@RequestMapping(value="/test", method = RequestMethod.POST)
  public String test() {
    return...;
  }
```

（2）@RequestParam 注解

@RequestParam 注解主要用于将请求参数赋给方法中的形参。当出现 HTTP 请求参数名称和映射不一致的情况时，如在 HTTP 中参数名为 user_name，而在 Controller 中设置接收参数值为 userName，二者存在不一致，需要使用 @RequestParam 注解来获取参数。该注解的常用属性见表 3-4。

表 3-4 @RequestParam 注解属性

属性	类型	说明
Name	String	请求的名称
Value	String	Name 的别名
Required	Boolean	参数是否有必要绑定
DefaultValue	String	未传入参数时的默认值

使用 @RequestParam 注解处理在 HTTP 中参数名为 user_name，而在 Controller 中设置接收参数值为 userName 的情况，示例代码如下。

```
// 使用 @RequestParam("user_name")指定映射 HTTP 参数名称
    public ModelAndView requestParam(@RequestParam("user_name") String user-
Name, String password) {
            System.out.println("userName=>"+ userName);
            System.out.println("password=>" + password);
            ModelAndView mv = new ModelAndView();
```

```
        mv.setViewName("index");
        return mv;
    }
```

（3）@PathVariable 注解

@PathVariable 注解主要用于将 URL 中的变量映射到功能处理方法的形参上，即将 URL 变量取出作为参数，它只支持一个 String 类型的属性，形式为：/ 路径 /{ 变量 }。

例如，当请求路径 URL 为"http://localhost:8080/Test/findById/1"时，将 URL 中的变量值绑定到 @PathVariable 注解的同名参数上，即将 productId 值赋值为 1，代码如下。

```
@Controller
public class TestController {
    @RequestMapping(value="/findById/{Id}")
    public String findById(@PathVariable("Id") Integer productId){
    System.out.println(" 通过 @PathVariable 获得数据 "+productId);
     return "hello";
   }
}
```

（4）@RequestBody 注解

@RequestBody 注解主要用来接收前端传递给后端的 JSON 字符串中的数据（请求体中的数据）。@RequestBody 可以将 JSON 的形式转换为 Java 的形式，并将传进来的数据装配到集合或实体类中。使用 @RequestBody 时，需遵循如下的规则。

① GET 方式无请求体，使用 @RequestBody 接收数据时，前端不能使用 GET 方式提交数据，而是使用 POST 方式进行提交。

②在后端 Controller 的同一个接收方法中，@RequestBody 与 @RequestParam()两种注解可以同时使用，但 @RequestBody 注解最多只能有一个，而 @RequestParam()可以有多个。

使用 @RequestBody 注解时，前端传递 JSON 到 Controller 中获取，示例代码如下。

```
@RequestMapping("/findRoles")
    public ModelAndView findRoles(@RequestBody RoleParams roleParams){
            List<Role> roleList = roleService.findRoles(roleParams);
            ModelAndView mv = new ModelAndView();
            // 绑定模型
            mv.addObject(roleList);
            // 设置为 JSON 视图
            mv.setView(new MappingJackson2JsonView());
            return mv;
    }
```

（5）@RestController 注解

@RestController 注解是 @Controller 和 @ResponseBody 两个注解的组成。

@Controller 将当前修饰的类注入 Spring Boot IOC 容器，当运行整体项目时，被标记的类就被实例化，即代表该类是作为控制器 Controller 使用的。

@ResponseBody 注解的基本作用是返回该类中所有的 API 接口数据，无论其下方法中返回的是 Map 集合或是 Object 对象等，都会以 JSON 字符串的形式返回给客户端。

例如，使用 @RestController 对 ControllerTest 类进行修饰时，编写对应的方法 test()和 testgetStr()的代码如下。

```
@RestController
public class ControllerTest {
  @GetMapping// 相当于 @RequestMapping（method =RequestMethod.GET）
  @RequestMapping（"test"）
  public Map test（）{
    HashMap<String，String> hashMap = new HashMap<>（）；
    hashMap.put（"name"，"zhang"）；
    return hashMap；
    }
  @RequestMapping（"test/str"）
  public String testStr（）{
    return "Strings"；
  }
}
```

当在地址栏输入 http：//localhost：8080/test 路径地址时，即获取了对应 hashMap 中的键值对信息，这里为“name”“zhang”，使用对应的 @RestController 注解，可将集合中的键值对信息转换为 JSON 字符串的形式，并返回至浏览器，效果如图 3-14 所示。

图 3-14 访问 test 方法

当在地址栏输入 http：//localhost：8080/test/test 路径地址时，由于使用了 @RestController 注解，return 中的字符串形式没有被编辑为页面显示的模板名称，而是直接将 String 类型的数据进行返回，并显示在浏览器上，效果如图 3-15 所示。

图 3-15　访问 test/str 方法

3. Controller 应用实例

【实例】使用登录(Login)页面,传递用户名和密码至 Controller 中,并跳转至登录成功(LoginSuccess)页面显示对应的用户名称。

第一步,编写 ControllerTest 控制器,编写 form()方法,获取 username 和 password 属性,并使用不同方式保存参数传递到其他页面中,如示例代码 3-11 所示。

示例代码 3-11:ControllerTest 控制器中的 form 方法

```
@RequestMapping(value = "/from",method = RequestMethod.GET)
    public ModelAndView form(@RequestParam("username") String name, @Request-
Param("password") String password,Model model)
    {
      DateFormat dateFormat = DateFormat.getDateInstance();// 日期格式,精确到日
      ModelAndView modelAndView = new ModelAndView();
// 三种方式获取属性值
      modelAndView.addObject("name",name);
      model.addAttribute("password",password);
      // 由于 ModelAndView 属于 Map 类型,可以使用 Map 的 put 方法
      modelAndView.getModel().put("time",dateFormat.format(new Date()));
      modelAndView.setViewName("LoginSuccess");
      return modelAndView;
    }
```

第二步,编写 LoginSuccess 页面,将 Login 页面中的参数传递到该页面中,显示数据,如示例代码 3-12 所示。

示例代码 3-12:LoginSuccess.html 页面

```
<! DOCTYPE html>
<html lang="en" xmlns:th="http://www.thymeleaf.org">
<head>
  <meta charset="UTF-8">
  <title> 登录成功 </title>
```

```
        <link  th: href="@{/css/bootstrap.css}"  href="../static/css/bootstrap.css"  rel="-
stylesheet"/>
    <link th:href="@{/css/bootstrap-theme.css}" href="../static/css/bootstrap-theme.css"
        rel="stylesheet"/>
    <script src="https://s3.pstatp.com/cdn/expire-1-M/jquery/3.1.1/jquery.min.js"></script>
    <script th:src="@{/js/bootstrap.js}" src="../static/js/bootstrap.js"></script>
</head>
<body>
<div class="container">
    <div class="row">
        <div class="col-md-4">
            <div class="panel panel-warning">
                <div class="panel-heading text-center">
                    <span class="panel-title"> 欢迎登录 </span>
                </div>
                <div class="panel-body">
                    用户名:<span th:text="${name}"></span><br/>
                    密码:<span th:text="${password}"></span><br/>
                    登录时间:<span th:text="${time}"></span><br/>
                </div>
            </div>
        </div>
    </div>
</div>
</body>
</html>
```

第三步，在 Login 页面中输入用户名和密码，点击"提交"跳转至 LoginSuccess 页面，显示用户信息以及登录时间，效果如图 3-16 所示。

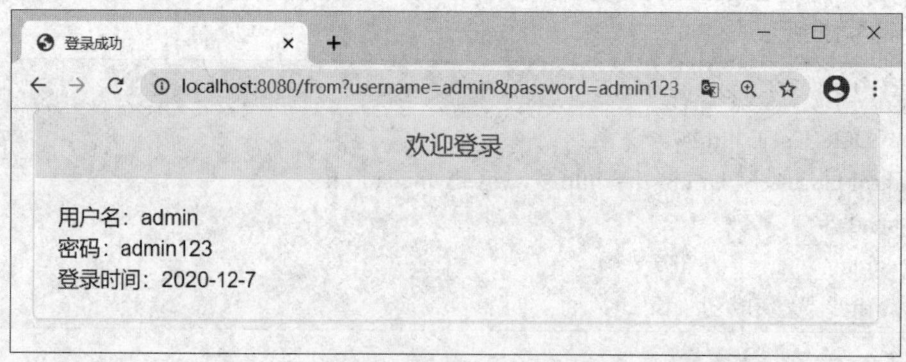

图 3-16 登录成功页面

技能点 4　Spring Boot 常用组件

在传统的 Web 开发中，对于 Servlet（小服务程序）、Filter（过滤器）和 Listener（监听器）都不会陌生，这三个组件被称为三大组件。Spring Boot 使用内嵌式 Servlet 容器，对这三大组件进行了整合，无须再进行 Web.xml 配置。

在 Spring Boot 中，实现三大组件有两种方式；第一种使用 @Component 注解，并设置自定义 @Configuration 配置将组件注入到容器中；第二种使用对应的注解，并在入口类上进行 @ServletComponentScan 注解的标记来完成装配。

具体组件的注解方法见表 3-5。

表 3-5　组件注解方法

注解	说明
Servlet	@WebServlet（name="",urlPatterns = ""）， name 指定 Servlet 的 name 属性， urlPatterns 指定一组 Servlet 的 URL 匹配模式
Listener	@WebListener
Filter	@WebFilter（filterName = "",urlPatterns = ""）， filterName 指定 Servlet 的 name 属性， urlPatterns 指定 Filter 的 URL 匹配模式

1. Servlet 组件

Java Servlet 是作为来自 Web 浏览器或其他 HTTP 客户端的请求和 HTTP 服务器上的数据库或应用程序之间的中间层的程序，其运行在 Web 服务器或应用服务器上。应用 Servlet，就可以创建网页，收集来自网页表单的用户信息，或在页面上呈现来自数据库的信息。

Spring Boot 作为快速启动框架，同样整合了 Servlet 组件，并提供了相应的对象让开发人员更方便地注册这些组件。引入 Servlet 主要是为扩展 Spring Boot 的功能、自定义用法，并且可以把它作为 Bean 加载到框架内，便于应用 Servlet 中的方法。

添加 Servlet 的具体方法如下。

在启动类上添加 @ServletComponentScan 注解的方向，可用来扫描本目录以及子目录中带有的 @WebServlet 注解的配置类。

@WebServlet 注解常用的属性和具体含义见表 3-6。

表 3-6　@WebServlet 注解的属性见含义

属性	详情
name	等价于 <servlet-name>，用于指定 Servlet 的 name 属性。如果没有显式指定，则 Servlet 的取值为类的全限定名

续表

属性	详情
value	等价于 urlPatterns 属性，但两个属性不能同时使用
urlPatterns	等价于 <url-pattern> 标签，用于指定一组 Servlet 的 URL 匹配模式
loadOnStartup	等价于 <load-on-startup> 标签，用于指定 Servlet 的加载顺序
initParams	等价于 <init-param> 标签，用于指定一组 Servlet 的初始化参数
asyncSupported	等价于 <async-supported> 标签，用于声明 Servlet 是否支持异步操作模式
description	等价于 <description> 标签，用于表示该 Servlet 的描述信息
displayName	等价于 <display-name> 标签，用于表示该 Servlet 的显示名，通常配合工具使用

Spring Boot 在项目中设置了 @WebServlet 注解，当请求其注解的 Servlet 时，服务器就会自动读取其中的信息。使用 @WebServlet（"/MyServlet"）时，若没有设置 @WebServlet 的 name 属性，默认值会是 Servlet 的类的完整名称，即表示该 Servlet 默认的请求路径为 .../MyServlet，这里省略了 urlPatterns 属性名，完整的写法应该是"@WebServlet（name="MyServlet"，urlPatterns = "/MyServlet"）"。如果在 @WebServlet 中需要设置多个属性，必须给属性值加上属性名称，并在中间用逗号隔开，否则会报错。

【案例】Spring Boot 整合 Servlet，实现在页面上显示信息。

第一步，编写 MyServlet 类，继承 HttpServlet，重写其 doGet（）和 doPost（）方法，如示例代码 3-13 所示。

示例代码 3-13：MyServlet 类

```
@WebServlet(name = "MyServlet", urlPatterns = "/MyServlet")
public class MyServlet extends HttpServlet {
    @Override
    protected void doGet(HttpServletRequest req, HttpServletResponse resp) throws ServletException, IOException {
        this.doPost(req, resp);
    }
    @Override
    protected void doPost(HttpServletRequest req, HttpServletResponse resp) throws ServletException, IOException {
        resp.setCharacterEncoding("utf-8");
        resp.setContentType("text/html; charset=utf-8");
        resp.getWriter().append("Spring Boot 整合 Servlet 成功 ");
    }
}
```

第二步，在入口类上编写 @ServletComponentScan 注解。启动项目时，会扫描本目录以及子目录中带有 @WebServlet 注解的类，并注入容器中，如示例代码 3-14 所示。

示例代码 3-14：DemoApplication 入口类

```
@ServletComponentScan
@SpringBootApplication
public class DemoApplication {
    public static void main(String[] args) {
        SpringApplication.run(DemoApplication.class, args);
    }
}
```

第三步，运行项目，在浏览器地址栏中输入地址 http://localhost:8080/MyServlet，访问该地址路径，效果如图 3-17 所示。

图 3-17　访问 MyServlet 方法

看到显示的中文提示词，说明使用 Spring Boot 整合 Servlet 成功，可根据需求修改示例代码 3-13 中 doPost()的内容，实现特殊的功能需求。

2. 监听器

监听器(Listener)是 Servlet 规范中定义的一种特殊类。用于监听 ServletContext、HttpSession 和 ServletRequest 等域对象的创建、销毁，变量的创建、销毁和修改等。可以在某些动作前后增加处理，实现监控。Listener 主要可用于监听 Servlet 上下文、用来初始化一些数据、监听 HTTP Session、用来获取当前在线的人数、监听客户端请求的 ServletRequest 对象并获取用户的访问信息等。通常，Listener 有如下的使用场景。

①统计在线人数和在线用户。

②在系统启动时加载初始化信息。

③统计网站的用户访问量。

④记录用户访问路径。

【案例】使用监听器，监听在线人数信息，并显示在页面上。

第一步，编写监听器类 MyListener，实现 HttpSessionListener 接口方法，创建 session 以及销毁 session。创建计数器 count，将在线人数进行记录，如示例代码 3-15 所示。

示例代码 3-15：MyListener 类

```
@Component    // 运用 @Component 注解将类引入容器，实现监听器
//@WebListener
// 应用路径扫描的方式也可整合监听器，同时要在入口类中增加 @ServletComponentScan 注解
```

```
public class MyListener implements HttpSessionListener {
    Integer count = 0;
    @Override
    public synchronized void sessionCreated(HttpSessionEvent httpSessionEvent) {
        System.out.println(" 新用户上线了 ");
        count++;
        httpSessionEvent.getSession().getServletContext().setAttribute("count", count);
    }
    @Override
    public synchronized void sessionDestroyed(HttpSessionEvent httpSessionEvent) {
        System.out.println(" 用户下线了 ");
        count--;
        httpSessionEvent.getSession().getServletContext().setAttribute("count", count);
    }
}
```

第二步，编写 ControllerTest 控制器 indexView 方法，通过 cookie 保存登录的 session，当用户再次登录时，将所保存的 sessionId 传递过去，服务器就不会再次创建新的 session 以此保证计数器的正确，如示例代码 3-16 所示。

示例代码 3-16：ControllerTest 类

```
@Controller
public class ControllerTest {
@RequestMapping("/index")
    public String IndexView(Model model, HttpServletRequest request, HttpServle-
tResponse response) {
    Cookie cookie;
    try {
        // 把 sessionId 记录在浏览器中，JSESSIONID 是一个 Cookie，Servlet 容器
(tomcat,jetty)用来记录用户 session
        cookie = new Cookie("JSESSIONID", URLEncoder.encode(request.getSession
().getId(), "utf-8"));
        cookie.setPath("/");
        // 设置 cookie 有效期为 5 分钟，可根据需求修改
        cookie.setMaxAge(5*60);
        response.addCookie(cookie);
    } catch (UnsupportedEncodingException e) {
        e.printStackTrace();
    }
```

```
        Integer count =(Integer)request.getSession().getServletContext().getAttribute
("count");
        model.addAttribute("count",count);
        return "index";
    }
}
```

第三步，依旧使用之前显示学生信息的前端 index.html 页面，添加对应的在线人数信息，如示例代码 3-17 所示。

示例代码 3-17：index.html 页面

```
<! DOCTYPE html>
<html lang="en" xmlns:th="http://www.thymeleaf.org">
<head>
  <title> 显示学生信息 </title>
  <meta charset="UTF-8"/>
 <! -- 省略引入 css 和 js 文件 -->
</head>
<body>
<div class="container">
  <div class="row">
    <div class="col-md-4">
      <div class="panel panel-primary">
        <div class="panel-heading text-center">
          <span class="panel-title"> 学生信息 </span>
        </div>
        <div class="panel-body">
          <h4 th:text="' 当前在线人数:'+${count}">0</h4>
          学号:<span th:text="#{student.id}">10</span><br/>
          姓名:<span th:text="#{student.name}">wang</span><br/>
          班级:<span th:text="#{student.className}">Class Two</span><br/>
          课程: <span th: text="#{student.course}">Mathemastics, Chinese, English
</span><br/>
          日期:<span th:text="#{student.elective.classTime}">2020/11/19</span><br/>
        </div>
      </div>
    </div>
  </div>
</div>
```

```
</body>
</html>
```

第四步，运行项目。在浏览器的地址栏中输入地址 http：//localhost：8080/index，访问该地址路径，效果如图 3-18 所示。

图 3-18　访问 index 方法

在该页面中，通过 Listener 成功显示了在线人数，实现了 Spring Boot 中整合监听器的需求。

3. 过滤器

过滤器（Filter）被 Web 开发人员用于管理服务器中的所有 Web 资源，实现对静态图片文件或静态 html 文件等进行拦截，对用户请求进行预处理，以及对 HttpServletResponse 进行后处理，从而实现一些特殊的功能，如实现权限访问控制、敏感词汇过滤、压缩响应信息等一些高级功能。

Filter 依靠 Servlet 容器，使用 AOP 思想。Filter 的完整流程：Filter 对用户请求进行预处理，接着将请求交给 Servlet 进行处理并生成响应，最后 Filter 再对服务器响应进行后处理，如图 3-19 所示。

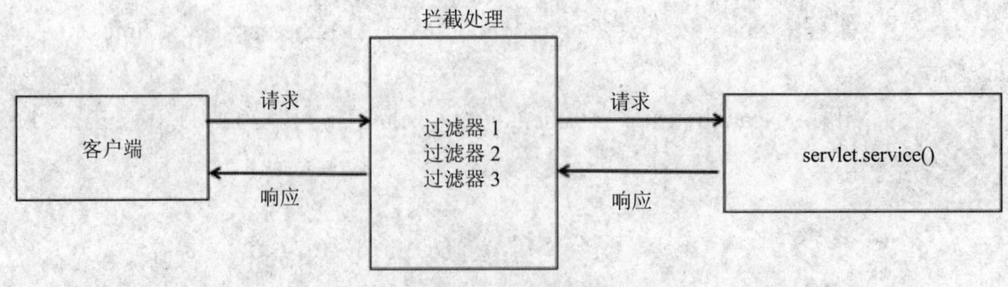

图 3-19　Filter 过滤器流程图

Filter 过滤器的使用方法见表 3-7，可通过重写这些方法达到实现特殊功能的目的。

表 3-7　Filter 过滤器方法

方法	说明
init（FilterConfig）	其为初始化方法。服务器启动时，会创建过滤器的实例化对象，并调用 init（）方法，完成初始化方法，过滤器对象只会创建一次，init（）方法也只会执行一次。在此方法中，可以读取 web.xml 文件中 Servlet 过滤器的初始化参数
doFilter（ServletRequest，ServletResponse，FilterChain）	其为完成具体过滤操作的方法。在访问过滤器关联的 URL 时，Serlvet 容器将先调用过滤器的 doFilter 方法，其中 FilterChain 参数用于访问后续的过滤器
destroy（）	Servlet 过滤器销毁前调用该方法，其可释放 Servlet 过滤器占用的资源

【案例】编写 Filter 过滤器，对错误的地址路径统一跳转至登录页面。

第一步，编写 MyFilter 类，实现 Filter 接口的三个方法，获取请求地址与“/index”进行对比，对于错误的地址路径，统一跳转至登录页面，如示例代码 3-18 所示。

示例代码 3-18：MyFilter 类

```
@WebFilter(filterName = "test", urlPatterns = "/index/*")
// 设置过滤器名称，设置其触发的路径
public class MyFilter implements Filter {
  @Override
  public void init(FilterConfig filterConfig) throws ServletException {
    System.out.println(" 启动过滤器 ");
  }
  @Override
  public void doFilter(ServletRequest servletRequest, ServletResponse servletResponse,
FilterChain filterChain) throws IOException, ServletException {
    HttpServletRequest request = (HttpServletRequest) servletRequest;
    HttpServletResponse response = (HttpServletResponse) servletResponse;
    String requestUrl = request.getRequestURI();
    System.out.println(" 请求地址为："+requestUrl);
    if(! requestUrl.equals("/index"))
    {
//servletRequest.getRequestDispatcher("/login").forward(servletRequest, servle-
tResponse);   // 应用转发进行页面跳转，所输入的地址不变
    response.sendRedirect("/login");
// 应用重定向对页面进行跳转，地址栏地址变化为 /login 结尾
    }
```

```
        else
        {
            filterChain.doFilter(servletRequest,servletResponse);
        }
    }
    @Override
    public void destroy() {
        System.out.println("销毁过滤器");
    }
}
```

第二步，使用 Forward 方法进行转发，在浏览器地址栏中输入 http://localhost:8080/index/success 并访问。因为在对应的项目中，该路径并没有页面，由于路径信息符合 urlPatterns =“/index/*”，会触发过滤器，对请求地址进行获取，并与“/index”进行对比，结果是地址路径错误，所以会跳转至登录页面进行登录，效果如图 3-20 所示。

图 3-20　过滤错误访问地址跳转至登录页面

第三步，使用 Redirect 方法进行重定向，访问地址同为 http://localhost:8080/index/success。经过 Redirect 方法跳转，地址栏信息改变为“login”，说明成功通过过滤器，将错误的地址路径跳转至登录页面，完成了 Spring Boot 对 Filter 的整合，效果如图 3-21 所示。

图 3-21　过滤错误访问地址重定向至登录页面

4. 拦截器

Spring MVC 拦截器（Interceptor）的功能是在每一个请求处理的前后，进行相关的业务处理。Spring MVC 拦截器可以在进入处理器之前，或是在处理器执行完成后，甚至是在渲染视图后进行拦截并处理相关数据。

类似于 Servlet 中的过滤器，但两者也有区别，区别如下。

①过滤器是依赖于 Servlet 容器，属于 Servlet 规范的一部分，而拦截器则是独立存在的，可以在任何情况下使用。

②过滤器的执行由 Servlet 容器回调完成，而拦截器通常通过动态代理的方式来执行。

③过滤器的生命周期由 Servlet 容器管理，而拦截器则可以通过控制反转（Inversion of Control，IoC）容器来管理。因此可以通过注入等方式来获取其他 Bean 的实例。

拦截器的本质就是面向切面编程即 AOP，所以符合切面编程思想的功能都可以在拦截器中进行实现，主要包括以下几个场景。

- 登录验证：判断用户是否已经登录了。
- 权限验证：判断该用户是否有该功能的访问权限。
- 日志记录：记录日志信息，用户的登录记录、操作记录、访问时间记录等。
- 性能监控：监控处理数据时长等。

【案例】使用拦截器实现页面的拦截操作。

第一步，编写 MyInterceptor 拦截器类，实现 HandlerInterceptor 接口，重写 preHandle 方法，并在请求发生前执行，如示例代码 3-19 所示。

示例代码 3-19：MyInterceptor 类

```
@Component
public class MyInterceptor implements HandlerInterceptor {
```

```
    public boolean preHandle(HttpServletRequest request, HttpServletResponse response,
Object handler) throws Exception {
        // 在业务处理器处理请求之前被调用
        try{
            System.out.println(" 执行 preHandle");
            String path = request.getRequestURI();
            System.out.println(path);
            if(path.equals("/index"))
            {
                return true;
            }
            response.sendRedirect(request.getContextPath()+"/login");// 重定向至 login
        }catch(IOException e)
        {
            e.printStackTrace();
        }
        return false;
    }
    public void postHandle(HttpServletRequest request, HttpServletResponse response,
Object handler, @Nullable ModelAndView modelAndView) throws Exception {
        // 在业务处理器处理请求执行完成后，生成视图之前执行
        System.out.println(" 执行 postHandle");
    }
    public void afterCompletion(HttpServletRequest request, HttpServletResponse re-
sponse, Object handler, @Nullable Exception ex) throws Exception {
        // 在 DispatcherServlet 完全处理完请求后被调用，可用于清理资源等。
        System.out.println(" 执行 afterCompletion");
    }
}
```

第二步，在 MyConfig 中注册拦截器，设置拦截路径和不拦截的路径，如示例代码 3-20 所示。

示例代码 3-20：addInterceptors 类

```
@Override
    public void addInterceptors(InterceptorRegistry registry) {
        // 注册 Interceptor 拦截器
            InterceptorRegistration registration = registry.addInterceptor(new MyInterceptor
());
```

```
        registration.addPathPatterns("/index/*");                  // 路径被拦截
        registration.excludePathPatterns(                       // 添加不拦截路径
            "/js/*.js",          //js 静态资源
            "/css/*.css"         //css 静态资源
        );
    }
```

第三步，在浏览器的地址栏中输入 http：//localhost：8080/index/success，通过 sendRedirect()方法跳转至登录页面，实现了拦截器的功能，效果如图 3-22 所示。

图 3-22　实现拦截器跳转页面

技能点 5　Spring Boot 文件上传与下载

在任何项目中，文件的上传和下载都是比较常见的。通常通过自定义上传下载的方法，来实现这一部分的功能。

1. 文件上传

由于浏览器本身的限制，Web 项目并不能直接操作文件系统。所以在进行文件上传时，需要调用统一的接口，并获得用户的主动授权。在获得授权之后将读取的文件内容放进指定内存里，然后在执行提交请求操作后将文件内容数据上传到服务端，最后服务端解析前端传来的数据信息后，储存文件。

前端页面利用 form 表单标签和类型为 file 的 input 标签来完成上传，需要将表单数据

编码格式设置为 multipart/form-data 类型，该编码类型会在上传时对文件内容进行处理，以便服务端处理程序解析文件类型与内容，完成上传操作。

【案例】实现文件上传功能。

第一步，编写 ControllerTest 控制器，创建 UploadFileLogin 方法，跳转至上传页面，如示例代码 3-21 所示。

示例代码 3-21：ControllerTest 控制器中 UploadFileLogin 方法

```
@RequestMapping("/upload")
public String UploadFileLogin(Model model) {
    return "FileUpload";
}
```

第二步，编写具体的上传逻辑代码，创建 UploadFile 方法，该方法使用 @ResponseBody 注解表示其返回值不再指向页面模板，而直接返回字符串，如示例代码 3-22 所示。

示例代码 3-22：UploadFile 方法

```
@ResponseBody
@RequestMapping(value = "/UploadFile")
public String UploadFile(@RequestParam("fileName") MultipartFile file){
    if(file.isEmpty()){
        return "文件为空，上传失败";
    }
    String fileName = file.getOriginalFilename();
    int size = (int) file.getSize();
    System.out.println(fileName + "-->" + size);

    String path = "D:/test";
    File dest = new File(path + "/" + fileName);
    if(! dest.getParentFile().exists()){ // 判断文件父目录是否存在
        dest.getParentFile().mkdirs();
    }
    try {
        file.transferTo(dest); // 保存文件
        return "上传成功";
    } catch (IllegalStateException | IOException e) {
        // TODO Auto-generated catch block
        e.printStackTrace();
        return "发生异常，上传失败";
    }
}
```

第三步，编写 FileUpload 前端页面，如示例代码 3-23 所示。

示例代码 3-23：FileUpload.html 前端页面

```
<! DOCTYPE html>
<html lang="en" xmlns:th="http://www.thymeleaf.org">
<head>
  <meta charset="UTF-8"/>
  <title> 文件上传 </title>
  <! -- 省略 css 和 js 文件的引入 -->
</head>
<body>
<form action="/UploadFile" method="post" enctype="multipart/form-data">
  <div class="col-md-4">
    <div class="form-group">
      <label for="InputFile" class="control-label"> 文件上传 </label>
      <input id="InputFile" type="file" name="fileName"/>
      <p class="help-block"> 上传文件的文件会显示在上方 </p>
    </div>
    <div class="form-group">
      <div>
        <input type="submit" class="btn btn-default" value=" 提交 "/>
      </div>
    </div>
  </div>
</form>
</body>
</html>
```

第四步，运行项目，在页面上点击“选择文件”，选择上传的文件，点击“提交”，即可完成上传，效果如图 3-23 所示。

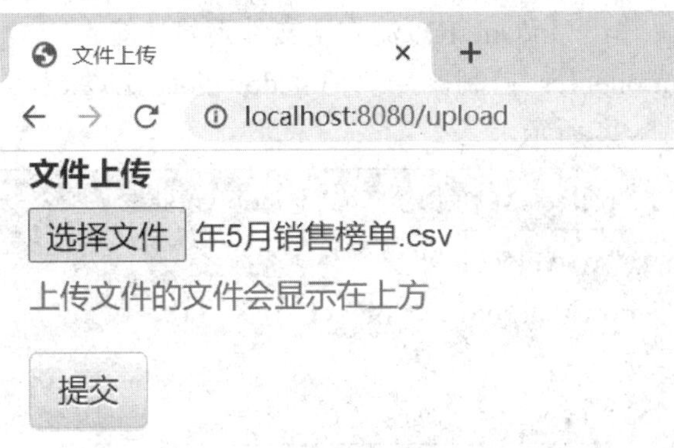

图 3-23　上传文件页面

在点击“提交”之后，可看到“上传成功”字样，完成上传，效果如图 3-24 所示。同时在控制台中可看到上传的文件名称以及文件大小的信息，效果如图 3-25 所示。

图 3-24　上传成功效果

```
2020-12-04 10:18:57.035  INFO 7156 --- [nio-8080-exec-1]
年5月销售榜单.csv-->19
```

图 3-25　控制台获取上传文件信息

打开对应地址“D: /test”可在目录中看到上传的文件，如图 3-26 所示。至此就完成了文件上传功能相关代码的编写。

› 此电脑 › 新加卷 (D:) › test

名称	修改日期	类型	大小
1605593509808学生缴费信息.xlsx	2020/12/3 16:08	XLSX 工作表	12 KB
ttteeaa.txt	2020/12/1 16:47	文本文档	2 KB
待修改的内容.txt	2020/11/6 16:37	文本文档	1 KB
年5月销售榜单.csv	2020/12/4 10:33	XLS 工作表	1 KB

图 3-26　上传文件位置

2. 文件下载

在实现文件下载功能时，页面显示可下载的文件，运用 FileInputStream 文件输入流和 OutputStream 输出流，获取对应的文件名称并判断路径是否存在。在构造流对象的时候，获取指定的文件对象，并连接程序内存与文件对象，来实现文件的下载功能。

【案例】通过输入 / 输出（In/Out，IO）流实现文件下载功能。

第一步，编写 ControllerTest 控制器，编写 Filedownload 方法，获取目标文件夹中所有文件的名称，如示例代码 3-24 所示。

示例代码 3-24：ControllerTest 控制器中的 Filedownload 方法

```
@RequestMapping("/downloadList")
  public String Filedownload(Model model)
  {
    List<String> list = new ArrayList<>();
```

```
        String filePath = "D:/test";  // 获取文件路径
        File file1 = new File(filePath);  // 获取文件列表
        File[] array = file1.listFiles();  // 判断文件是否存在
        assert array != null;
        for (File f : array)
        {
          String fileName = f.getName();// 获取文件名称
          list.add(fileName);// 放入 list 集合中
        }
        model.addAttribute("fileList", list);// 传递至前端显示
        return "FileDownload";
    }
```

第二步，编写下载逻辑代码，编写 DownloadFile 方法，获取对应文件的名称，使用 IO 流下载文件，如示例代码 3-25 所示。

示例代码 3-25：DownloadFile 方法

```
    @RequestMapping("/download")
        public String DownloadFile(@RequestParam("file") String filename, HttpServle-
tResponse response) throws UnsupportedEncodingException {
        String filePath = "D:/test";
        File file = new File(filePath + "/" + filename);
        if(file.exists()){ // 判断文件父目录是否存在
          response.setContentType("application/vnd.ms-excel; charset=UTF-8");
          response.setCharacterEncoding("UTF-8");
            response.setHeader("Content-Disposition", "attachment; fileName=" +java.net.
URLEncoder.encode(filename,"UTF-8"));
          byte[] buffer = new byte[1024];
          FileInputStream fis = null; // 文件输入流
          BufferedInputStream bis = null;
          OutputStream os = null; // 输出流
          try {
            os = response.getOutputStream();
            fis = new FileInputStream(file);
            bis = new BufferedInputStream(fis);
            int i = bis.read(buffer);
            while(i != -1){
              os.write(buffer);
              i = bis.read(buffer);
```

```
            }
        } catch (Exception e) {
            // TODO Auto-generated catch block
            e.printStackTrace();
        }
        System.out.println(" 下载文件："+ filename);
        try {
            assert bis != null;
            bis.close();
            fis.close();
        } catch (IOException e) {
            // TODO Auto-generated catch block
            e.printStackTrace();
        }
    }
    return null;
}
```

第三步，编写 FileDownload 前端页面，将 ControllerTest 传递的文件名称列表使用 th:each 进行变量化并显示在页面上。如果没有显示设置状态变量，则 Thymeleaf 将创建一个默认的迭代变量，该状态迭代变量名称为迭代变量 +"Stat"。Thymeleaf 在此情况下创建的迭代变量名称为 fileStat，对该迭代变量使用 count 方法即可完成序号的编写。使用标签 th:href="@{/download(file=${file})}" 跳转至对应的 ControllerTest 方法，即完成下载操作，如示例代码 3-26 所示。

示例代码 3-26：FileDownload.html 页面

```
<!DOCTYPE html>
<html lang="en" xmlns:th="http://www.thymeleaf.org">
<head>
    <meta charset="UTF-8">
    <title> 下载列表 </title>
    <!-- 省略 css 和 js 文件的引入 -->
</head>
<body>
<table class="table table-striped table-hover">
    <tr class="warning">
        <th>#</th>
        <th> 文件名称 </th>
        <th> 操作 </th>
```

```
        </tr>
        <tr th:each="file:${fileList}" class="active">
            <td th:text="${fileStat.count}">11</td>
            <td th:text="${file}">11</td>
            <td><a th:href="@{/download(file=${file})}"> 下载文件 </a></td>
        </tr>
    </table>
    </body>
    </html>
```

第四步，运行项目，在浏览器的地址栏输入 http://localhost:8080/downloadList，访问文件下载列表页面，效果如图 3-27 所示。

下载列表　localhost:8080/downloadList

#	文件名称	操作
1	1605593509808学生缴费信息.xlsx	下载文件
2	ttteeaa.txt	下载文件
3	年5月销售榜单.csv	下载文件
4	待修改的内容.txt	下载文件

图 3-27　下载列表页面

第五步，点击文件后的“下载文件”即可下载对应的文件。同时在控制台中可看到下载的文件名称，效果如图 3-28 所示。在用户电脑上默认的文件下载文件夹中，可以看到下载的文件，效果如图 3-29 所示。至此，完成了文件下载功能相关代码的编写。

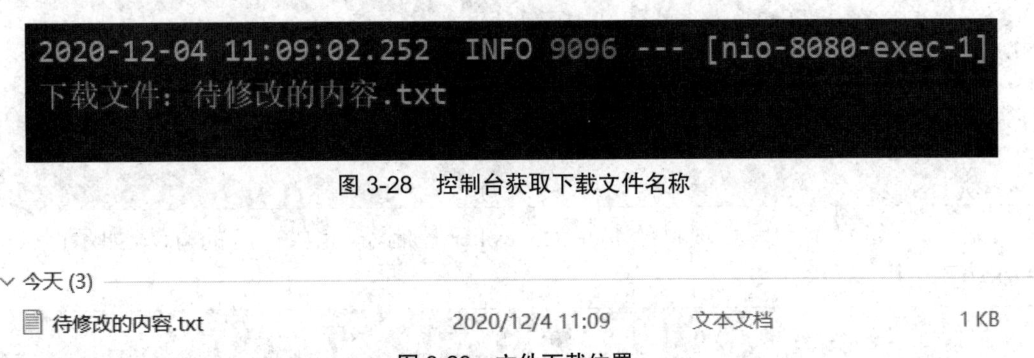

图 3-28　控制台获取下载文件名称

图 3-29　文件下载位置

完成个人博客首页展示部分，内容包括获取热度前十的文章信息、查询文章详情、发布文章、更新文章和删除文章。根据功能编写 Service 层接口和控制器部分。

第一步，编写个人博客项目的 IArticleService 接口，如示例代码 3-27 所示。该接口实现方法将在后面的任务中进行讲解。

示例代码 3-27：IArticleService 接口

```
public interface IArticleService {
    // 分页查询文章列表
    public PageInfo<Article> selectArticleWithPage(Integer page, Integer count);

    // 统计热度前十的文章信息
    public List<Article> getHeatArticles();

    // 查询文章详情
    public Article selectArticleWithId(Integer id);

    // 发布文章
    public void publish(Article article);

    // 更新文章
    public void updateArticleWithId(Article article);

    // 删除文章
    public void deleteArticleWithId(int id);
}
```

第二步，编写个人博客项目的 IndexController 控制器，完成相应的方法，如示例代码 3-28 所示。

示例代码 3-28：IndexController 控制器类

```
@Controller
public class IndexController {
// 打印日志信息
    private static final Logger logger = LoggerFactory.getLogger(IndexController.class);
```

```
    //文章相关实体类
    @Autowired
  private IArticleService articleServiceImpl;
    //评论相关实体类
    @Autowired
  private ICommentService commentServiceImpl;
  //文章动态信息统计实体类
    @Autowired
    private ISiteService siteServiceImpl;

    //博客首页，会自动跳转到文章页
    @GetMapping(value = "/")
    private String index(HttpServletRequest request) {
      return this.index(request, 1, 5);
    }

    //文章页
    @GetMapping(value = "/page/{p}")
    public String index(HttpServletRequest request, @PathVariable("p") int page, @Re-
questParam(value = "count", defaultValue = "5") int count) {
           PageInfo<Article> articles = articleServiceImpl.selectArticleWithPage(page,
count);
      //获取文章热度统计信息
      List<Article> articleList = articleServiceImpl.getHeatArticles();
      //信息保存并传递至前端页面
      request.setAttribute("articles", articles);
      request.setAttribute("articleList", articleList);
  //跳转至index页面
      return "client/index";
    }

    //文章详情查询
    @GetMapping(value = "/article/{id}")
    public String getArticleById(@PathVariable("id") Integer id, HttpServletRequest re-
quest){
      Article article = articleServiceImpl.selectArticleWithId(id);
      if(article! =null){
        //查询封装评论相关数据
```

```
            getArticleComments(request, article);
            // 更新文章点击量
            siteServiceImpl.updateStatistics(article);
            request.setAttribute("article",article);
            return "client/articleDetails";
        }else {
            logger.warn("查询文章详情结果为空,查询文章 id: "+id);
            // 未找到对应文章页面,跳转到提示页
            return "comm/error_404";
        }
    }

    // 查询文章的评论信息,并补充到文章详情里面
    private void getArticleComments(HttpServletRequest request, Article article) {
        if(article.getAllowComment()) {
            // cp 表示评论页码,commentPage
            String cp = request.getParameter("cp");
            cp = StringUtils.isBlank(cp) ? "1" : cp;
            request.setAttribute("cp", cp);
                PageInfo<Comment> comments = commentServiceImpl.getComments(article.
getId(),Integer.parseInt(cp),3);
            request.setAttribute("cp", cp);
            request.setAttribute("comments", comments);

        }
    }
}
```

第三步,编写前端页面的 index.html,应用 Thymeleaf 引入相应的信息,如示例代码 3-29 所示。

示例代码 3-29:index.html 页面

```
<!DOCTYPE html>
<html lang="en" xmlns:th="http://www.thymeleaf.org">
<div th:replace="/client/header::header(null,null)"/>
<body style="background-color:#f9ffff">
<header class="am-g am-g-fixed blog-fixed index-page">
    <div class="masthead" th:style="'background:url('+ @{/assets/img/about-bg.jpg}+');
background-size:cover;'" >
```

```
    <div class="overlay">
      <div class="container">
        <div class="row">
          <div class="site-heading">
            <h1 align="center"> 个人博客站 </h1>
            <h2 align="center"> 随笔 </h2>
          </div>
        </div>
      </div>
    </div>
  </div>
  <div>
    <! -- 信息描述 -->
    <div class="am-u-md-4 am-u-sm-12 blog-sidebar">
      <div class="blog-sidebar-widget blog-bor">
        <h2 class="blog-text-center blog-title"><span>Pote</span></h2>
        <img th:src="@{/assets/img/cat.jpg}" alt="touxiang" class="blog-entry-img"/>
        <p>
          Java 后台开发
        </p>
        <p>Pote 个人博客，主要发表 Java、Spring 等相关文章 </p>
      </div>
      <div class="blog-sidebar-widget blog-bor">
        <h2 class="blog-text-center blog-title"><span> 阅读排行榜 </span></h2>
        <div style="text-align: left">
          <th:block th:each="article : ${articleList}">
            <a style="font-size: 15px; " th:href="@{'/article/'+${article.id}}"
             th:text="${articleStat.index+1}+'、'+${article.title}+'('+${article.hits}+')'">
            </a>
            <hr style="margin-top: 0.6rem; margin-bottom: 0.4rem"/>
          </th:block>
        </div>
      </div>
    </div>
    <div class="am-u-md-8 am-u-sm-12">
      <! -- 文章遍历并分页展示 -->
      <div th:each="article: ${articles.list}">
        <article class="am-g blog-entry-article">
```

```
                <div class="am-u-lg-6 am-u-md-12 am-u-sm-12 blog-entry-img">
                    <img width="100%" class="am-u-sm-12" th: src="@{${commons.show_
thumb(article)}}"/>
                </div>
                <div class="am-u-lg-6 am-u-md-12 am-u-sm-12 blog-entry-text">
                  <span>      </span>
                  <! -- 发布时间 -->
                  <span style="font-size: 15px; " th: text="' 发布时间: '+ ${commons.date-
Format(article.created)}"/>
                  <h2>
                    <div><a style="color: #0f9ae0; font-size: 20px; "th: href="${commons.
permalink(article.id)}" th: text="${article.title}"/>
                    </div>
                  </h2>
                  <! -- 文章摘要 -->
                  <div style="font-size: 16px; "th: utext="${commons.intro(article, 75)}"/>
                </div>
              </article>

            </div>
            <! -- 文章分页信息 -->
            <div class="am-pagination">
                <div th: replace="/comm/paging: : pageNav(${articles}, '上一页', '下一
页','page')"/>
            </div>
          </div>
        </div>
    </header>
    </body>
    <! -- 载入文章尾部页面,位置在 /client 文件夹下的 footer 模板页面,模板名称 th:
fragment 为 footer -->
    <div th: replace="/client/footer: : footer"/>
    </html>
```

通过本任务的学习，读者能够了解 Controller 控制器的编写方式，学习 Thymeleaf 模板获取信息的方法，加深对于 Spring Boot 的理解，从而通过代码编写，实现整体的控制器模块功能，完成个人博客主页。

Mapping	映射	Resources	资源
Model	模型	Params	参数
Variable	变量		

一、填空题

1. 应用 ____________ 注解全面接管完成 Spring MVC 自动配置。

2. 在全局配置文件中使用 __________________________ 注解开启对 Thymeleaf 模板的参数配置。

3. @RestController 注解是 ____________ 注解和 _______________ 注解的组合使用。

4. 在 WebMvcConfigurer 接口中，可制定视图管理的方法为 ____________________。

5. 在启动项目时，_______________ 注解会扫描本目录以及子目录中带有 @WebServlet 注解的类。

二、简答题

简述 Controller 控制器的工作流程。

项目四 个人博客项目数据访问

通过学习 Spring Boot 整合 MyBatis、JPA、Redis 的相关知识，了解 Spring Boot 缓存的知识，熟悉 Spring Boot 的数据访问方法和应用场景，编写完善的 Spring Boot 个人博客项目。

- 掌握 Spring Boot 数据访问基本概念。
- 掌握 Spring Boot 整合 MyBatis。
- 掌握 Spring Boot 整合 JPA。
- 掌握 Spring Boot 整合 Redis。
- 掌握 Spring Boot 缓存管理。

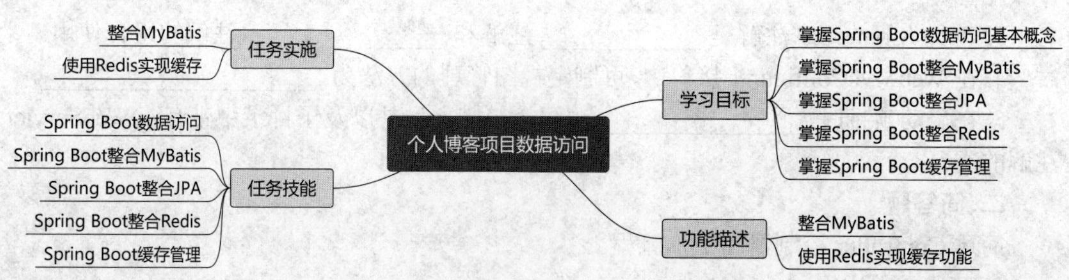

【情境导入】

在任何的项目中，都需要跟数据进行交互，否则该项目是没有意义的。Spring Boot 可以整合相关的框架，如持久层框架 MyBatis、JPA 框架、非关系型数据库 Redis 等。使用缓存的主要目的是减小数据库的访问压力、提高用户体验。开发人员在进行 Web 开发时，可以使用 Spring Boot 整合 Redis 实现缓存功能。

【功能描述】

- 整合 MayBatis。
- 使用 Redis 实现缓存功能。

技能点 1　Spring Boot 数据访问

数据访问是系统连接到数据源来实现数据访问的一种行为，现有的大部分系统都离不开数据访问。数据来源有很多，如本地文件、网络数据源或数据库。Web 应用一般使用数据库进行数据储存，其中数据库主要包括 SQL 和 NOSQL。SQL 指关系型数据库，常见的有 SQL Server、Oracle、MySQL（开源）；NOSQL 泛指非关系型数据库，常见的有 MongoDB、Redis。

在进行 Spring 项目开发时，一般常用的对象关系映射（Object Relational Mapping，ORM）框架有 JDBC、MyBatis 和 Hibernate。现在比较常用的是 MyBatis，其常见的数据库依赖项见表 4-1。

表 4-1 MyBatis 数据库依赖项表

名称	解释
spring-boot-start-data-redis	Redis 数据库与 Spring Data Redis 和 Redis 客户端的启动器
spring-boot-start-data-jpa	Spring Data JPA 和 Hibernate 的启动器
spring-boot-start-data-mongodb	Spring Data MongoDB 和 MongoDB 的启动器
spring-boot-start-data-neo4j	Spring Data Neo4j 和 Neo4j 数据库的启动器

技能点 2 Spring Boot 整合 MyBatis

1. MyBatis 简介

（1）MyBatis 概述及发展历史

MyBatis 源自 Apache 软件基金会的一个开源项目 iBatis。2010 年 6 月，这个项目由 Apache 转到了谷歌公司。由于开发人员转移到谷歌公司，iBatis 3.x 正式更名为 MyBatis，相关代码也于 2013 年 11 月迁移到 Github。当前，MyBatis 的最新版本是 MyBatis 3.5.6，其 logo 如图 4-1 所示。

图 4-1 MyBatis 的 logo

MyBatis 是持久层框架，它支持存储过程、高级映射。MyBatis 可以使用 XML 文件或注解来配置和映射原生信息，将接口和 Java 的 POJO 映射到数据库记录。MyBatis 的优点是减少了 Java 数据库连接代码、设置参数和获取结果集的工作。

（2）MyBatis 的特点

① MyBatis 框架非常小，而且简单易学。它没有任何的第三方依赖，只需要配置几个 SQL 映射文件和两个 JAR 包即可完成框架相关的配置。

② MyBatis 框架十分灵活，通过 SQL 语句可以满足操作数据库的任何需求。SQL 语句写在 XML 文件里，便于开发人员管理和优化。

③ MyBatis 框架解除 SQL 与程序代码之间的耦合，通过提供 DAO 层，将数据访问逻辑和业务逻辑分离，使系统的设计更加清晰，单元测试更加容易。SQL 和代码的分离，提高了开发人员的维护效率。

④ MyBatis 框架提供 XML 标签，可以编写动态 SQL，这是该框架的一大特点。

2. MyBatis 的功能架构

MyBatis 的功能架构主要分为三层，如图 4-2 所示。

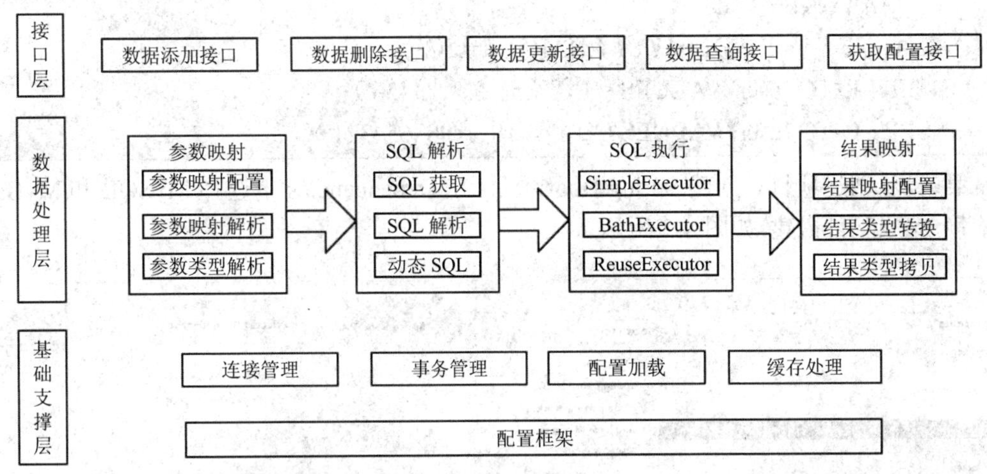

图 4-2　MyBatis 功能架构图

①接口层：负责提供外部使用的接口，开发人员通过这些接口来操纵数据库。当接收到调用请求时，它就会调用数据处理层去完成具体的数据处理。

②数据处理层：负责参数映射、SQL 解析、SQL 执行和结果映射。它的主要作用是根据请求完成数据库操作。

③基础支撑层：负责连接管理、事务管理、配置加载和缓存处理。它作为整个框架中最为基础的组件，为数据处理层提供了支撑。

3. Spring Boot 整合 MyBatis

虽然 Spring Boot 官方没有对 MyBatis 进行整合，但是 MyBatis 自行适配了对应的启动器，进一步简化了 MyBatis 对于数据的操作。具体整合 MyBatis 的步骤以下面的案例进行说明。

【案例】Spring Boot 整合 MyBatis。

第一步，新建数据库。创建一个名为 SpringBoot 的数据库，并在该数据库中创建名为 user 的表，并插入数据，如示例代码 4-1 所示。

示例代码 4-1：创建 user

```
// 创建数据库
CREATE DATABASE SpringBoot;
// 选择数据库
USER SpringBoot;
```

```
// 创建 user 表并插入数据
DROP TABLE IF EXISTS `user`;
CREATE TABLE `user` (
  `id` int(20) NOT NULL,
  `name` varchar(255) NOT NULL,
  `password` varchar(255) NOT NULL,
  PRIMARY KEY (`id`)
) ENGINE=InnoDB DEFAULT CHARSET=utf8;
INSERT INTO `user` VALUES ('1', '张三', '123456');
INSERT INTO `user` VALUES ('2', '李四', '987654321');
```

第二步，创建项目。创建 Spring Boot 项目，在 Dependencies 中选择 MySQL 和 MyBatis 的依赖，完成项目搭建，如图 4-3 所示。

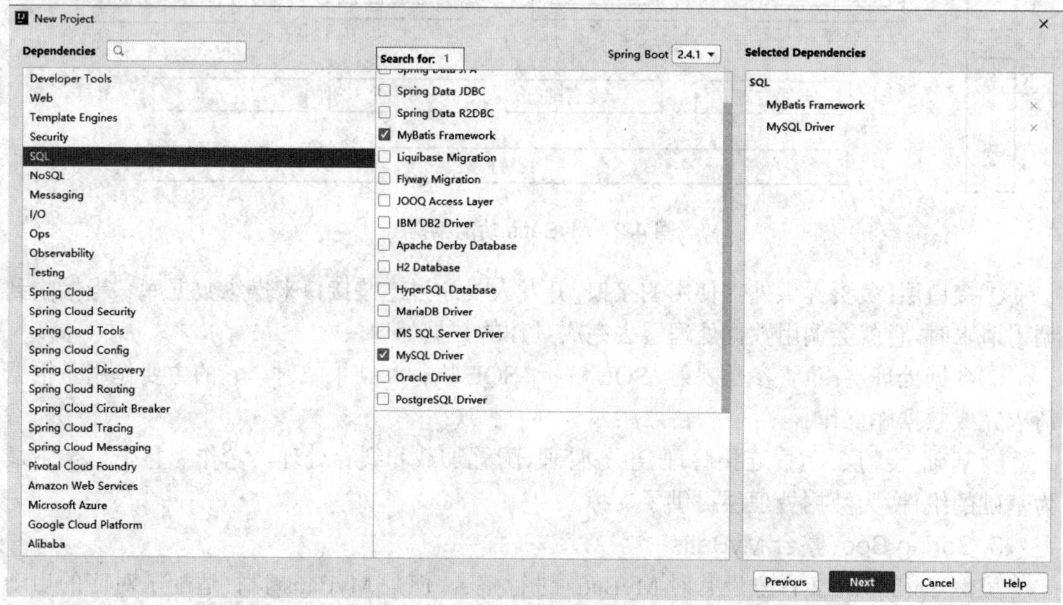

图 4-3 配置 Dependencies

第三步，编写 User 实体类。创建 POJO 包，在 POJO 包中创建与 user 表对应的 User 类，如示例代码 4-2 所示。

示例代码 4-2：User 实体类

```
public class User {
    private Integer id;
    private  String name;
    private  String password;
// 忽略 getter、setter 和 toString 方法。
}
```

第四步，在 pom.xml 中添加数据源依赖，如 Druid 数据源。Druid 是一个目前常用的数据库连接池，其在功能、性能、扩展性方面都超过其他数据库连接池，包括 DBCP、C3P0、BoneCP、Proxool、JBoss DataSource。添加数据源依赖的代码如示例代码 4-3 所示。

示例代码 4-3：pom.xml 文件

```
//Druid 数据源版本号
<properties>
<druid.version>1.1.16</druid.version>
</properties>
//Druid 数据源依赖
<dependency>
  <groupId>com.alibaba</groupId>
  <artifactId>druid</artifactId>
  <version>${druid.version}</version>
</dependency>
```

第五步，在 application.properties 配置文件中配置数据库连接并配置第三方数据源，需要修改第三方数据源的运行参数，可以在配置文件中修改，如示例代码 4-4 所示。

示例代码 4-4：application.properties 配置文件

```
server.port=8088
// 添加并配置 Druid 数据源
spring.datasource.type=com.alibaba.druid.pool.DruidDataSource
//MySQL 数据库连接配置
spring.datasource.url=jdbc：mysql：//localhost/springboot？ useSSL=true&serverTime-
zone=UTC&characterEncoding=UTF8
 spring.datasource.driver-class-name=com.mysql.cj.jdbc.Driver
 spring.datasource.username=root
 spring.datasource.password=root
```

第六步，整合 MyBatis。

（1）使用注解的方式整合 MyBatis

Spring Boot 与 MyBatis 整合，相比于与 Spring 整合来说更加便捷，省去了大量配置文件，并且支持 XML 与注解两种配置方法。

①创建 Mapper 接口，创建一个 mapper 包，在包下创建一个 UserMapper 接口，在接口类上添加了 @Mapper 注解，在编译之后会生成相应的接口实现类，能够被 Spring Boot 自动扫描到 Spring 容器中。如果想把每个接口都变成实现类，可以在 Spring Boot 启动类上添加 @MapperScan（"xxx"），就不需要将每个接口都添加上 @Mapper 注解；在接口的内部添加 @Select、@Delect、@Insert、@Update 注解完成增－查－更－删（Creat-Retrieve-Update-Delete，CRUD）操作，如示例代码 4-5 所示。

示例代码 4-5：UserMapper 接口

```
@Mapper
public interface UserMapper {
    @Insert("insert into user(id, name, password) values (#{id}, #{name}, {pass-
word})")
    public int insertUser(User user);
    @Select("select * from user where id=#{id}")
    public User finduserById(Integer id);
}
```

②编写单元测试，打开项目一使用 Initializr 方式构建的 demo 项目的代码，使用 Demo2ApplicationTests 测试类，通过 @Autowired 注解将 UserMapper 接口装配成 Bean，如示例代码 4-6 所示。

示例代码 4-6：Demo2ApplicationTests 测试类

```
@RunWith(SpringRunner.class)
@SpringBootTest
public class Demo2ApplicationTests {
    @Autowired
    private UserMapper userMapper;
    @Test
    public void selecctByID(){
        System.out.println(userMapper.finduserById(1));
    }
}
```

③选中所需测试方法，用鼠标右键点击"Run selecctByID()"启动测试方法，控制台上会显示相关信息，如图 4-4 所示。

```
2020-12-14 15:22:47.252  INFO 13472 --- [           main] com.alibaba.druid.pool.DruidDataSource   : {dataSource-1} inited
User{id=1, name='张三', password='123456'}

2020-12-14 15:22:47.885  INFO 13472 --- [extShutdownHook] com.alibaba.druid.pool.DruidDataSource   : {dataSource-1} closing ...
2020-12-14 15:22:47.918  INFO 13472 --- [extShutdownHook] com.alibaba.druid.pool.DruidDataSource   : {dataSource-1} closed

Process finished with exit code 0
```

图 4-4　测试整合图

（2）使用配置文件的方式整合 MyBatis

①创建一个 Mapper 接口文件，在 mapper 包下创建一个名为 ConsumerMapper 的接口文件，创建一个查询方法和一个添加方法，如示例代码 4-7 所示。

示例代码 4-7：ConsumerMapper 接口

```
@Mapper
public interface ConsumerMapper {
   int insertUser（User user）;
   User finduserById（Integer id）;
}
```

②创建 XML 映射文件。在 resources 目录下，创建一个名为 mapper 的包，在该包下创建一个名字与 ConsumerMapper 接口相同的 XML 文件。namespace 关键字在 XML 里为命名空间，通常是一个统一资源识别符（URI）的名字。而 URI 只当名字用，主要目的是为了避免名字发生冲突，在该 XML 文件中使用接口文件的全路径名称，编写两个方法所对应的 SQL 语句。在配置数据类型映射时，使用类的别名，如示例代码 4-8 所示。

示例代码 4-8：ConsumerMapper.xml 文件

```
<? xml version="1.0" encoding="UTF-8"? >
<! DOCTYPE mapper
      PUBLIC "-//mybatis.org//DTD Mapper 3.0//EN"
      "http://mybatis.org/dtd/mybatis-3-mapper.dtd">
<mapper namespace="com.aaa.demo2.mapper.UserMapper">
   <insert id="insertUser" parameterType="com.aaa.demo2.pojo.User">
      insert into user
      values（#{id},#{name}, {password}）;
   </insert>
   <select id="findUserById" resultType="com.aaa.demo2.pojo.User">
      select * from user where id=#{id}
   </select>
</mapper>
```

③配置 XML 映射文件的路径。需要在 application.properties 全局配置文件中添加 XML 映射文件的路径，因为在配置数据类型映射时使用了类的别名，所以需要配置 XML 映射文件中实体类的别名，如示例代码 4-9 所示。

示例代码 4-9：application.properties 配置文件

```
// 配置 XML 映射文件的路径
mybatis.mapper-Locations=classpath: mapper/*.xml
// 配置 XML 映射文件中实体类的别名
mybatis.type-aliases-package=com.aaa.demo2.pojo
```

④编写单元测试，在测试类中引入 ConsumerMapper 接口，对方法进行测试，如示例代码 4-10 所示。

示例代码 4-10：selecctByID1 测试方法

```
@Autowired
private ConsumerMapper consumerMapper;

@Test
public void selecctByID1(){
System.out.println(consumerMapper.finduserById(1));
}
```

⑤选中所需测试方法，用鼠标右键点击“Run finduserById()”启动测试方法，控制台上会显示相关信息，如图 4-5 所示。

```
2020-12-14 15:22:47.252  INFO 13472 --- [           main] com.alibaba.druid.pool.DruidDataSource   : {dataSource-1} inited
User{id=1, name='张三', password='123456'}

2020-12-14 15:22:47.885  INFO 13472 --- [extShutdownHook] com.alibaba.druid.pool.DruidDataSource   : {dataSource-1} closing ...
2020-12-14 15:22:47.918  INFO 13472 --- [extShutdownHook] com.alibaba.druid.pool.DruidDataSource   : {dataSource-1} closed

Process finished with exit code 0
```

图 4-5 测试结果

技能点 3 Spring Boot 整合 JPA

JPA 标准是 Java 社区流程（Java Community Process，JCP）组织于 2005 年推出的三个 JavaEE 系列标准之一。所有声称符合新的 JPA 系列标准的应用框架均必须遵循相同的标准架构，提供相同的访问 API，这可以保证基于 JPA 标准开发的大型企业应用，经过少量的修改就能够在不同的 JPA 标准框架下运行。

推出 JPA 标准的主要目的之一是提供更简单的程序模型。例如，在 JPA 框架中创建实体，就如同创建 Java 类一样，没有任何的约束和限制，只需使用 javax.persistence.entity 进行注解。JPA 框架的界面也非常简单，没有过多的特殊规则。开发人员很容易就能掌握其设计模式和要求。 JPA 是基于非侵入原则设计的，因此它可以与其他框架或容器相结合。JPA 的实现原理如图 4-6 所示。

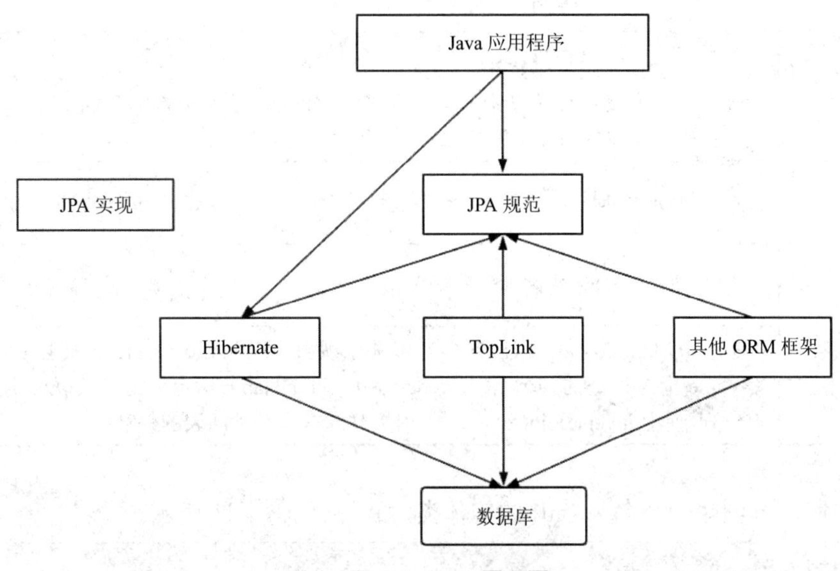

图 4-6　JPA 原理图

1. JPA 的基本使用

Spring Data JPA 提供了 CRUD 的功能，可以使开发人员用较少的代码实现数据操作。在讲解 Spring Boot 整合 JPA 之前，先对 Spring Data JPA 的基本使用进行介绍，具体内容如下。

第一步，编写 ORM 实体类。

① @Entity 表明该类（staff）为一个实体类，它默认对应数据库中的表名是 staff。可以写成 @Entity（name = "staff "），其中的 name，表示对应的数据库中的表名。

② @Id 的属性为主键。该属性值可以通过自身创建，推荐通过 Hibernate 生成。

③ @GeneratedValue 用于指定主键的生成策略。其包括四个值，见表 4-2。

表 4-2　@GeneratedValue 的参数表

值	解释
TABLE	使用表保存 id 值
IDENTITY	主键自增
SEQUENCR	不支持主键自增的数据库主键生成
AUTO	根据数据库的类型自动使用上面三个值

④关系映射所使用的注解，见表 4-3。

表 4-3 关系映射注解表

名称	解释
@ManyToOne 单向多对一	使用 @ManyToOne 来映射多对一关系映射，可以使用 @ManyToOne 的 fetch 来修改默认的关联属性的加载策略
@OneToMany 单向一对多	使用 @OneToMany 来映射一对多关系映射，使用 @JoinColumn 来映射外键
@OneToOne 双向一对一	使用 @OneToOne 来映射一对一关系映射
@ManyToMany 双向多对多	使用 @ManyToMany 来映射多对多关系映射，使用 @JoinTable 来映射中间表，使用 name 指定外键的列名，使用 rederenedColumnName 指定外键列关联的当期表的那一列，使用 inverseJoinColumns 映射关联的类所在中间表的外键

根据上述注解，创建一个名为 staff 的实体类，如示例代码 4-11 所示。

示例代码 4-11：Staff 实体类

```
@Entity（name = "staff"）
public class Staff {
    @Id
    @GeneratedValue（strategy = GenerationType.IDENTITY）
    private Integer id;
    private String name;
    private String email;
// 忽略 getter、setter 和 toString 方法。
}
```

第二步，编写 Repository 接口。

创建 StaffRepository 接口，并编写所需的数据操作方法。

StaffRepository 接口继承 JpaRepository<T, ID> 接口，其中 T 代表需要操作的实体类，ID 代表该实体类主键的数据类型。

StaffRepository 接口继承 JpaRepository 接口，默认包含了常用的 CRUD 操作。在 StaffRepository 接口中，可以使用方法名的关键字进行查询，各关键字见表 4-4。

表 4-4 方法名的关键字表

关键字	方法名	SQL 语句
Is，Equals	findByFirstname，findByFirstnameIs	where x.firstname = ？1
Between	findByStartDateBetween	where x.startDate between ？1 and ？2
LessThan	findByAgeLessThan	where x.age < ？1
LessThanEqual	findByAgeLessThanEqual	where x.age <= ？1
GreaterThan	findByAgeGreaterThan	where x.age > ？1

续表

关键字	方法名	SQL 语句
GreaterThanEqual	findByAgeGreaterThanEqual	where x.age >= ? 1
After	findByStartDateAfter	where x.startDate > ? 1
Before	findByStartDateBefore	where x.startDate < ? 1
Not	findByLastnameNot	where x.lastname <> ? 1
In	findByAgeIn（Collection<Age> ages）	where x.age in ? 1
NotIn	findByAgeNotIn（Collection<Age> age）	where x.age not in ? 1
True	findByActiveTrue（）	where x.active = true
False	findByActiveFalse（）	where x.active = false
IsNull	findByAgeIsNull	where x.age is null
IsNotNull，NotNull	findByAge（Is）NotNull	where x.age not null
Like	findByFirstnameLike	where x.firstname like ? 1
NotLike	findByFirstnameNotLike	where x.firstname not like ? 1
StartingWith	findByFirstnameStartingWith	where x.firstname like ? 1（%）
EndingWith	findByFirstnameEndingWith	where x.firstname like ? 1（%）
Containing	findByFirstnameContaining	where x.firstname like ? 1（%）
IgnoreCase	findByFirstnameIgnoreCase	where UPPER（x.firstame） = UPPER（? 1）
Or	findByLastnameOrFirstname	where x.lastname = ? 1 or x.firstname = ? 2
And	findByLastnameAndFirstname	where x.lastname = ? 1 and x.firstname = ? 2
OrderBy	findByAgeOrderByLastnameDesc	where x.age = ? 1 order by x.lastname desc

@Query 注解适用于所查询的数据无法通过关键字查询得到结果时的查询。

在编写实现删除或修改操作代码的时候，必须加上 @Modifying 注解，用于通知 Spring Data 这是一个删除或修改操作。

对于修改、删除的数据操作，必须在方法上添加 @Transactional 注解，进行事务管理，如果在 Service 层上已经添加了 @Transactional 注解，那么在 StaffRepository 接口中便不需要添加。

根据上述内容，创建一个名为 StaffRepository 的接口，如示例代码 4-12 所示。

示例代码 4-12：StaffRepository 接口

```
public interface StaffRepository extends JpaRepository<Staff, Integer> {
  List<Staff> findByEmail(String email);
@Transactional
```

```
    @Modifying
    @Query("update staff s set s.name = ? 1 where s.id = ? 2")
    int updateStaff(String name, Integer id);
}
```

JpaRepository 接口的继承结构如图 4-7 所示。

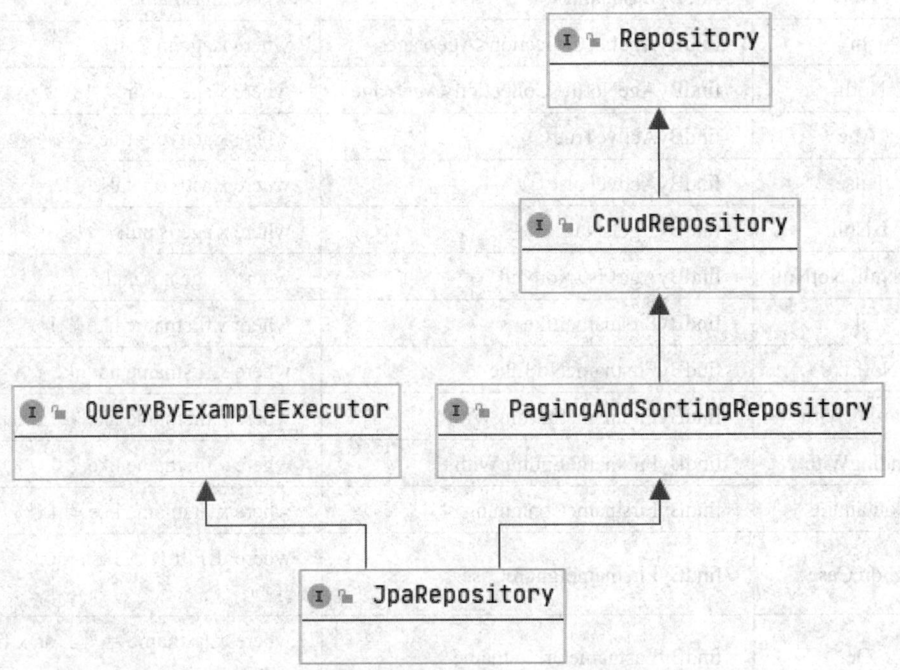

图 4-7 JpaRepository 接口的继承结构

① JpaRepository 接口继承了 PagingAndSortingRepository 接口和 QueryByExampleExecutor 并且提供了一些数据操作方法。其中，PagingAndSortingRepository 接口继承 CrudRepository 接口，并提供了排序和分页方法；QueryByExampleExecutor 接口，允许使用 Example 实例进行复杂的条件查询。

② CrudRepository 接口继承了 Repository 接口，并提供了一些简单的 CRUD 方法。

③ Repository 接口是顶级的父类接口，该接口中没有任何的方法。

2. Spring Boot 整合 JPA

前文中讲解了如何使用 JPA，现在将讲解如何实现 Spring Data 整合 JPA，具体步骤如以下案例所示。

【案例】Spring Boot 整合 JPA。

第一步，添加 JPA 依赖。在项目中引入 JPA 的 Maven 依赖，如示例代码 4-13 所示。

示例代码 4-13：pom.xml 文件

```
<dependency>
   <groupId>org.springframework.boot</groupId>
```

```
    <artifactId>spring-boot-starter-data-jpa</artifactId>
</dependency>
```

第二步，创建实体类。创建 Users 实体类，并创建 t_users 数据表，根据上文所述添加 JPA 对应的注解，如示例代码 4-14 所示。

示例代码 4-14：Users 实体类

```
@Entity(name = "t_users")
public class Users {
    @Id
    @GeneratedValue(strategy = GenerationType.IDENTITY)
    private Integer id;
    private String name;
    private Integer age;
    private String address;
// 忽略 getter、setter 和 toString 方法。
}
```

第三步，创建 UsersRepository 接口。创建一个 com.aaa.demo2.repository 包，并在该包下创建一个 UsersRepository 接口，根据之前所学习的知识，编写数据操作的方法，如示例代码 4-15 所示。

示例代码 4-15：UsersRepository 接口

```
public interface UsersRepository extends JpaRepository<Users, Integer> {
    // 根据姓名查询
List<Users> findByName(String name);
// 根据 id 修改客户姓名
    @Query("update t_users u set u.name=? 1 where u.id=? 2")
    @Modifying
    void updateUsersNameById(String name, Integer id);
}
```

第四步，编写单元测试。编写多个测试方法对 UsersRepository 接口的方法进行测试，分别对 JpaRepository 内部方法、方法名关键字、@Query 注解进行数据操作，如示例代码 4-16 所示。

示例代码 4-16：Demo2ApplicationTests 测试类

```
@RunWith(SpringRunner.class)
@SpringBootTest
public class Demo2ApplicationTests {
@Autowired
    private UsersRepository usersRepository;
```

```
// 使用 JpaRepository 内部方法进行数据操作
    @Test
    public void save(){
        Users users=new Users(null," 李四 ",24," 天津市 ");
        System.out.println(usersRepository.save(users));
    }
// 使用方法名关键字进行数据操作
    @Test
    public void findByName(){
        System.out.println(usersRepository.findByName(" 张三 "));
    }
// 使用 @Query 注解进行数据操作
    @Test
    @Transactional //@Transactional 与 @Test 一起使用时,事务是自动回滚的
    @Rollback(false)// 取消自动回滚
    public void testUpdateUsersNameById() {
        this.usersRepository.updateUsersNameById(" 张三三 ", 1);
    }
}
```

第五步,选中所需测试方法。用鼠标右键点击"Run finduserById()"启动测试方法,控制台上会显示如下信息,如图 4-8 所示。

图 4-8　测试图

技能点 4　Spring Boot 整合 Redis

1. Redis 简介

Redis 是一种开放源代码的内存数据存储结构，可用作数据库、缓存和信息代理等。Redis 支持五种数据类型：string（字符串）、hash（哈希）、list（列表）、set（集合）及 zset（sorted set：有序集合）。 Redis 具有内置的复制、Lua 脚本、最近最少使用（Least Recently Used，LRU）驱逐算法、事务和不同级别的磁盘持久性，并通过 Redis Sentinel 和 Redis Cluster 自动分区，提供了具有高可用性的功能，可用于缓存、事件发布或订阅、高速队列等场景。

Redis 有以下优点。

①性能极好：读取速度与写入速度都非常快。

②拥有 6 种数据结构：可以满足存储各种数据结构体的需要，数据类型少，则所需的逻辑和判断就少，提高了读 / 写速度。

③原子性：要求操作要么成功执行要么完全失败（不执行）。单个操作是原子性的；多个操作则支持多个事务，通过 multi 和 exec 指令包来实现。

④功能丰富：其堆栈支持 publish / subscribe 、通知和 key 过期等特征。

Redis 拥有良好的性能和丰富的功能，但仍有一些缺点，如对持久性软件支持得不够好。一般情况下，在 Redis 中，孤立的业务数据不作为主业务数据库中的数据进行存储，这些数据的存储是与一些传统的关系型业务数据库互相配合完成。

2. Redis 安装使用

第一步，在 Windows 操作系统下安装。Redis 支持 Windows、Linux、Docker 镜像安装等安装方式。

Redis 下载地址为 https: //github.com/tporadowski/redis/releases。在浏览器中输入网址，进入 gethub，选择 Redis 版本，下载 Redis-x64-xxx.zip 压缩包，如图 4-9 所示。

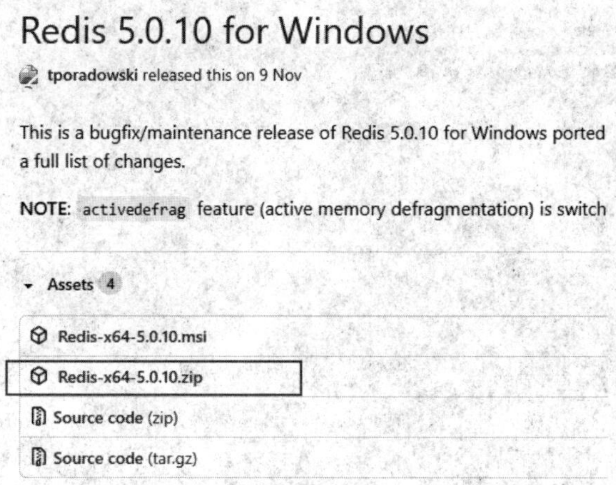

图 4-9　下载图

第二步，将压缩包解压到文件夹中。解压后，将文件夹重新命名为“redis”，不需要任何配置，Redis 就安装完成，如图 4-10 所示。

图 4-10　解压图

第三步，在安装完成之后，开启 Redis 服务。

Redis 的安装包在解压后，会产生有多个目录和文件，其中有两个重要的可执行文件：redis-cli.exe 和 redis-server.exe。redis-server.exe 用于开启 Redis 服务，redis-cli.exe 用于开启客户端工具。

用鼠标左键双击 redis-server.exe，在终端窗口会显示 Redis 的版本和默认启动端口号 6379，如图 4-11 所示。

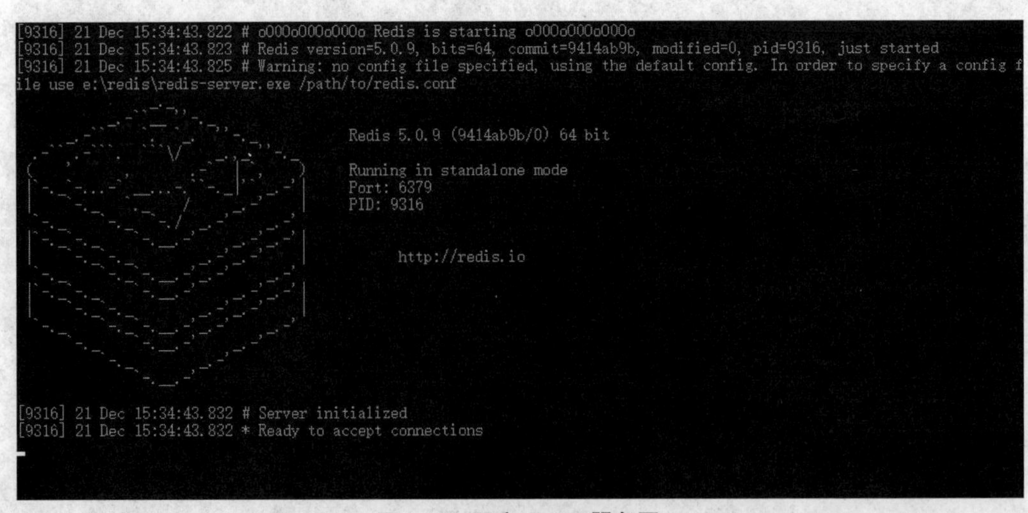

图 4-11　开启 Redis 服务图

第四步，下载安装 Redis 可视化客户端工具。访问官网下载地址：https：//redisdesktop.com/pricing，将该工具下载，如图 4-12 所示。

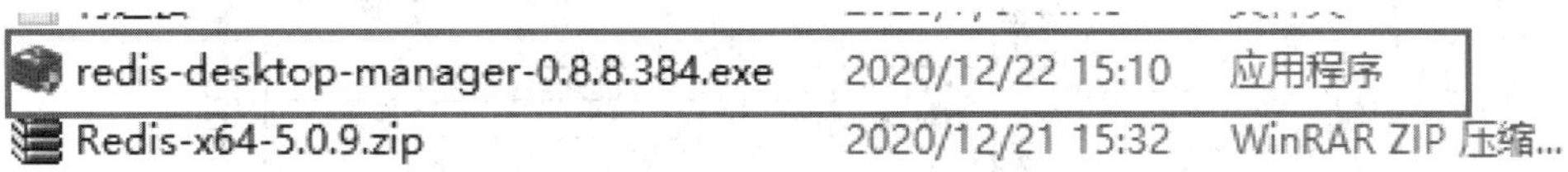

redis-desktop-manager-0.8.8.384.exe 2020/12/22 15:10 应用程序
Redis-x64-5.0.9.zip 2020/12/21 15:32 WinRAR ZIP 压缩...

图 4-12 下载可视化客户端工具图

第五步，下载完成之后，用鼠标左键双击文件进行安装，效果如图 4-13 所示。

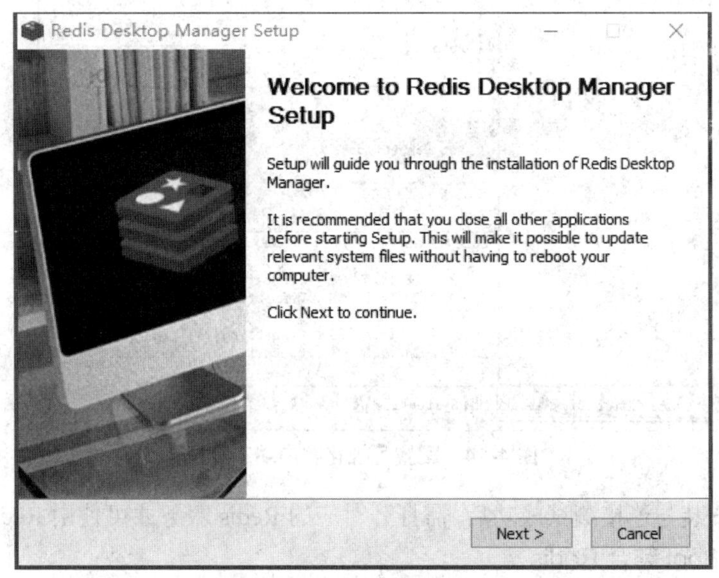

图 4-13 安装可视化客户端工具图

第六步，依次打开 Redis 服务和可视化工具 Redis Desktop。

点击“Connect to Redis Server”打开 Redis 连接配置，在窗口中填写对应的连接名称（Name），连接主机（Host，默认地址是 127.0.0.1），连接端口（Port，默认端口号为 6397）。然后点击“Test Connection”进行连接测试，连接成功，点击“OK”。至此，Redis 服务开启与连接配置全部完成，如图 4-14 所示。

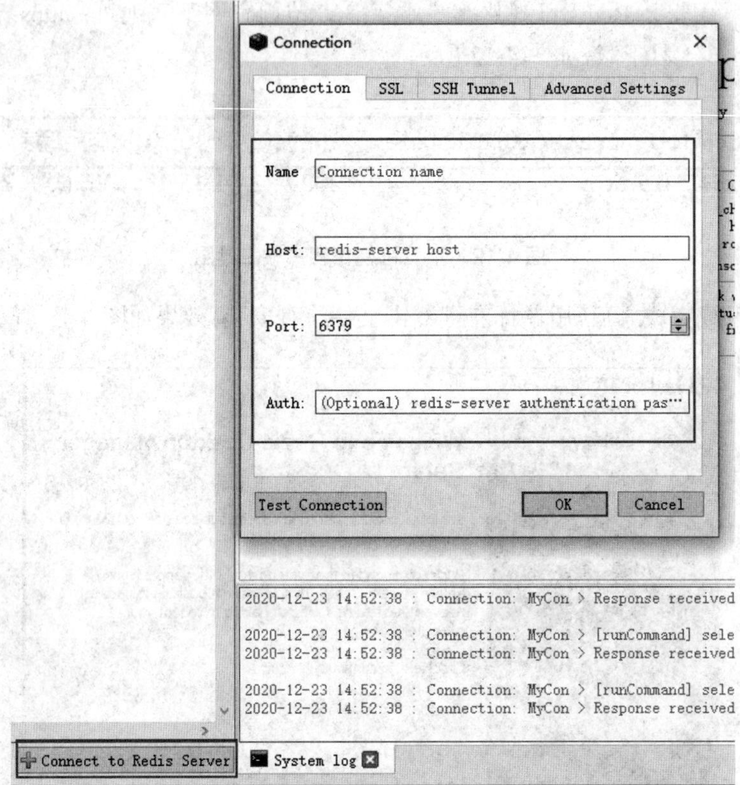

图 4-14　配置可视化客户端工具图

在 Redis 安装与连接测试成功后，需打开对应的 Redis 服务才可使用 Redis。

3. Spring Boot 整合 Redis

Spring Boot 除了支持关系型数据库的整合，对非关系型数据库的整合同样支持。关于 Spring Boot 与 Redis 的整合使用，如下面案例所示。

【案例】Spring Boot 整合 Redis。

第一步，添加依赖。在 pom.xml 文件中，添加 spring-boot-starter-data-redis 依赖，如示例代码 4-17 所示。

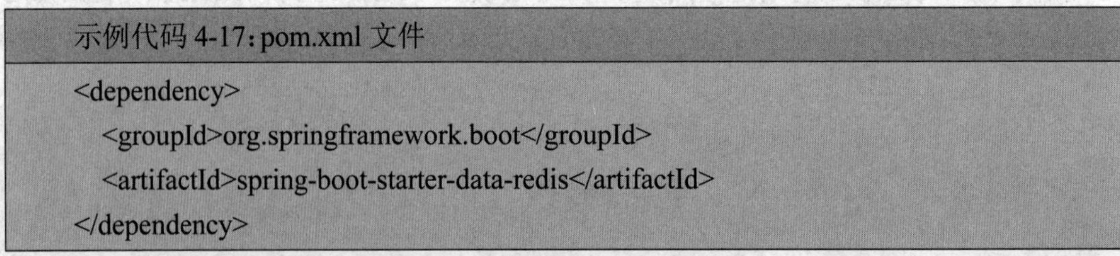
示例代码 4-17：pom.xml 文件

```
<dependency>
  <groupId>org.springframework.boot</groupId>
  <artifactId>spring-boot-starter-data-redis</artifactId>
</dependency>
```

第二步，Redis 数据库连接配置。在全局配置文件 application.properties 中添加 Redis 数据库连接，主要配置 Redis 数据库的服务地址（默认为 127.0.0.1）、连接端口（默认为 6379）、连接密码（默认为空），如示例代码 4-18 所示。

示例代码 4-18：application.properties 配置文件

```
#Redis 服务地址，默认地址为 127.0.0.1
spring.redis.host=127.0.0.1
#Redis 服务器连接端口，默认端口号为 6379
spring.redis.port=6379
#Redis 服务器连接密码，默认密码为空
spring.redis.password=
```

第三步，编写实体类。在之前所创建的项目中，创建三个实体类，即在 com.example.demo2.pojo 包下创建名为 Department（部门）、Staffs（员工）、Role（角色）的三个实体类，分别如示例代码 4-19、4-20、4-21 所示。

其中，@Id 声明此属性为主键，Redis 数据库会默认生成字符串形式的 HashKey 表示 ID；@Indexed 用于在 Redis 数据库中生成二级索引，以便方便地进行数据查询；@RedisHash（"staffs"）用于操作实体类对象在 Redis 数据库的存储空间。

示例代码 4-19：Deparment 实体类

```
public class Department {
    @Indexed
    private String departmentname;
}
```

示例代码 4-20：Staffs 实体类

```
@RedisHash("staffs")
public class Staffs {
    @Id
    private Integer id;
    @Indexed
    private String name;
    private Department department;
    private List<Role> role;
}
```

示例代码 4-21：Role 实体类

```
public class Role {
    @Indexed
    private String rolename;
}
```

第四步，编写 StaffsRepository 接口。spring-data 是对于数据库的 CRUD 操作的中央存储库，通过它可以对数据库（Redis）进行各种操作，基本的实现方法是用 CrudRepository，其

与 Spring Data JPA 操作数据的方法基本相同。

创建一个名为 StaffsRepository 的接口，如示例代码 4-22 所示。

示例代码 4-22：StaffsRepository 接口

```
public interface StaffsRepository  extends CrudRepository<Staffs，Integer> {
    List<Staffs> findByName（String name）；
    List<Staffs> findByRoleList_Rolename（Role role_rolename）；
}
```

第五步，编写单元测试。编写多个测试方法对 StaffsRepository 接口的方法进行测试，分别对 Redis 数据库增加和查询进行数据操作，如示例代码 4-23 所示。

示例代码 4-23：RedisTest 测试类

```
@RunWith（SpringRunner.class）
@SpringBootTest
public class RedisTest {
    @Autowired
    private StaffsRepository staffsRepository；

    @Test
    public void saveStaff（）{
        Staffs staffs =new Staffs（null，" 张三 "）；
        Department department =new Department（" 研发部 "）；
        staffs.setDepartment（department）；
        List<Role> list=new ArrayList<>（）；
        Role aaa=new Role（" 管理员 "）；
        list.add（aaa）；
        staffs.setRole（list）；
        Staffs save= staffsRepository.save（staffs）；
        System.out.println（save）；
    }
@Test
public void findStaff（）{
    List<Staffs> list=staffsRepository.findByDepartment_Departmentname（" 研发部 "）；
    System.out.println（list）；
}
}
```

第六步，选中所需的测试方法，用鼠标右键点击“Run saveStaff（）”启动测试方法，控制台上会显示如下信息，如图 4-15 所示。

```
2020-12-24 08:54:38.546  INFO 5252 --- [           main] com.aaa.demo2.RedisTest
Staffs{id=1, name='张三',department=Department{,departmentname='研发部'},role=[com.aaa.demo2.pojo.R
2020-12-24 08:54:47.517  INFO 5252 --- [extShutdownHook] o.s.s.concurrent.ThreadPoolTaskExecutor
```

图 4-15　测试图

打开 Redis Desktop，可以看到 saveStaff()方法执行成功，数据已经添加到 Redis 数据库中，如图 4-16 所示。如图，在窗口的左侧还生成了一些二级索引，可以通过二级索引来进行数据查询。

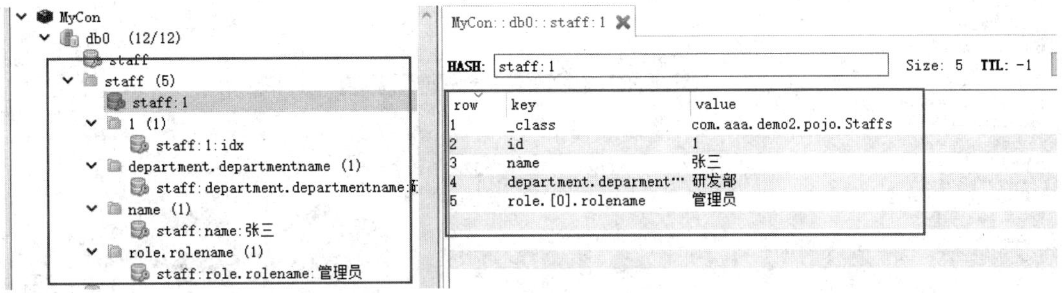

图 4-16　客户端测试图

测试 findStaff 方法，显示结果如图 4-17 所示。

```
2020-12-24 09:29:42.463  INFO 14164 --- [           main] com.aaa.demo2.RedisTest                  :
[Staffs{id=1, name='张三',department=Department{,departmentname='研发部'},role=[com.aaa.demo2.pojo.Ro
2020-12-24 09:29:43.736  INFO 14164 --- [extShutdownHook] o.s.s.concurrent.ThreadPoolTaskExecutor  :
```

图 4-17　测试图

技能点 5　Spring Boot 缓存管理

缓存是用于提升系统的性能、加速系统的访问、降低成本的一种技术。可以将一些高频、热点信息放入缓存中，避免直接从数据库中查询，如网上商城中的商品首页这种经常被访问的数据。

Spring Boot 支持向应用程序透明地添加缓存。开发人员可以自由地选择缓存的具体实现方式。其核心是将缓存应用于方法，从而减少了缓存中可用信息的执行次数。缓存逻辑对于应用程序来说是透明的，对调用程序没有干扰。

1. 基础环境搭建

使用缓存的主要目的是减小数据库的访问压力、提高用户体验。对于 Spring Boot 的缓存管理，其具体步骤如以下案例所示。

【案例】基础环境搭建。

第一步，使用 Spring Initializr 创建一个名为 demo3 的 Spring Boot 项目，需要在“Dependencies”选项中选择 Spring Web 依赖、SQL 模块中的 Spring Data JPA 依赖和 MySQL Driver 依赖，如图 4-18 所示。

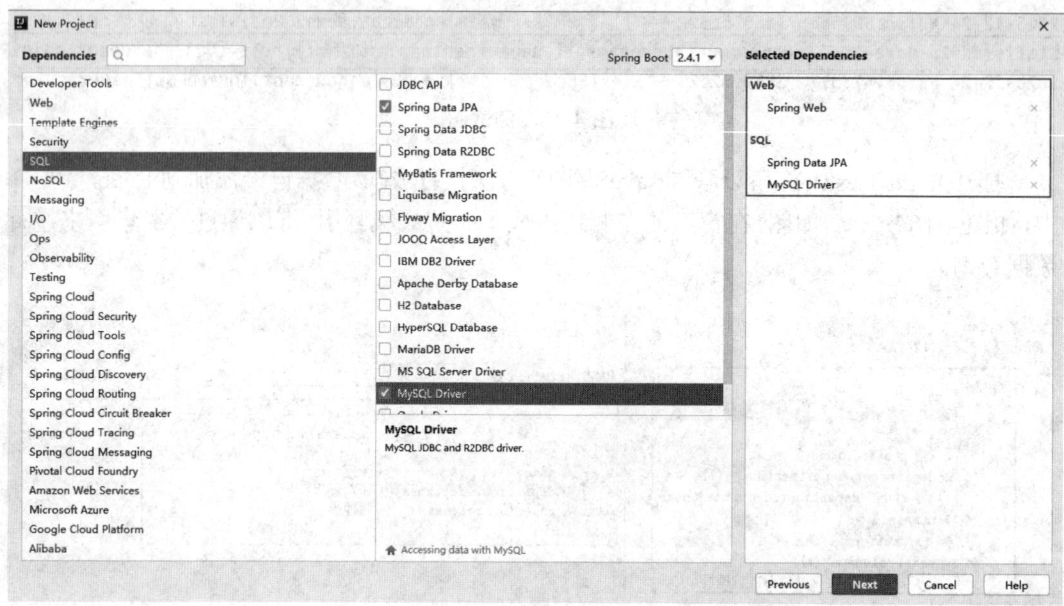

图 4-18　创建项目图

第二步，编写全局配置文件。在全局配置文件 application.properties 中配置 MySQL 数据库连接池，并为之后开启缓存后进行效果演示，添加“spring.jpa.show-sql=true”语句，用于显示 SQL 语句，如示例代码 4-24 所示。

示例代码 4-24：application.properties 配置文件

```
server.port=8088
spring.datasource.type=com.alibaba.druid.pool.DruidDataSource
spring.datasource.url=jdbc：mysql：//localhost/springboot？ useSSL=true&serverTime-
zone=UTC&characterEncoding=UTF8
spring.datasource.driver-class-name=com.mysql.cj.jdbc.Driver
spring.datasource.username=root
spring.jpa.show-sql=true
```

第三步，编写实体类。在 com.example.demo3 包下创建 pojo 包，在 pojo 包中创建 Users 类，并使用 JPA 相关注解配置。@JsonIgnoreProperties（value = {“hibernateLazyInitializer”}）的作用是在 JSON 序列化时忽略 Bean 中的一些不需要转化的属性，如示例代码 4-25 所示。

示例代码 4-25：Users 实体类

```
@JsonIgnoreProperties（value = { "hibernateLazyInitializer"}）
@Entity（name = "t_users"）
public class Users {
```

```
    @Id
    @GeneratedValue(strategy = GenerationType.IDENTITY)
    private Integer id;
    private String name;
    private Integer age;
    private String address;
}
```

第四步，创建一个 com.example.demo3.repository 包，并在该包下创建一个 UsersRepository 接口，根据学习 Spring Boot 整合 JPA 时的知识点，编写数据操作的方法，如示例代码 4-26 所示。

示例代码 4-26：UsersRepository 接口

```
public interface UsersRepository extends JpaRepository<Users,Integer> {
    List<Users> findByName(String name);
//修改用户名
    @Query("update t_users u set u.name=? 1 where u.id=? 2")
    @Modifying
    int updateUsersNameById(Users uers);
}
```

第五步，创建一个 com.example.demo3.service 包，并在该包下创建一个 UserService 接口，用于相关业务的操作，如示例代码 4-27 所示。

示例代码 4-27：UserService 接口

```
public interface UserService {
    boolean update(Users users);
    Users getById(Integer id);
}
```

第六步，在 com.example.demo3.service 包下创建一个 UserServiceImpl 实现类，注入 usersRepository 实例，完成查询、修改数据的操作，如示例代码 4-28 所示。

示例代码 4-28：UserServiceImpl 实现类

```
@Service
public class UserServiceImpl implements UserService {
    @Autowired
    private  UsersRepository usersRepository;
    @Override
    public boolean update(@RequestBody Users users) {
        if(null == users ){
```

```
            System.out.println(users);
            return false;
        }
        return usersRepository.updateUsersNameById(users)>0;
    }
    @Override
        public Users getById(Integer id) {
        if(null == id || id < 1) {
            return null;
        }
        return usersRepository.getOne(id);
    }
}
```

第七步，创建一个 com.example.demo3.controller 包，并在该包下创建一个 UserController 类，用于访问控制，如示例代码 4-29 所示。

示例代码 4-29：UserController 类

```
@RestController
public class UserController {
    @Autowired
    private UserServiceImpl userServiceImpl;
    @GetMapping("/get/{id}")
    public Users findByID(@PathVariable("id") int id){
        return userServiceimpl.getById(id);
    }
}
```

第八步，项目测试。启动项目，在浏览器的地址栏中输入访问地址“localhost：8088/get/1”，显示效果如图 4-19 所示。

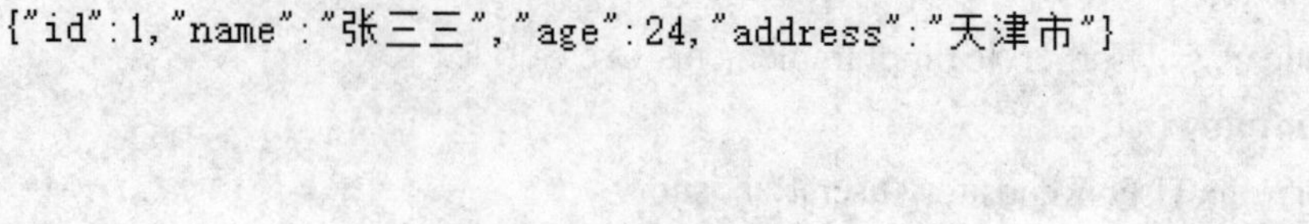

```
localhost:8088/get/1
{"id":1,"name":"张三三","age":24,"address":"天津市"}
```

图 4-19 测试图

由于没有开启缓存，虽然数据并没有发生变化，但是每一次数据查询，都会执行一次 SQL 语句，这会使数据规模变得越来越大，影响系统的性能、增加成本，影响用户的使用。

因此，可以将一些高频、热点信息放入缓存中，如图 4-20 所示。

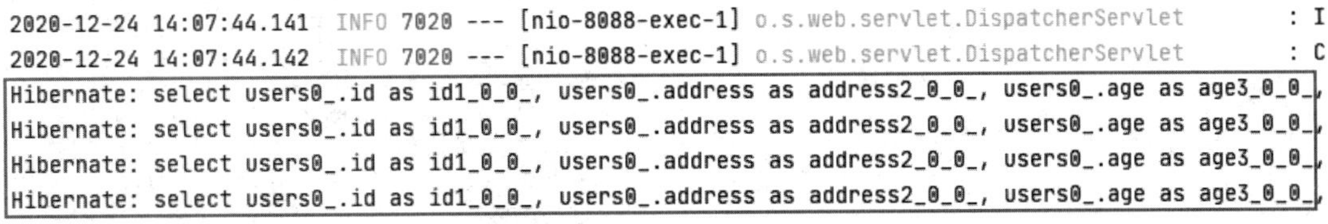

```
2020-12-24 14:07:44.141  INFO 7020 --- [nio-8088-exec-1] o.s.web.servlet.DispatcherServlet        : I
2020-12-24 14:07:44.142  INFO 7020 --- [nio-8088-exec-1] o.s.web.servlet.DispatcherServlet        : C
Hibernate: select users0_.id as id1_0_0_, users0_.address as address2_0_0_, users0_.age as age3_0_0_,
Hibernate: select users0_.id as id1_0_0_, users0_.address as address2_0_0_, users0_.age as age3_0_0_,
Hibernate: select users0_.id as id1_0_0_, users0_.address as address2_0_0_, users0_.age as age3_0_0_,
Hibernate: select users0_.id as id1_0_0_, users0_.address as address2_0_0_, users0_.age as age3_0_0_,
```

图 4-20　测试图

2. Spring Boot 默认缓存

（1）JSR-107 缓存规范

Java Caching 是 JSR-107 的最终定制规范，开发人员共同遵守这一规范以便简化沟通，让开发更加轻松，其主要的目标如下。

- 为应用程序提供缓存 Java 对象的功能。
- 定义了一套通用的缓存概念和工具。
- 最小化开发人员使用缓存的学习成本。
- 最大化应用程序在使用不同缓存实现之间的可移植性。
- 支持进程内和分布式的缓存实现。
- 支持 by-value 和 by-reference 的缓存对象。
- 定义了一套 Java 编程语言的元素。

Java Caching 定义了 5 个核心接口，分别是 CachingProvider、CacheManager、Cache、Expiry 和 Entry，见表 4-5。

表 4-5　Java Caching 的核心接口表

核心接口	解释
CachingProvider	定义了创建、配置、获取、管理和控制多个 CacheManager；一个应用可以在运行期访问多个 CachingProvider
CacheManager	定义了创建、配置、获取、管理和控制多个唯一命名的 Cache，这些 Cache 存在于 CacheManager 的上下文中
Cache	是一个类似 Map 的数据结构并临时存储以 Key 为索引的值；一个 Cache 仅被一个 CacheManager 所拥有
Expiry	每一个存储在 Cache 中的条目有一个定义的有效期，一旦超过这个有效期，条目便为过期状态
Entry	是一个存储在 Cache 中的 key-value 键值对

ConfigurableEnvironment 类如图 4-21 所示。

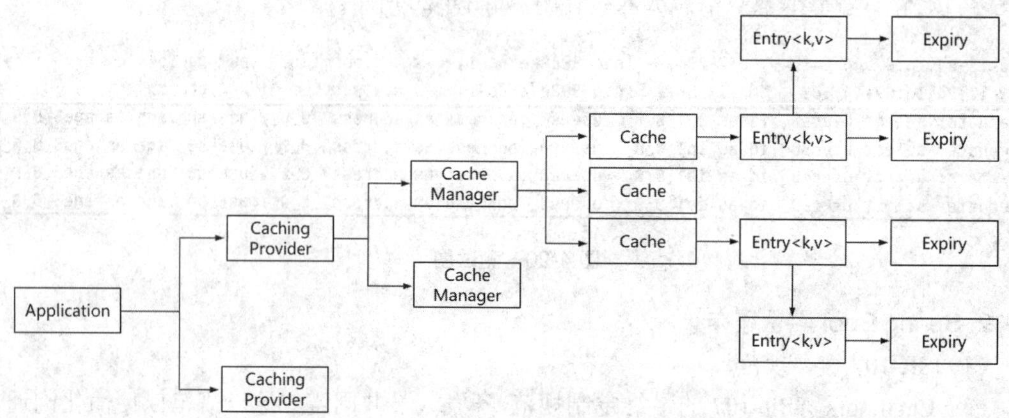

图 4-21 ConfigurableEnvironment 类图

（2）Spring 缓存抽象

Spring 缓存抽象的核心是将缓存应用于 Java 方法，这样就减少了基于缓存中可用信息的执行次数，即每次在调用目标方法时都会检查该方法是否已经给定参数并执行过。若有，则直接返回缓存的结果，如果没有，则执行方法，缓存结果，以便在下次访问时返回缓存的结果。从 Spring 3.1 版起，系统中便定义了 org.springframework.cache.Cache 和 org.springframework.cache.CacheManager 接口，来统一不同的缓存技术并支持使用 JCache（JSR-107）注解来简化开发，如图 4-22 所示。

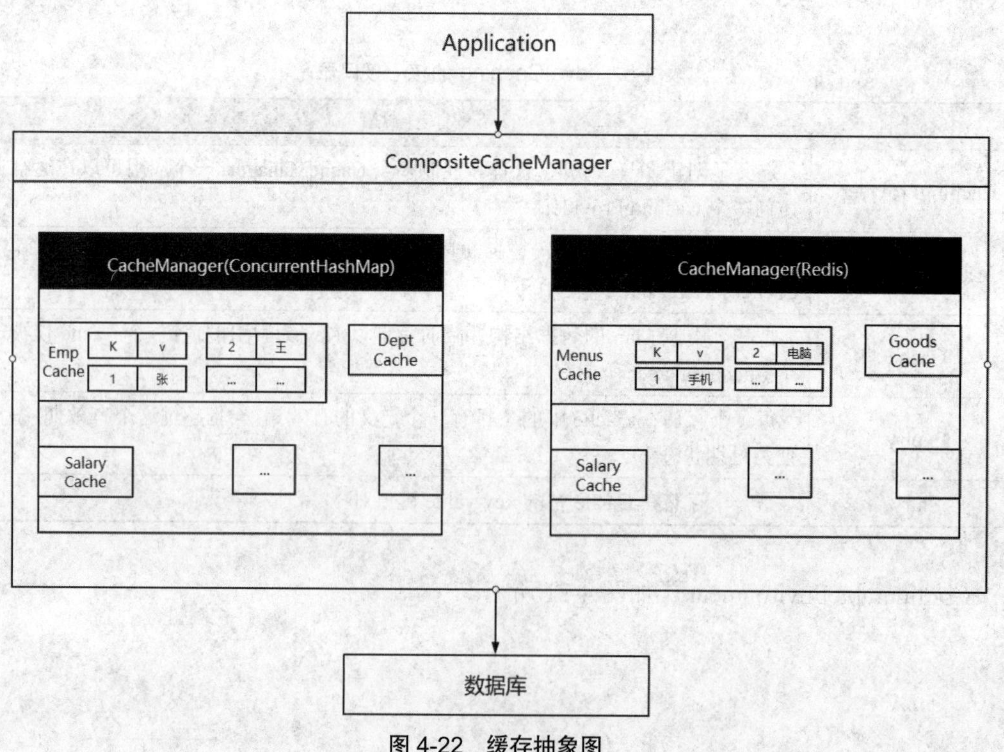

图 4-22 缓存抽象图

【案例】开启 Spring Boot 缓存。

第一步，需要在项目的启动类上添加 @EnableCaching 注解，用于开启缓存支持，如示例代码 4-30 所示。

示例代码 4-30：Demo3Application 启动类

```
@EnableCaching
@SpringBootApplication
public class Demo3Application {
    public static void main(String[] args) {
        SpringApplication.run(Demo3Application.class, args);
    }
}
```

第二步，需要在 UserSevice 类上添加 @Cacheable(cacheNames = "users") 注解，该注解的作用是将查询结果存放在名为 users 的名称空间中，如示例代码 4-31 所示。

示例代码 4-31：UserServiceImpl 实现类

```
@Service
public class UserServiceImpl implements UserService {
    @Autowired
    private  UsersRepository usersRepository;
    @Override
    public boolean update(@RequestBody Users users) {
        if(null == users ){
            System.out.println(users);
            return false;
        }
        return usersRepository.updateUsersNameById(users)>0;
    }
    @Override
    @Cacheable(cacheNames = "users")
    public Users getById(Integer id) {
        if(null == id || id < 1) {
            return null;
        }
        return usersRepository.getOne(id);
    }
}
```

第三步，测试开启缓存后的项目。重新运行项目，在浏览器中访问“localhost: 8088/get/1”地址，无论刷新多少次，页面的查询结果只有一个，并且控制台打印的 SQL 语句只有一条，说明缓存开启成功，如图 4-23 所示。

```
2020-12-24 14:18:23.049  INFO 8384 --- [nio-8088-exec-1] o.s.web.servlet.DispatcherServlet        :
2020-12-24 14:18:23.050  INFO 8384 --- [nio-8088-exec-1] o.s.web.servlet.DispatcherServlet        :
Hibernate: select users0_.id as id1_0_0_, users0_.address as address2_0_0_, users0_.age as age3_0_0
```

图 4-23　测试图

3. Spring Boot 缓存注解

（1）@Cacheable 注解

@Cacheable 注解可以用在类或方法上，通常应用在查询数据的方法上。该方法的返回结果会放在缓存中，以后使用相同的参数调用该方法时，会返回缓存中的值，而不会实际执行该方法。@Cacheable 注解的属性具体使用方法如下。

① cacheName 和 value。这两种属性的作用相同，均用于指定缓存名，还可以同时指定多个名称空间，如 @Cacheable（{"menu", "menuById"}）。

② cacheResolver 和 cacheManager。这两种参数分别用于指定缓存解析器和缓存管理器，cacheManager 和 cacheResolver 是互斥参数，同时使用时可能会导致异常。

③ keyGenerator 自动生成。如果在缓存数据时，没有指定 key 参数，Spring 有一个默认的 SimpleKeyGenerator。在 Spring Boot 自动化配置中，该属性会被默认注入，该缓存名下的所有 key 会使用 KeyGenerator 根据参数自动生成。默认情况下，如果该缓存方法没有参数，返回 SimpleKey.EMPTY；如果该缓存方法有一个参数，返回该参数的实例；如果该缓存方法有多个参数，返回一个包含所有参数的 SimpleKey。

④显式指定 key。Spring 官方更推荐指定 key 的方式，即指定 @Cacheable 的 key 参数。虽然这属于显式指定，但是 key 的值还是需要根据具体参数来生成，可以使用 SpEL（Spring Expression Language，Spring 表达式语言）来生成，SpEL 表达式级见表 4-6。

表 4-6　SpEL 表达式级表

名称	位置	描述
metnodName	root object	当前被调用的方法名
method	root object	当前被调用的方法
target	root object	当前被调用的目标对象
targetClass	root object	当前被调用的目标对象的类
args	root object	当前被调用的方法的参数列表
caches	root object	当前方法调用使用的缓存列表
result	执行上下文	方法执行后的返回值
argument name	执行上下文	当前被调用的方法参数

⑤ condition。它接收一个结果为 true 或 false 的表达式，如果表达式结果为 true，方法的返回值会被缓存，否则在调用方法时就好像该方法没有声明缓存一样。该表达式支持

SpEL。

⑥ unless。用于在执行后判断不缓存的条件。unless 接收一个结果为 true 或 false 的表达式，当结果为 true 时，不缓存。该表达式支持 SpEL。

⑦ sync。它用于判断数据存储过程中是否同步，值为 true 或 false。sync = true 可以有效地避免缓存击穿问题，默认值为 false。

（2）@CacheEvict 注解

@CacheEvict 的作用主要是针对方法配置，其能够根据一定的条件对缓存进行清空，执行顺序是先进行缓存，然后再清理缓存。@CacheEvict 提供的属性与 @Cacheable 注解所提供的属性基本相同，但 @CacheEvict 提供了两个额外的属性。

① allEntries。该属性表示是否清空所有缓存内容，默认值为 false。如果指定为 true，则方法调用后将立即清空所有缓存。

② beforeInvocation。该属性表示是否在方法执行前就清空，默认值为 false，如果指定为 true，则在方法还没有执行的时候就清空缓存。

（3）@CachePut 注解

@CachePut 注解可以用在类或方法上，通常用在更新数据的方法上。该注解的作用是实现缓存与数据库的同步更新，执行顺序是先进行方法调用，然后再更新缓存。@CachePut 注解提供的属性与 @Cacheable 注解提供的属性完全相同。

（4）@Caching 注解

当进行复杂的数据缓存时，可以使用 @Caching 注解应用在类或方法上。@Caching 注解有三个属性，cacheable、put、evict，这三个属性分别相当于 @Cacheable 注解、@CachePut 注解、@CacheEvict 注解，如示例代码 4-32 所示。

示例代码 4-32

```
@Caching（cacheable = {@Cacheable（value = "users"，key ="#id" ）},
put = {@CachePut（value = "users",key = "#result.id"，condition = "#result ！ =null" ）}
）
  public Users getById（Integer id）{
      return usersRepository.getOne（id）;
  }
```

（5）@CacheConfig 注解

@CacheConfig 注解可以用在类上，其作用是管理 @Cacheable、@CachePut、@CacheEvict 注解标注的公共属性，包括 cacheName、KeyGenerator、CacheManager 和 CacheResolver，如示例代码 4-33 所示。

示例代码 4-33：UserServiceImpl 实现类

```
@CacheConfig（cacheNames = "users"）
@Service
public class UserServiceImpl implements UserService {
  @Autowired
```

```
    private UsersRepository usersRepository;
    @Override
    @Cacheable
    public Users getById(Integer id) {
        if(null == id || id < 1) {
            return null;
        }
        return usersRepository.getOne(id);
    }
}
```

4. Spring Boot 整合 Redis 实现缓存

在 Spring Boot 默认缓存管理的基础上可通过引入 Redis 缓存组件实现缓存，整合 Redis 的具体步骤如以下案例所示。

【案例】Spring Boot 整合 Redis 实现缓存。

第一步，添加 Spring Data Redis 依赖，在上一知识点中创建的 demo3 项目的 pom.xml 文件中，添加所需要的依赖，如示例代码 4-34 所示。

示例代码 4-34：pom.xml 文件

```
<dependency>
    <groupId>org.springframework.boot</groupId>
    <artifactId>spring-boot-starter-data-redis</artifactId>
</dependency>
```

第二步，Redis 服务连接配置。使用 Redis 第三方组件进行缓存管理时，需要搭建数据库进行缓存存储，而不是像 Spring Boot 默认缓存管理将数据直接存储在内存中，所以必须开启 Redis 服务。在项目的全局配置文件 application.properties 中添加相关配置，并在文件中配置日志，打印 SQL 以便查看缓存情况，如示例代码 4-35 所示。

示例代码 4-35：application.properties 配置文件

```
// Redis 服务地址，默认地址为 127.0.0.1
spring.redis.host=127.0.0.1
// Redis 服务器连接端口，默认端口号为 6379
spring.redis.port=6379
// Redis 服务器连接密码，默认密码为空
spring.redis.password=
logging.level.com.example.demo3.repository=debug
```

第三步，编写实体类。使用之前项目创建的 Users 实体类，由于对实体类对象进行缓存时必须先序列化，所以 Users 实体类必须要实现 JDK 自带的 Serializable 接口，避免出现异常，如示例代码 4-36 所示。

示例代码 4-36：Users 实体类

```
@Entity(name ="t_users")
public class Users implements Serializable {
    @Id
    @GeneratedValue(strategy = GenerationType.IDENTITY)
    private Integer id;
    private String name;
    private Integer age;
    private String address;
}
```

第四步，自定义 RedisTemlate 序列化机制。如果使用自定义序列化方式的 RedisTemlate 进行数据缓存，需要在 com.example.demo3 包下创建一个名为 Config 的包，并在该包下创建一个名为 RedisConfig 的自定义配置类。

在类上添加 @Configuration 注解，将 RedisConfig 标注为一个配置类，使用 @Bean 注解注入一个默认名称为 RedisTemplate 的组件，使用自定义的 Jackson2JsonRedisSerialize 替换默认序列化，自定义一个 RedisTemlate，如示例代码 4-37 所示。

示例代码 4-37：RedisConfig 配置类

```
@Configuration
public class RedisConfig {
@Bean
public RedisTemplate<String,Object> redisTemplate(RedisConnectionFactory factory) {
RedisTemplate<String, Object> template = new RedisTemplate<String, Object>();
        template.setConnectionFactory(factory);
      // 使用 jackson2JsonRedisSerialize 替换默认序列化(默认采用的是 JDK 序列化)
            GenericJackson2JsonRedisSerializer jackson2JsonRedisSerializer = new Generic-
Jackson2JsonRedisSerializer();
        StringRedisSerializer stringRedisSerializer = new StringRedisSerializer();
        // key 采用 String 的序列化方式
        template.setKeySerializer(stringRedisSerializer);
        // value 序列化方式采用 jackson
        template.setValueSerializer(jackson2JsonRedisSerializer);
        // hash 的 key 也采用 String 的序列化方式
        template.setHashKeySerializer(stringRedisSerializer);
        // hash 的 value 序列化方式采用 jackson
        template.setHashValueSerializer(jackson2JsonRedisSerializer);
        template.afterPropertiesSet();
        return template;
```

```
    }
}
```

第五步，创建 Spring Redis 工具类。由于 Redis 进行缓存数据时，其存取过程均使用 RedisTemplate 过于烦琐，所以需要将 RedisTemplate 进一步进行封装。创建一个 RedisUtil 工具类，加入 @Component 注解自动生成对象，@Autowired 自动注入 RedisTemplate，如示例代码 4-38 所示。

示例代码 4-38：RedisUtil 工具类

```
@Component
public class RedisUtil {
    @Autowired
    public RedisTemplate redisTemplate;
    /**
     * 缓存基本的对象，Integer、String、实体类等
     *
     * @param key 缓存的键值
     * @param value 缓存的值
     * @return 缓存的对象
     */
    public <T> ValueOperations<String, T> setCacheObject(String key, T value)
    {
        ValueOperations<String, T> operation = redisTemplate.opsForValue();
        operation.set(key, value);
        return operation;
    }
    /**
     * 缓存基本的对象，Integer、String、实体类等
     *
     * @param key 缓存的键值
     * @param value 缓存的值
     * @param timeout 时间
     * @return 缓存的对象
     */
     public <T> ValueOperations<String, T> setCacheObject(String key, T value, Long
timeout)
  {
        ValueOperations<String, T> operation = redisTemplate.opsForValue();
        operation.set(key, value, timeout);
```

```
        return operation;
    }
    /**
     * 获得缓存的基本对象
     *
     * @param key 缓存键值
     * @return 缓存键值对应的数据
     */
    public <T> T getCacheObject(String key)
    {
        ValueOperations<String, T> operation = redisTemplate.opsForValue();
        return operation.get(key);
    }
```

第六步，编写 Mapper 接口。创建 UsersMapper 接口，在接口类上添加 @Mapper 注解，在编译之后会生成相应的接口实现类，能够被 Spring Boot 自动扫描到 Spring 容器中。在接口的内部添加 @Select、@Update 注解完成相关操作，如示例代码 4-39 所示。

示例代码 4-39：UsersMapper 接口

```
@Mapper
public interface UsersMapper {
    @Update("update t_users set name=#{name} where id=#{id}")
    int updateName(@Param("id") int id  , @Param("name") String name);
    @Select("select * from t_users where id=#{id}")
    Users finduserById(int id);
}
```

第七步，创建 Service 类。在 com.example.demo3.service 包下创建一个 UsersService 实现类，注入 usersMapper 实例完成查询、修改数据的操作，读取用户的缓存逻辑是先从缓存中查找，如果找到了则返回客户端，如果没有找到则连接数据库读取用户。读取成功后，将用户数据存入缓存中，如示例代码 4-40 所示。

示例代码 4-40：UsersService 类

```
@Service
public class UsersService {
    @Autowired
    private UsersMapper usersMapper;
    @Autowired
    RedisUtil redisUtil ;
    public Users getById(Integer id) {
```

```
        if(null == id || id < 1) {
            return null ;
        }
        Users n = null ;
        // Redis 中缓存用户 key 是 users_id
        String key = "users_" + id;
        if(redisUtil.hasKey(key)) {
            n = redisUtil.getCacheObject(key) ;
            if(null ! = n) {
                return n ;
            }
        }
        n = usersMapper.finduserById(id);
        // 放入缓存
        redisUtil.setCacheObject(key , n);
        return n ;
    }
    public Users updateNews(Integer id, String name) {
        usersMapper.updateName(id , name) ;
        Users n = usersMapper.finduserById(id) ;
        redisUtil.setCacheObject("users_" + id , n) ;
        return n ;
    }
}
```

第八步，在 com.example.demo3.controller 包下，创建一个 UserController2 类，用于 User 访问控制，如示例代码 4-41 所示。

示例代码 4-41：UserController2 类

```
@RestController
public class UserController2 {
    @Autowired
    private UsersService usersService
    @GetMapping("/select/{id}")
    public Users findByID(@PathVariable("id") Integer id) {
        System.out.println("11");
        return usersService.getById(id);
    }
```

```
    @GetMapping("/update/{id}/{name}")
    public Users update(@PathVariable("id") Integer id, @PathVariable("name") String name){
        return usersService.updateNews(id,name);
    }
}
```

第九步，测试缓存是否成功。

①在浏览器中访问“localhost: 8088/select/1”地址，查询 id 为 1 的用户信息，重复刷新浏览器查询同一条数据信息，查询结果如图 4-24 所示。

图 4-24　查询结果图

控制台打印的 SQL 语句，如图 4-25 所示。

```
c.e.d.r.UsersRepository.finduserById      : ==>  Preparing: select * from t_users where id=?
c.e.d.r.UsersRepository.finduserById      : ==> Parameters: 1(Integer)
c.e.d.r.UsersRepository.finduserById      : <==      Total: 1
```

图 4-25　SQL 语句打印图

打开 Redis 客户端可视化工具 Redis Desktop 连接 Redis 服务，查看缓存效果，如图 4-26 所示。可以看出用户信息被正确地存储到了 Redis 缓存中，且存储到 Redis 缓存中的数据已以 JSON 格式进行存储。

```
{
  "@class": "com.example.demo3.pojo.Users",
  "id": 1,
  "name": "李",
  "age": 24,
  "address": "天津市"
}
```

图 4-26　Redis 客户端可视化工具查看效果图

② Redis 缓存更新测试。在浏览器中访问“localhost: 8088/update/1/ 李四”地址，更新 id 为 1 的用户姓名，然后访问“localhost: 8088/select/1”，查看 id 为 1 的用户信息，如图 4-27 所示。

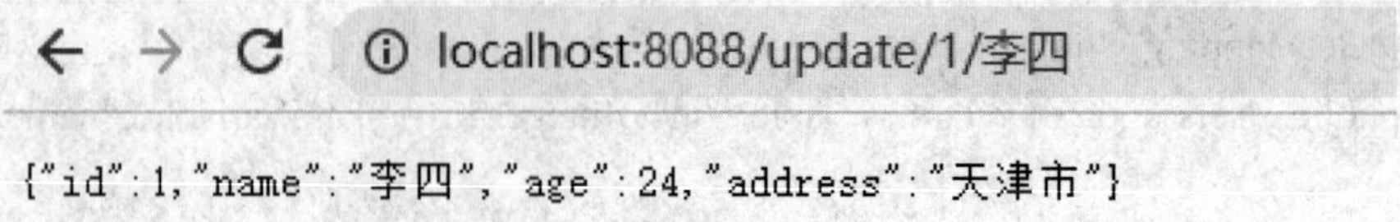

localhost:8088/update/1/李四

```
{"id":1,"name":"李四","age":24,"address":"天津市"}
```

图 4-27 更新缓存结果图

控制台打印的 SQL 语句,如图 4-28 所示。

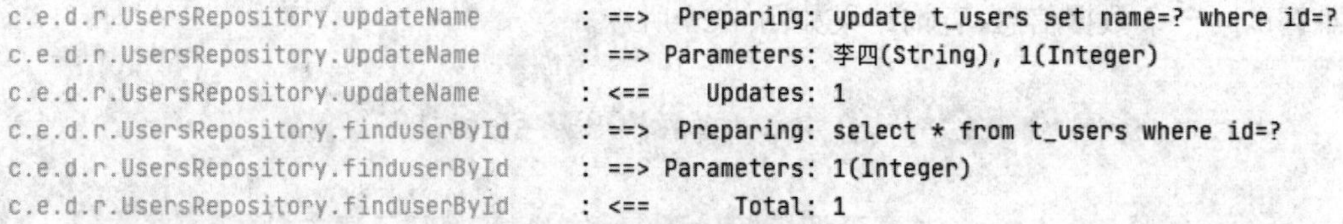

```
c.e.d.r.UsersRepository.updateName      : ==>  Preparing: update t_users set name=? where id=?
c.e.d.r.UsersRepository.updateName      : ==> Parameters: 李四(String), 1(Integer)
c.e.d.r.UsersRepository.updateName      : <==    Updates: 1
c.e.d.r.UsersRepository.finduserById    : ==>  Preparing: select * from t_users where id=?
c.e.d.r.UsersRepository.finduserById    : ==> Parameters: 1(Integer)
c.e.d.r.UsersRepository.finduserById    : <==      Total: 1
```

图 4-28 输出 sql 语句图

打开 Redis 客户端可视化工具 Redis Desktop,连接 Redis 服务,查看缓存效果,如图 4-29 所示。

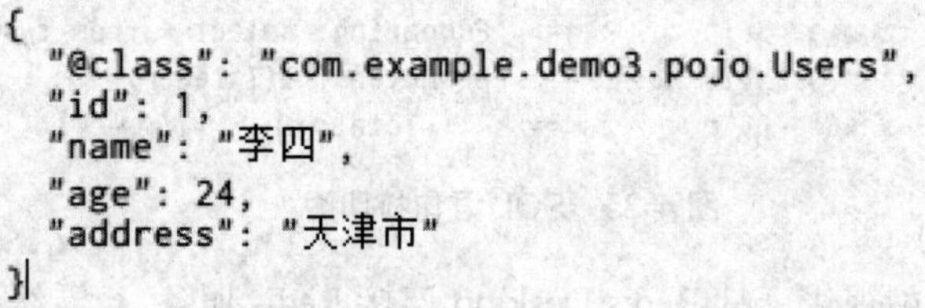

```
{
  "@class": "com.example.demo3.pojo.Users",
  "id": 1,
  "name": "李四",
  "age": 24,
  "address": "天津市"
}
```

图 4-29 Redis 客户端可视化工具查看效果图

1. 整合 MyBatis 框架

第一步,在 pom.xml 中添加数据源依赖、MyBatis 依赖、MySQL 依赖。在 resources 目录下加入静态资源,如示例代码 4-42 所示。

示例代码 4-42:pom.xml 文件

```
<! -- Druid 数据源依赖启动器 -->
<dependency>
    <groupId>com.alibaba</groupId>
    <artifactId>druid-spring-boot-starter</artifactId>
```

```
    <version>1.1.10</version>
  </dependency>
  <! -- MyBatis 依赖启动器 -->
  <dependency>
    <groupId>org.mybatis.spring.boot</groupId>
    <artifactId>mybatis-spring-boot-starter</artifactId>
    <version>2.0.0</version>
  </dependency>
  <! -- MySQL 数据库连接驱动 -->
  <dependency>
    <groupId>mysql</groupId>
    <artifactId>mysql-connector-java</artifactId>
    <scope>runtime</scope>
  </dependency>
  <! -- MyBatis 分页插件 -->
  <dependency>
    <groupId>com.github.pagehelper</groupId>
    <artifactId>pagehelper-spring-boot-starter</artifactId>
    <version>1.2.8</version>
  </dependency>
```

第二步，在 application-jdbc.properties 配置文件中配置数据库连接并且配置第三方数据源，如示例代码 4-43 所示。

示例代码 4-43：application-jdbc.properties 配置文件

```
// 添加并配置第三方数据库连接池 Druid
spring.datasource.type = com.alibaba.druid.pool.DruidDataSource
spring.datasource.initialSize=20
spring.datasource.minIdle=10
spring.datasource.maxActive=100

// 数据源连接配置
spring.datasource.url=jdbc：mysql：//localhost：3306/blog_system？ serverTime-
zone=UTC&useSSL=false
spring.datasource.username = root
// spring.datasource.password = 251125
// driver-class-name 可以省略
// spring.datasource.driver-class-name = com.mysql.jdbc.Driver
```

在 application.yml 配置文件中配置 MyBatis 和 pagehelper 分页设置，如示例代码 4-44

所示。

示例代码 4-44：application.yml 配置文件

```
// MyBatis 配置
mybatis：
  configuration：
    // 开启驼峰命名匹配映射
    map-underscore-to-camel-case：true
  // 配置 MyBatis 的 XML 映射文件路径
  mapper-locations：classpath：mapper/*.xml
  // 配置 XML 映射文件中指定的实体类别名路径
  type-aliases-package：com.xtgj.model.domain
// pagehelper 分页设置

pagehelper：
  helper-dialect：mysql
  reasonable：true
  support-methods-arguments：true
  params：count=countSql
```

第三步，创建 Mapper 接口，创建一个 DAO 包，在包下创建一个 ArticleMapper 接口，在接口类上添加 @Mapper。在编译之后，会生成相应的接口实现类，能够被 Spring Boot 自动扫描到 Spring 容器中，如示例代码 4-45 所示。

示例代码 4-45：ArticleMapper 接口

```
@Mapper
public interface ArticleMapper {
  // 根据 id 查询文章信息
  @Select（"SELECT * FROM t_article WHERE id=#{id}"）
  public Article selectArticleWithId（Integer id）；
  // 发表文章，同时使用 @Options 注解获取自动生成的主键 id
  @Insert（"INSERT INTO t_article（title，created，modified，tags，categories，" +
        " allow_comment，thumbnail，content）"+
        " VALUES（#{title}，#{created}，#{modified}，#{tags}，#{categories}，"+
        " #{allowComment}，#{thumbnail}，#{content}）"）
  @Options（useGeneratedKeys=true，keyProperty="id"，keyColumn="id"）
  public Integer publishArticle（Article article）；

  // 文章分页查询
  @Select（"SELECT * FROM t_article ORDER BY id DESC"）
```

```
    public List<Article> selectArticleWithPage();
    // 通过 id 删除文章
    @Delete("DELETE FROM t_article WHERE id=#{id}")
    public void deleteArticleWithId(int id);
    // 站点服务统计,统计文章数量
    @Select("SELECT COUNT(1) FROM t_article")
    public Integer countArticle();
    // 通过 id 更新文章
    public Integer updateArticleWithId(Article article);
}
```

创建 XML 映射文件,在 resources 目录下,创建一个名为 mapper 的包,在该包下创建一个名字与 ArticleMapper 接口相同的 XML 文件,如示例代码 4-46 所示。

示例代码 4-46: ArticleMapper .xml 文件

```
<? xml version="1.0" encoding="UTF-8"? >
<! DOCTYPE mapper PUBLIC"-//mybatis.org//DTD Mapper 3.0//EN" "http: //mybatis.org/dtd/mybatis-3-mapper.dtd">
<mapper namespace="com.xtgj.dao.ArticleMapper">
  <update id="updateArticleWithId" parameterType="com.xtgj.model.domain.Article">
    update t_article
    <set>
      <if test="title ! = null">
        title = #{title},
      </if>
      <if test="created ! = null">
        created = #{created},
      </if>
      <if test="modified ! = null">
        modified = #{modified},
      </if>
      <if test="tags ! = null">
        tags = #{tags},
      </if>
      <if test="categories ! = null">
        categories = #{categories},
      </if>
      <if test="hits ! = null">
        hits = #{hits},
```

```
            </if>
            <if test="commentsNum ! = null">
              comments_num = #{commentsNum},
            </if>
            <if test="allowComment ! = null">
              allow_comment = #{allowComment},
            </if>
            <if test="thumbnail ! = null">
              thumbnail = #{thumbnail},
            </if>
            <if test="content ! = null">
              content = #{content},
            </if>
          </set>
          where id = #{id}
      </update>
  </mapper>
```

第四步,进行整合测试。在 AdminController 编写代码打印数据,运行项目,测试是否整合完成,如示例代码 4-47 所示。

示例代码 4-47:AdminController 类

```
// 跳转到后台文章列表页面
@GetMapping(value = "/article")
public String index(@RequestParam(value = "page", defaultValue ="1") int page,
@RequestParam(value = "count", defaultValue ="10") int count, HttpServletRequest re-
quest) {
 PageInfo<Article> pageInfo = articleServiceImpl.selectArticleWithPage(page, count);
   request.setAttribute("articles", pageInfo);
   System.out.println(pageInfo);
   return "back/article_list";
}
```

测试成功后,控制台上打印的 SQL 语句,如图 4-30 所示。

```
2021-01-05 09:37:21.740  INFO 4284 --- [nio-8085-exec-1] o.s.web.servlet.DispatcherServlet        :
PageInfo{pageNum=1, pageSize=10, size=3, startRow=1, endRow=3, total=3, pages=1, list=Page{count=tr
```

图 4-30 输出 SQL 语句图

2. 使用 Redis 实现缓存

第一步,在个人博客项目中,引入 Spring Data Redis 依赖,如示例代码 4-48 所示。

示例代码 4-48：pom.xml 文件

```
<! -- Redis 服务启动器 -->
<dependency>
  <groupId>org.springframework.boot</groupId>
  <artifactId>spring-boot-starter-data-redis</artifactId>
</dependency>
```

第二步，Redis 服务连接配置，在项目的全局配置文件 application-redis.properties 中添加相关配置，如示例代码 4-49 所示。

示例代码 4-49：application-redis.properties 配置文件

```
// Redis 服务器地址，另外注意要开启 Redis 服务
spring.redis.host=127.0.0.1
// Redis 服务器连接端口
spring.redis.port=6379
// Redis 服务器连接密码（默认为空）
spring.redis.password=
// 连接池最大连接数（使用负值表示没有限制）
spring.redis.jedis.pool.max-active=8
// 连接池最大阻塞等待时间（使用负值表示没有限制）
spring.redis.jedis.pool.max-wait=-1
// 连接池中的最大空闲连接
spring.redis.jedis.pool.max-idle=8
```

第三步，自定义 Redis 配置类，进行序列化以及 RedisTemplate 设置，如示例代码 4-50 所示。

示例代码 4-50：RedisConfig 配置类

```
@Configuration
public class RedisConfig extends CachingConfigurerSupport {
  /**
   * 定制 Redis API 模板 RedisTemplate
   * @param redisConnectionFactory
   * @return
   */
  @Bean
  public RedisTemplate<Object, Object> redisTemplate(RedisConnectionFactory redis-
ConnectionFactory) {
    RedisTemplate<Object, Object> template = new RedisTemplate();
    template.setConnectionFactory(redisConnectionFactory);
```

```
        // 使用 JSON 格式序列化对象,对缓存数据 key 和 value 进行转换
        Jackson2JsonRedisSerializer jacksonSeial = new Jackson2JsonRedisSerializer(Ob-
ject.class);
        // 解决查询缓存转换异常的问题
        ObjectMapper om = new ObjectMapper();
        // 指定要序列化的域，field、get 和 set,以及修饰符范围，ANY 是都有,包括 pri-
vate 和 public
        om.setVisibility(PropertyAccessor.ALL, JsonAutoDetect.Visibility.ANY);
        // 指定序列化输入的类型,类必须是非 final 修饰的，final 修饰的类,比如 String,
Integer 等会跑出异常
        om.enableDefaultTyping(ObjectMapper.DefaultTyping.NON_FINAL);
        jacksonSeial.setObjectMapper(om);
        // 设置 RedisTemplate 模板 API 的序列化方式为 JSON
        template.setDefaultSerializer(jacksonSeial);
        return template;
    }
    /**
     * 定制 Redis 缓存管理器 RedisCacheManager,实现自定义序列化并设置缓存时效
     * @param redisConnectionFactory
     * @return
     */
    @Bean
    public RedisCacheManager cacheManager(RedisConnectionFactory redisConnection-
Factory) {
        // 分别创建 String 和 JSON 格式序列化对象,对缓存数据 key 和 value 进行转换
        RedisSerializer<String> strSerializer = new StringRedisSerializer();
        Jackson2JsonRedisSerializer jacksonSeial = new Jackson2JsonRedisSerializer(Ob-
ject.class);
        // 解决查询缓存转换异常的问题
        ObjectMapper om = new ObjectMapper();
        om.setVisibility(PropertyAccessor.ALL, JsonAutoDetect.Visibility.ANY);
        om.enableDefaultTyping(ObjectMapper.DefaultTyping.NON_FINAL);
        jacksonSeial.setObjectMapper(om);
        // 定制缓存数据序列化方式及时效
        RedisCacheConfiguration config = RedisCacheConfiguration.defaultCacheConfig()
            .entryTtl(Duration.ofDays(7)) // 设置缓存有效期为 1 天
              .serializeKeysWith(RedisSerializationContext.SerializationPair.fromSerializer
(strSerializer))
```

```
            .serializeValuesWith(RedisSerializationContext.SerializationPair.fromSerializer
(jacksonSeial))
            .disableCachingNullValues();  // 对空数据不进行缓存
        RedisCacheManager cacheManager = RedisCacheManager.builder(redisConnection-
Factory).cacheDefaults(config).build();
        return cacheManager;
    }
}
```

第四步，编写 ArticleServiceImpl，注入 redisTemplate 实例，编写方法使用 Redis 进行缓存，如示例代码 4-50 所示。

示例代码 4-51：ArticleServiceImpl 类

```
@Autowired
private RedisTemplate redisTemplate;

// 根据 id 查询单个文章详情，并使用 Redis 进行缓存管理
public Article selectArticleWithId(Integer id){
    Article article = null;
    Object o = redisTemplate.opsForValue().get("article_" + id);
    if(o! =null){
        article=(Article)o;
    }else{
        article = articleMapper.selectArticleWithId(id);
        if(article! =null){
            redisTemplate.opsForValue().set("article_" + id, article);
        }
    }
    return article;
}
```

第五步，测试缓存。在浏览器访问“http: //localhost: 8085/article/1”地址，查询 id 为 1 的用户信息，重复刷新浏览器查询同一条数据信息，查询结果如图 4-31 所示。

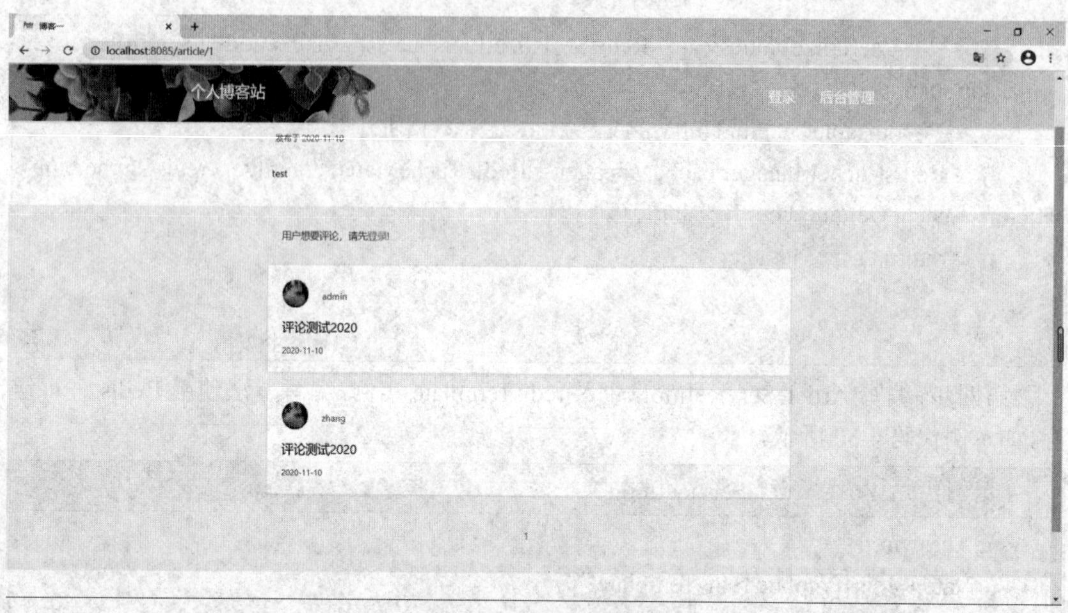

图 4-31 查询结果图

打开 Redis 客户端可视化工具 Redis Desktop 连接 Redis 服务，查看缓存效果，如图 4-32 所示。可以看出，用户信息被正确地存储到了 Redis 缓存中，且存储到 Redis 中的数据已以 JSON 格式进行存储。

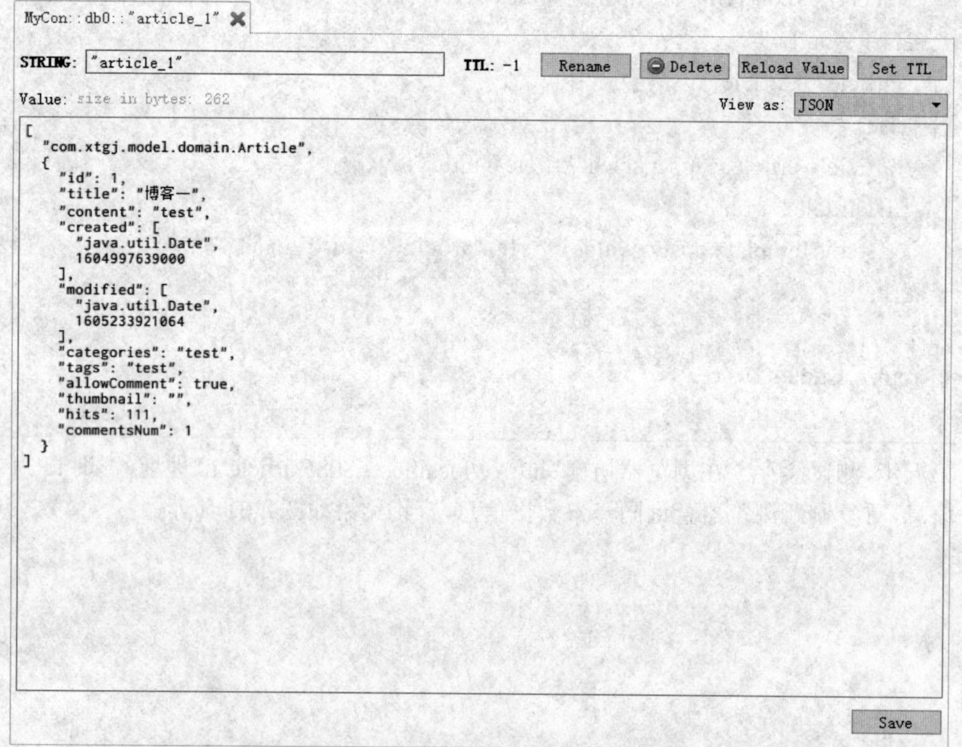

图 4-32 Redis 客户端可视化工具查看效果图

本任务讲解了 Spring Boot 整合 MyBatis 和使用 Redis 实现缓存的方法。在实现了基本功能的基础上，应用了 MyBatis 分页插件 pagehelper 的知识。本任务使读者加深了对于 Spring Boot 的理解，掌握了基本的 Spring Boot 技术。

Dependencies	依赖	Initializr	初始化
Caching	缓存	Expression	表达

一、选择题

1. 下列关于 Spring Boot 整合 MyBatis 实现的说法不正确的是（　　）。

A. Sprina Boot 整合 MyBatis 时，必须要在 pom.xml 中添加数据源依赖

B. @MaoperScan（"xxx"）注解的作用和 @Mapper 注解不一样

C. 创建 XML 映射文件时，文件名与相对应的接口名相同

D. 使用注解方式编写 MyBatis 接口文件数据修改方法时，还需要使用 @Transactional 注解

2. 下列说法错误的是（　　）。

A. @Entity 表明该类为一个实体类

B. @Id 注解标注在类属性上，表示主键对应的属性

C. @GeneratedValue 指定主键的生成策略

D. @Transient 注解表示指定属性不是到数据库表的字段的映射，ORM 框架将忽略该属性

3. 当 Redis 作为数据库时，下列与 Spring Boot 整合使用的相关说明，正确的是（　　）。

A. @RedisHash（"persons"）用于指定操作实体类对象在 Redis 数据库中的存储空间

B. @Id 用于标识实体类主键，需要手动指定 id 生成策略

C. 使用 Redis 数据库，必须为实体类属性添加 @Indexed 属性生成二级索引

D. 编写操作 Redis 数据库的 Repository 接口文件时，需要继承 JpaRepository 接口

4. 下列关于 Spring Boot 提供的缓存管理相关注解说法不正确的是（　　）。

A. @Cacheable 注解可以用在类或方法上，通常应用在查询数据的方法上

B. @CachEvict 的作用主要针对方法配置，该注解的作用是根据一定的条件对缓存进行清空，执行顺序是先清理缓存，然后再进行缓存

C. @CachePut 注解的作用是实现缓存与数据库的同步更新，执行顺序是先进行方法调用，然后再更新缓存

D. 当进行复杂的数据缓存时，可以使用 @Caching 注解应用在类或者方法上

5. 下列关于 Spring Boot 定制 Redis 缓存序列化机制的说法，错误的是（　　）。

A. 在定制序列化方式中，要定义一个 ObjectMapper 用于进行数据转换设置

B. 自定义 redisTemplate 组件时，方法名必须为 redisTemplate

C. Spring Boot 1.X 版本中，定制 RedisTemplate 组件序列化配置后，就完成了基于 API 和注解方式 Redis 序列化的定制

D. 使用自定义 Aedislemplaie 序列化机制缓存实体类数据，实体类不用再实现序列化

二、简答题

1. Redis 的优点有哪些？

2. MyBatis 的功能架构有哪几层，请具体解释？

项目五　个人博客项目安全管理

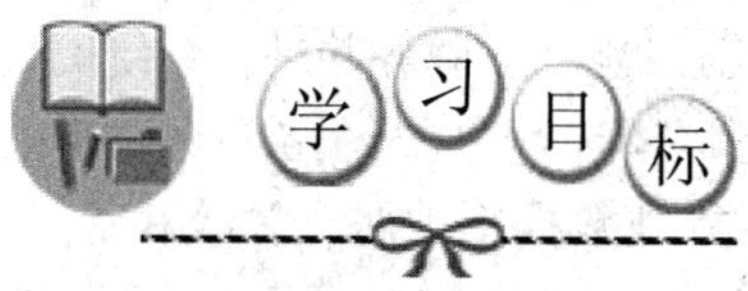

通过学习 Spring Security 的相关知识，掌握使用 Spring Security 实现用户登录认证、应用权限管理、自定义认证页面以及记住我功能。能编写个人博客项目安全管理服务，完成自定义登录、登出过程以及设置项目各模块权限等操作。

- 掌握使用 Spring Security 实现用户登录认证功能。
- 掌握使用 Spring Security 实现应用权限管理功能。
- 掌握使用 Spring Security 实现自定义认证页面功能。
- 掌握使用 Spring Security 实现记住我功能。

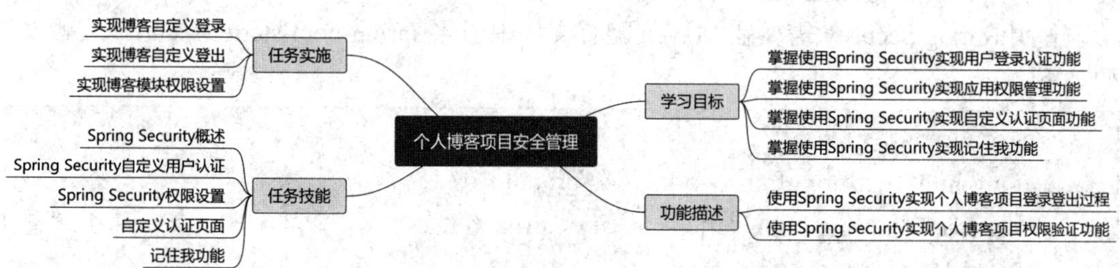

【情境导入】

在构建网站的过程中，必须要考虑数据安全问题。例如，对于一些涉及重要数据变更操作的请求，需要进行身份验证后才可以执行；或者在调用第三方公司服务并进行系统间数据交互时，也需要验证调用方身份才能进行业务处理。在互联网环境中，存在各种各样的恶意攻击，而数据往往是一个互联网公司最重要的资产，为了保证数据不被恶意篡改和泄露，为应用编写安全机制是开发中必不可少的一环。

【功能描述】

- 使用 Spring Security 实现个人博客项目的登录登出过程。
- 使用 Spring Security 实现个人博客项目的权限验证功能。

技能点 1　Spring Security 概述

1. Spring Security 快速入门

Spring Security 是 Spring 框架，采用 AOP 思想，基于 Servlet 过滤器实现的安全框架。它提供了完善的认证机制和方法级的授权功能，是一款优秀的权限管理框架。它可以在 Spring 应用上下文中进行配置，充分利用了 Spring 框架的强大特性，为应用系统提供声明式的安全访问控制功能，使企业应用安全机制的编写更加方便快捷。

使用 Spring Security 需要在 Maven 配置文件中引入 spring-boot-starter-security 依赖，代码如下。

```
<dependency>
   <groupId>org.springframework.boot</groupId>
   <artifactId>spring-boot-starter-security</artifactId>
</dependency>
```

在 Spring Boot 项目中引入 Spring Security 依赖后，直接启动项目即可启用 Spring Security，为了验证项目的各种认证功能，首先创建两个处理请求的方法，这两个方法响应的 URL 地址分别为“admin/visit”和“user/hello”，方法的代码如下。

```
@Controller
@RequestMapping("admin")
public class AdminController {
    @RequestMapping("visit")
    @ResponseBody
    public String hello(){
        return " 访问成功 ";
    }
}
```

```
@Controller
@RequestMapping("user")
public class UserController {
    @RequestMapping("hello")
    @ResponseBody
    public String hello(){
        return "hello world";
    }
}
```

启动项目后，因为未进行任何配置，所以 Spring Security 会自动生成默认登录用户，用户名为 user，密码是一串自动生成的字符串，会随着项目启动输出在控制台上，登录密码生成日志如下。

```
2020-12-07 09:33:21.640 INFO 12200 --- [main] .s.s.UserDetailsServiceAutoConfiguration:

Using generated security password: 7357d7f9-3032-4703-aa51-43aaf87ddf3a

2020-12-07 09: 33: 21.770 INFO 12200 --- [main] o.s.s.web.DefaultSecurityFilterChain: Creating filter chain: any request, [org.springframework.security.web.context.request.async.WebAsyncManagerIntegrationFilter@7f973a14, org.springframework.security.web.context.SecurityContextPersistenceFilter@5a49af50, org.springframework.security.web.header.HeaderWriterFilter@4833eff3, org.springframework.security.web.csrf.CsrfFilter@2489e84a, org.springframework.security.web.authentication.logout.LogoutFilter@421def93, org.springframework.security.web.authentication.UsernamePasswordAuthenticationFilter@118dcbbd, org.springframework.security.web.authentication.ui.DefaultLoginPageGeneratingFilter@31142d58, org.springframework.security.web.authentication.ui.DefaultLogoutPageGeneratingFilter@76130a29, org.springframework.security.web.authentication.www.BasicAuthenticationFilter@3c380bd8, org.springframework.security.web.savedrequest.RequestCacheAwareFilter@793d163b, org.springframework.security.web.servletapi.SecurityContextHolderAwareRequestFilter@7fc420b8, org.springframework.security.web.authentication.AnonymousAuthenticationFilter@124d02b2, org.springframework.security.web.session.SessionManagementFilter@56928e17, org.springframework.security.web.access.ExceptionTranslationFilter@2676dc05, org.springframework.security.web.access.intercept.FilterSecurityInterceptor@6a9cd0f8]
```

日志中的黑体字符串就是项目启动时随机生成的密码，在浏览器中输入项目请求 URL 后，会出现默认登录界面，如图 5-1 所示。

图 5-1 默认登录界面

在登录界面文本框中输入默认用户名 user，密码为控制台上显示的随机密码，然后点击“Sign in”按钮，即可完成登录。登录后，界面跳转到请求路径，如图 5-2 所示。

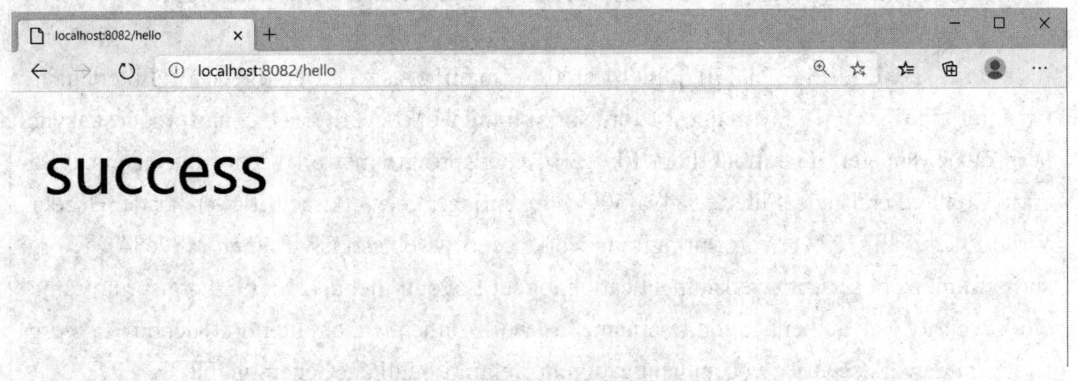

图 5-2 登录后跳转成功界面

2. Spring Security 基本原理

传统的 Java Web 项目使用 Servlet 过滤器来动态地拦截请求，然后通过编写的用户验证逻辑代码来判断是否放行请求。Spring Security 框架登录验证的核心是一条由过滤器组成的验证链，当一个请求到达时，按照验证链的顺序依次进行处理。通过所有过滤器的验证后，就可以访问 API 接口了。Spring Security 过滤器链如图 5-3 所示。

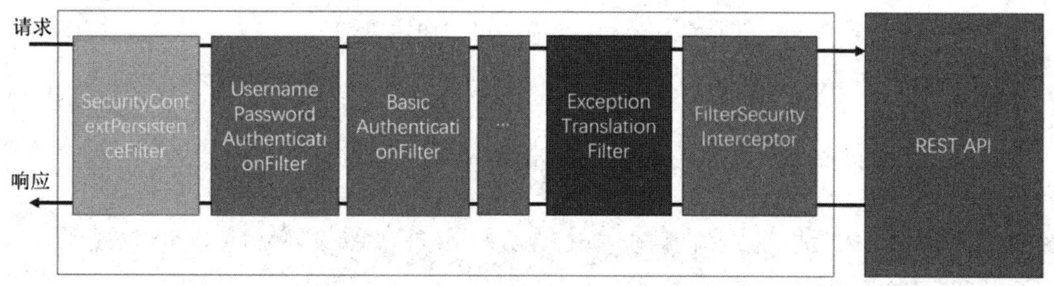

图 5-3　Spring Security 过滤器链

用户发送请求到服务器时，首先要经过 SecurityContextPersistenceFilter，该过滤器会在 session 中保存或更新一个 SecurityContext 对象，以供后续的过滤器使用，该对象中存储了当前用户的认证和权限信息。

接着该请求经过一组过滤器，其中最主要的是 UsernamePasswordAuthenticationFilter 和 BasicAuthenticationFilter，前者用于处理表单登录，后者用于处理 HttpBasic 登录，这组过滤器会检查请求中是否附带有本过滤器需要验证的信息，如包含所需信息则进行验证。每个过滤器成功完成验证之后，都会在这个请求上做标记，表明这个用户已经认证成功了，该请求经过这些过滤器后，会到达 FilterSecurityInterceptor 拦截器。

该拦截器中保存了用户自定义设置的认证规则，它会验证该请求用户在前面过滤器链中保存的过滤结果是否符合身份认证或权限认证的规则。如结果符合则放行该请求，使其调用请求的服务；如果不符合，会使用 ExceptionTranslationFilter 来抛出并返回相应的异常，这样就结束了一次对访问请求的处理。

技能点 2　Spring Security 自定义用户认证

Spring Security 提供了 SecurityConfigurer 接口来实现对 Spring Security 的配置，但该接口只提供了对配置的简单定义，为了让实际项目更方便使用，Spring 为 Web 工程提供了专门的接口 WebSecurityConfigurer，并且为这个接口提供了一个抽象实现类 WebSecurityConfigurerAdapter。开发人员通过继承该抽象类，并在类上添加 @EnableWebSecurity 注解就可以使用 Spring Security 的默认设置来开启安全功能，也可以通过覆盖该类的方法，自定义安全拦截的逻辑。WebSecurityConfigurerAdapter 中提供了三种默认方法，列举如下。

```
/**
 * 主要用于配置用户的认证服务，以及为用户添加角色
 * @param auth 身份验证管理器构造器，用于构建具体的权限认证控制
 */
protected void configure（AuthenticationManagerBuilder auth）；
/**
 * 主要用于配置拦截保护的请求，如控制哪些请求需要验证，哪些请求可以放行
 * @param http http 安全请求对象
 */
protected void configure（HttpSecurity http）；
/**
 * 主要用于配置用户的认证服务，以及为用户添加角色
 * @param auth 身份验证管理器构造器，用于构建具体的权限认证控制
 */
protected void configure（WebSecurity web）；
```

以上三个 configure 方法用于配置用户对于 Spring Security 的自定义设置，当 Spring Security 整合进 Spring Boot 时，如果没有配置任何自定义设置，框架会自动拦截所有请求，将请求跳转到 Spring Security 默认的登录验证页面，并生成一个名称为 user、密码为随机密码的用户账号。该用户的访问密码会输出在程序的日志中，只有使用该账号进行登录才能继续访问请求的 URL。而重写 WebSecurityConfigurerAdapter 类中的 configure 方法可以实现自定义用户身份认证，Spring Security 还提供了内存身份验证和数据库身份验证来实现应用的安全管理。

1. 内存身份验证

内存身份验证是将用户信息保存在内存当中，此方法编写方便、设置简单，一般用于需要快速搭建的测试环境，此类验证的使用方法如示例代码 5-1 所示。

示例代码 5-1：SecurityDemo1.java

```
Import org.springframework.security.config.annotation.authentication.builders.AuthenticationManagerBuilder;
import org.springframework.security.config.annotation.web.configuration.EnableWebSecurity;
import org.springframework.security.config.annotation.web.configuration.WebSecurityConfigurerAdapter;
import org.springframework.security.crypto.bcrypt.BCryptPasswordEncoder;
import org.springframework.security.crypto.password.PasswordEncoder;
@EnableWebSecurity
public class SecurityDemo1 extends WebSecurityConfigurerAdapter {
  @Override
```

```
protected void configure(AuthenticationManagerBuilder auth) throws Exception {
  // 实例化 Spring Security 密码编码器
    PasswordEncoder passwordEncoder = new BCryptPasswordEncoder();
    // 开启内存身份验证
    auth.inMemoryAuthentication()
        // 验证过程中启用 Spring Security 密码编码器解析密码
        .passwordEncoder(passwordEncoder)
        // 创建用户 testUser,用户密码为 123456,并赋予用户 USER 权限
        .withUser("testUser").password(passwordEncoder.encode("123456"))
        .roles("USER")
        //and 连接方法
        .and()
        // 创建用户 admin,用户密码为 abcdef,并赋予用户 ADMIN 权限
        .withUser("admin").password(passwordEncoder.encode("abcdef"))
        .roles("ADMIN");
  }
}
```

在 Spring 5 的 Security 中，自定义用户验证设置都要求使用 Spring Security 提供的密码编码器，否则会出现异常，所以代码中首先创建了一个 BCryptPasswordEncoder 实例，它实现了 PasswordEncoder 接口，该类可以使用单向不可逆的密码加密方式对密码进行加密，类中常用方法见表 5-1。

表 5-1　PasswordEncoder 接口方法

方法名称	详情说明
encode(CharSequence)	将原始密码编码加密
matches(CharSequence, String)	验证提交的密码与编码加密后的密码密文是否匹配

configure 方法的 AuthenticationManagerBuilder 参数可以构建内存身份验证管理器配置器和 JDBC 身份验证管理器配置器。在示例代码 5-1 中使用 inMemoryAuthentication() 方法返回内存身份验证管理器配置器 InMemoryUserDetailsManagerConfigurer，通过该配置器可以配置并实例化用于管理内存身份验证的 InMemoryUserDetailsManager 类。该管理类用于定义登录时所使用的用户（user）、密码（password）和角色（role）等属性，在示例代码 5-1 中使用的配置方法如下。

①使用 passwordEncoder() 方法以实例化 BCryptPasswordEncoder 类为参数配置密码编辑器。

②使用 withUser() 方法来注册用户名称，该方法返回用户详情构造器 UserDetailsBuilder 对象，通过该对象可以配置用户详细信息；使用 password() 方法来配置用户密码，采

用的是以 BCrypt 方式进行加密后的密码字符串；使用 roles()方法为用户赋予权限角色，该方法会在角色字符串前自动添加 ROLE_ 前缀，如 roles("USER")则赋予的角色为 ROLE_USER。在该示例中，通过 withUser()方法注册了两个用户，一个是 testUser 用户，其密码为 123456，权限角色为 ROLE_ USER；另一个是 admin 用户，其密码为 abcdef，权限角色为 ROLE_ADMIN。UserDetailsBuilder 对象的其他常用方法见表 5-2。

表 5-2 UserDetailsBuilder 常用方法

方法名称	详情说明
username(String)	设置用户名
password(String)	设置用户密码
authorities(GrantedAuthority...)	赋予用户权限
authorities(List<? extends GrantedAuthority>)	使用 List 赋予用户一个或多个权限
roles(Sting..)	赋予用户权限，会自动加入前缀"ROLE_"
accountExpired(boolean)	设置账号是否过期
accountLocked(boolean)	是否锁定账号
credentialsExpired(boolean)	定义凭证是否过期
disabled(boolean)	是否禁用用户

③使用 and()连接符来开启另一个用户构造。

实例编写完成后，启动服务测试，此时程序控制台中已经不会生成随机访问密码，通过浏览器访问"http://localhost:8082/admin/visit"地址，请求被 Spring Security 拦截跳转到登录页面，如图 5-4 所示。

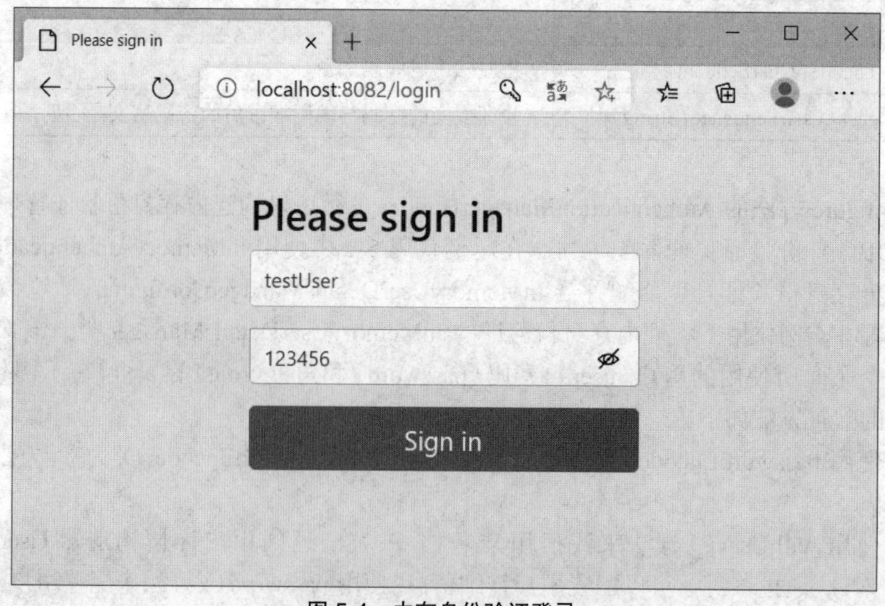

图 5-4 内存身份验证登录

在登录页面上使用 testUser 进行登录，登录成功后页面自动跳转到“http://localhost:8082/admin/visit”页面，效果如图 5-5 所示。

图 5-5 内存身份验证登录成功

创建用户时，除了使用 and()方法连接，还可以使用 InMemoryUserDetailsManagerConfigurer 类来进行逐个创建，如示例代码 5-2 所示。

示例代码 5-2：使用 userDetailsConfig 方法创建用户

```
@EnableWebSecurity
public class SecurityDemo2 extends WebSecurityConfigurerAdapter {
  @Override
  protected void configure(AuthenticationManagerBuilder auth) throws Exception {
    PasswordEncoder encoder = new BCryptPasswordEncoder();
     InMemoryUserDetailsManagerConfigurer<AuthenticationManagerBuilder> userDe-
tailsConfig= auth.inMemoryAuthentication().passwordEncoder(encoder);
    userDetailsConfig.withUser("testUser")
        .password(encoder.encode("123456"))
        .roles("USER");
    userDetailsConfig.withUser("admin").
        password(encoder.encode("abcdef"))
        .roles("ADMIN");
  }
}
```

除了在测试阶段，一般不使用内存身份验证，因为服务器内存空间有限，且该方式会占用 Java 虚拟机(Java Virtual Machine，JVM)的内存空间，所以只在需要快速启动测试时才考虑使用。

2.JDBC 身份验证

在实际项目中，用户的信息会存放在数据库中，在需要进行身份验证时，通过用户名或其他身份 ID 来对数据库中存放的用户数据进行查询比对。对于此种方式，Spring Security 提供了 JdbcUserDetailsManagerConfigurer 类来对数据查询过程进行配置，该类也提供了默认的 SQL 语句用于查询,但在日常使用中,一般是基于业务逻辑来自定义编写 SQL 语句进行查询。

(1)准备数据

为了学习数据库身份验证,首先需要准备 MySQL 数据库表和表内数据,如示例代码 5-3 所示。

示例代码 5-3:数据库创建语句

```
/** 创建权限角色表 **/
create table t_role(
id int(11) not null auto_increment primary key,
role varchar(255),
valid tinyint(1)
);
/** 创建用户表 **/
create table t_user(
id int(11) not null auto_increment primary key,
user_name varchar(255) not null,
password varchar(255) not null
);
/** 创建用户权限表 **/
create table t_user_role(
id int(11) not null auto_increment primary key,
role_id int(11) not null,
user_id int(11) not null,
constraint Reference_1 foreign key(role_id) references t_role(id)
on delete restrict on update restrict,
constraint Reference_2 foreign key(user_id) references t_user(id)
on delete restrict on update restrict);
```

上述代码创建了三个表:t_role、t_user 和 t_user_role。其中,t_role 表是权限角色表,用于保存用户的权限标识;t_user 表中用户详情表,用于保存用户的详细信息,包括用户名和密码等;t_user_role 表是用户权限表,用于保存每个用户所具有的权限角色。t_user_role 表中的 role_id 关联着权限角色表的主键 id,user_id 关联着用户详情表的主键 id。建表完成后,需在三个表中插入自定义的测试数据,以供后续学习使用。

(2)编写身份验证

重写 configure(AuthenticationManagerBuilder auth)方法,在方法中使用数据库身份验证的方式进行自定义用户验证,如示例代码 5-4 所示。

示例代码 5-4:SecurityDemo3.java

```
import org.springframework.beans.factory.annotation.Autowired;
import org.springframework.security.config.annotation.authentication.builders.AuthenticationManagerBuilder;
```

```
    import  org.springframework.security.config.annotation.web.configuration.EnableWebSe-
curity;
    import   org.springframework.security.config.annotation.web.configuration.WebSecurity-
ConfigurerAdapter;
    import org.springframework.security.crypto.bcrypt.BCryptPasswordEncoder;
    import javax.sql.DataSource;
    @EnableWebSecurity
    public class SecurityDemo3 extends WebSecurityConfigurerAdapter {
      // 注入 DataSource 数据源
      @Autowired
    private DataSource dataSource;
     // 通过用户名查询用户详情 SQL 语句
    private String userSQL = "select username, password, valid from t_user where username
= ? ";
     // 通过用户名查询用户权限 SQL 语句
      private String roleSQL = "select u.username,r.role "
            +"from t_user u,t_role r,t_user_role ur "
            +"where u.id = ur.user_id and r.id = ur.role_id and u.username = ? ";

      protected void configure(AuthenticationManagerBuilder auth) throws Exception {
        BCryptPasswordEncoder encoder = new BCryptPasswordEncoder();
        // 使用 JDBC 进行身份认证
        auth.jdbcAuthentication().passwordEncoder(encoder)
        // 绑定 JDBC 数据源
            .dataSource(dataSource)
        // 绑定通过用户名查询用户密码 SQL 语句
            .usersByUsernameQuery(userSQL)
        // 绑定通过用户名查询用户权限 SQL 语句
            .authoritiesByUsernameQuery(roleSQL);
      }
    }
```

代码 5-4 中使用的配置说明如下。

① 使用 @Autowired 注解注入 JDBC 数据源，注入完成后即可使用 SQL 语句对数据库中的数据进行查询。

② 定义两条 SQL 语句：userSQL 和 roleSQL。userSQL 语句使用 username 字段作为变量在数据库中查询用户详情，该语句中返回的值必须与用户名 username、密码 password、是否为有效用户 valid 三个值对应；roleSQL 语句使用 username 字段作为变量在数据库中查询用户所含有的权限角色，该语句中返回的值必须与用户名 username、权限 role 两个值对应。

③ 在 configure()方法中使用 AuthenticationManagerBuilder 对象的 JdbcAuthentication()方法构建 JDBC 身份验证管理器配置器 JdbcUserDetailsManagerConfigurer 对象。通过该配置器,可以配置并实例化用于管理 JDBC 身份验证的 JdbcUserDetailsManager 类,该类通过 dataSource()方法绑定 JDBC 数据源来进行数据查询,使用 passwordEncoder()方法来设置密码解码器。

④ 在使用 usersByUsernameQuery()方法进行登录身份验证查询时,方法需传入通过用户名查询用户详情的 SQL 语句,查询后须返回三列,分别是用户名、密码和验证是否为有效用户的布尔值。当返回值为 1 时,布尔值为 true,用户被标记为有效用户;返回值为 0 时,布尔值被赋值为 false,用户被标记为无效用户,无法成功登录。

⑤ 使用 authoritiesByUsernameQuery()方法进行用户权限验证查询,方法需传入通过用户名查询用户所含权限角色的 SQL 语句,查询后须返回两列,分别是用户名和用户权限。返回结果后, Spring Security 就会根据查询的结果赋予权限。当查询语句返回多条结果时,会赋予用户多个权限角色。

(3)启动服务测试

在测试服务前,先启动 MySQL 数据库服务器,然后启动 Spring Boot 项目,通过浏览器访问“http://localhost:8082/admin/visit”地址,请求被 Spring Security 拦截跳转到登录页面,如图 5-6 所示。

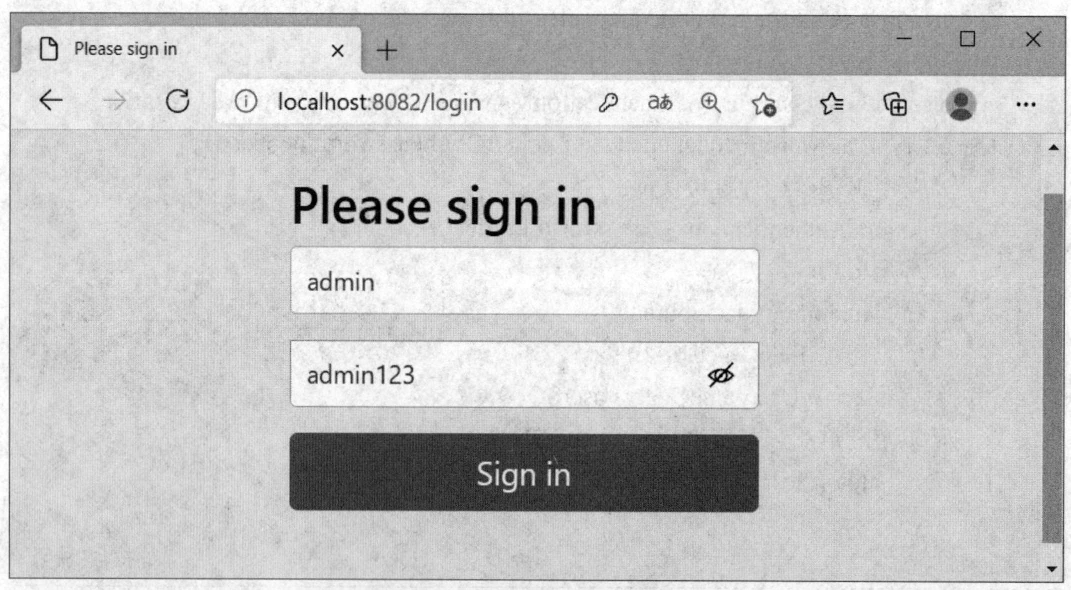

图 5-6 JDBC 身份验证登录页面

在页面中使用保存在 MySQL 数据库中的用户 admin 进行登录,登录成功后自动跳转到 http://localhost:8082/admin/visit 页面。

(4)Bcrypt 加密密码

在实际使用中,用 BCrypt 加密的密码虽然很难破译,但是仍旧不能避免用户使用类似“123456”“abcdefg”这样的简单密码,如果这些简单密码的密文被人截取匹配,那么可能会造成用户账户被盗用。为了克服此问题,可以在加密时添加密钥进行加密,而密钥存在于企业服务器中，这样即使密文被人截取，别人也无法得到密钥破解密文，这样就能够提高网站的安全性。Spring Security 提供了可使用密钥的密码编码器 Pbkdf2PasswordEncoder，在该类的构造函数中可以添加一个密钥作为参数，协助加密，如示例代码 5-5 所示。

示例代码 5-5：Pbkdf2PasswordEncoder 密码编辑器

```
// 密码加密密钥
private String key = "qwert098";
protected void configure(AuthenticationManagerBuilder auth) throws Exception {
    // 使用 Pbkdf2PasswordEncoder 密码编码器类加载加密密钥
    PasswordEncoder encoder = new Pbkdf2PasswordEncoder(key);
    // 使用 JDBC 进行身份认证，并加载密码编码器
    auth.jdbcAuthentication().passwordEncoder(encoder)
        .dataSource(dataSource)
        .usersByUsernameQuery(userSQL)
        .authoritiesByUsernameQuery(roleSQL);
}
```

Spring Security 还拥有 SCryptPasswordEncoder 和 DelegatingPasswordEncoder 等密码加载器，用户可以根据自己的需求去使用不同的密码编码器，或者可以通过实现密码编码器 PasswordEncoder 接口，来根据自己的需求自定义编码器。

3. 自定义身份验证

在一些用户流量较大或业务繁杂的项目中，频繁地使用 JDBC 进行身份验证会受到 JDBC 本身性能的限制，从而降低网站的响应速度。因此，实际项目一般会对数据层统一进行开发管理，这时可以使用自定义身份验证，在已有的用户信息查询代码上进行开发。

（1）MyBatis 实现用户详情及用户权限查询服务

首先，创建用户详情及用户权限 POJO 对象。用户详情类 SysUser 含有用户名 username、用户 id 和用户密码 password 三个属性；用户权限类 SysUserRole 含有用户名 username 和用户权限角色 role 两个属性。这两个 POJO 对象如示例代码 5-6 所示。

示例代码 5-6：POJO 对象代码

```
/**
 * 用户详情类
 */
public class SysUser {
    private String username;
    private String password;
    private long id;
    //get、set 方法
}
```

```
/**
 * 用户权限类
 */
public class SysUserRole {
    private String username;
    private String role;
//get、set 方法
}
```

接下来，编写 MyBatis 的 Mapper 接口以及相应的 XML 文件，在 XML 文件中编写根据用户名查询用户详情的 selectPasswordByUsername 代码，以及根据用户名查询用户权限的 selectUserRoleByUsername 代码，如示例代码 5-7 所示。

示例代码 5-7：sysUserMapper.xml

```
<? xml version="1.0" encoding="UTF-8" ? >
<! DOCTYPE mapper PUBLIC "-//mybatis.org//DTD Mapper 3.0//EN"
                        "http://mybatis.org/dtd/mybatis-3-mapper.dtd">
<mapper namespace="com.xtgj.mapper.SysUserMapper">
<! -- 定义 SysUser 以及 SysUserRole 类的结果映射 -->
 <resultMap id="SysUser" type="com.xtgj.dao.SysUser">
    <id property="id" column="id"/>
    <result property="username" column="username"/>
    <result property="password" column="password"/>
 </resultMap>
 <resultMap id="SysUserRole" type="com.xtgj.dao.SysUserRole">
    <result property="username" column="username"/>
    <result property="role" column="role"/>
 </resultMap>
 <! -- 定义查询 SQL 语句 -->
  <select id="selectPasswordByUsername" parameterType="String" resultMap="SysUser">
select username, password
    from t_user
    where username = #{username}
 </select>
 <select id="selectUserRoleByUsername" parameterType="String" resultMap="SysUser-
Role">
select u.username, r.role
    from t_user u, t_role r, t_user_role ur
    where u.id = ur.user_id and r.id = ur.role_id
```

```
  and u.username = #{username}
 </select>
</mapper>
```

（2）封装用户认证信息

UserDetailsService 是 Spring Security 提供的用于封装用户认证信息的接口，该接口只有一个 loadUserByUsername（String username）方法，在该方法中需要实现通过传入用户名字符串来查询用户详细信息和权限的功能，返回 UserDetails 对象。UserDetailsService 接口实现类的代码如示例代码 5-8 所示。

示例代码 5-8：UserDetailsServiceImpl.java

```
import com.xtgj.dao.SysUser;
import com.xtgj.dao.SysUserRole;
import com.xtgj.service.SysUserService;
import org.springframework.beans.factory.annotation.Autowired;
import org.springframework.security.core.GrantedAuthority;
import org.springframework.security.core.authority.SimpleGrantedAuthority;
import org.springframework.security.core.userdetails.User;
import org.springframework.security.core.userdetails.UserDetails;
import org.springframework.security.core.userdetails.UserDetailsService;
import org.springframework.security.core.userdetails.UsernameNotFoundException;
import org.springframework.stereotype.Service;
import java.util.ArrayList;
import java.util.List;

@Service
public class UserDetailsServiceImpl implements UserDetailsService {
  @Autowired
  private SysUserService sysUserService;
  @Override
  public UserDetails loadUserByUsername(String username) throws UsernameNotFoundException {
    SysUser sysUser = sysUserService.selectPasswordByUsername(username);
    List<SysUserRole> sysUserRoleList = sysUserService.selectUserRoleByUsername(username);
    // 创建 Spring Security 用户权限列表对象
    List<GrantedAuthority> grantedAuthorityList = new ArrayList<GrantedAuthority>();
    // 将查询到的用户权限赋值给用户权限列表
```

```
        for(SysUserRole sysUserRole: sysUserRoleList){
            GrantedAuthority grantedAuthority = new SimpleGrantedAuthority(sysUserRole.
getRole());
        }
        if(sysUser! =null){
    // 创建 Spring Security 用户详情对象，将用户名、用户密码以及用户权限赋值给对
象
            UserDetails userDetails =
    new User(sysUser.getUsername(), sysUser.getPassword(), grantedAuthorityList);
            return userDetails;
        }else {
            throw new UsernameNotFoundException(" 当前用户不存在 ");
        }
    }
}
```

对示例代码 5-8 中使用的配置说明如下。

① 首先，其将类注释为 @Service，这样 Spring 就可以扫描并将其自动装配为 Bean；然后，注入 UserRoleService 接口；最后，在类中使用 @Autowired 注入用户认证数据层的 SysUserService 对象。

② 重写接口的 loadUserByUsername(string username)方法。方法中通过 SysUserService 接口查询到用户详情以及用户权限列表，将查询结果通过构造方法传入 UserDetails 接口的实现类 User 中。需要注意的是，在实例化 User 对象前，必须对用户详情对象进行非空判断，如查询到用户为空，则抛出 UsernameNotFoundException 异常。

（3）使用 UserDetailsService 实现身份验证

重写 configure(AuthenticationManagerBuilder auth)方法，并在方法中使用 UserDetailsService 实现身份验证的代码，如示例代码 5-9 所示。

示例代码 5-9：SecurityDemo4.java

```
import org.springframework.security.config.annotation.authentication.builders.Authentica-
tionManagerBuilder;
import org.springframework.security.config.annotation.web.configuration.EnableWebSe-
curity;
import org.springframework.security.config.annotation.web.configuration.WebSecurity-
ConfigurerAdapter;
import org.springframework.security.core.userdetails.UserDetailsService;
import org.springframework.security.crypto.bcrypt.BCryptPasswordEncoder;
import javax.annotation.Resource;
```

```
@EnableWebSecurity
public class SecurityDemo4 extends WebSecurityConfigurerAdapter {
    @Resource(name ="userDetailsServiceImpl")
    private UserDetailsService userDetailsService;
    @Override
    protected void configure(AuthenticationManagerBuilder auth) throws Exception {
        BCryptPasswordEncoder encoder = new BCryptPasswordEncoder();
        auth.userDetailsService(userDetailsService)
                .passwordEncoder(encoder);
    }
}
```

在代码 5-9 中，首先使用 @Resource 注解注入 UserDetailsService 实现类 UserDetailsServiceImpl，然后在 configure 方法中使用 AuthenticationManagerBuilder 实例的 userDetailsService()方法调用 UserDetailsServiceImpl 对象进行自定义身份验证，同时绑定密码编码器来对验证过程中的密码进行解码。

（4）启动服务测试

在测试服务前，需先启动 MySQL 数据库服务器，然后启动 Spring Boot 项目，通过浏览器访问“http://localhost:8082/admin/visit”地址，与前例相似，请求被 Spring Security 拦截并跳转到登录页面，如图 5-7 所示。

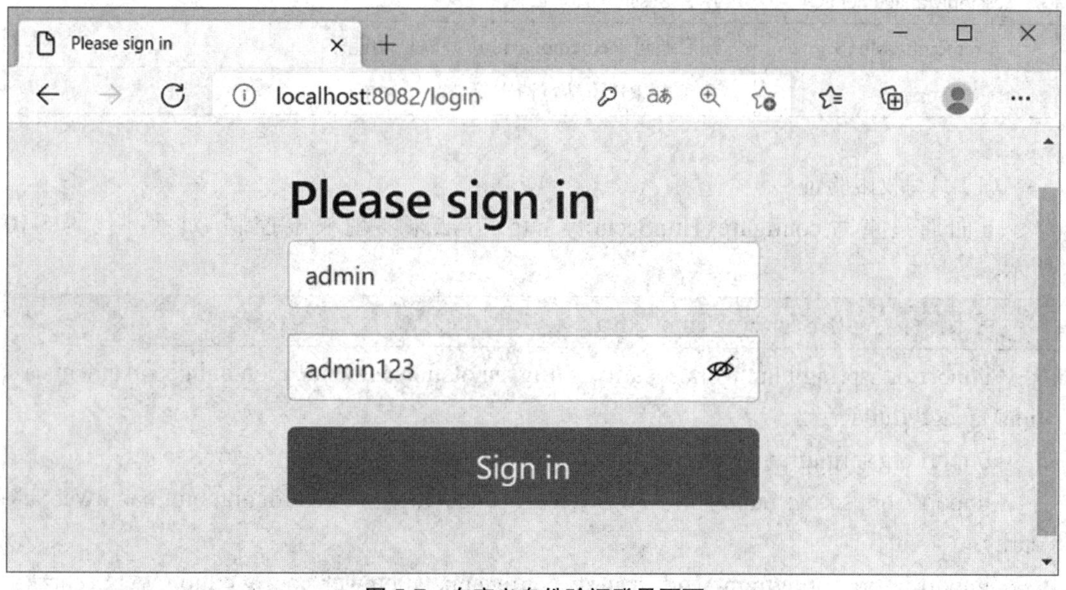

图 5-7　自定义身份验证登录页面

在页面中使用保存在 MySQL 数据库中的用户 admin 进行登录，登录成功后页面自动跳回到 http://localhost:8082/admin/visit 地址对应的页面。

技能点 3 Spring Security 权限设置

在一个应用中,常常将不同的权限赋予不同的用户,以此来保证系统日常的使用以及数据的安全。例如,一个网站可能存在普通用户和管理员用户,普通用户一般只拥有对自身数据操作的权限,而管理员用户所拥有的数据操作权限范围会比普通用户大得多。下面将对实际应用中常见的自定义用户权限设置进行讲解。

1. 配置请求访问权限

Spring Security 的权限控制是通过重写 WebSecurityConfigurerAdapter 类中的 configure (HttpSecurity http)方法来实现的, HttpSecurity 类提供了 HTTP 请求权限限制、Session 管理配置、跨站点请求伪造 Cross-Site Request Forgery, CSRF) 防护等功能,其常用方法见表 5-3。

表 5-3 HttpSecurity 类常用方法

方法名称	详情说明
authorizeRequests()	开启基于 HttpServletRequest 的请求访问权限设置
formLogin()	启用基于表单的用户登录
httpBasic()	启用基于 HTTP 请求的 Basic 认证登录
logout()	启用退出登录支持
sessionManagement()	开启 Session 管理配置
rememberMe()	开启通过 RememberMe 访问验证功能
csrf()	开启 CSRF 跨站点请求伪造防护功能

(1)编写权限验证

下面通过重写 configure(HttpSecurity http)方法来实现权限控制,如示例代码 5-10 所示。

示例代码 5-10: SecurityDemo6.java

```
import org.springframework.security.config.annotation.authentication.builders.AuthenticationManagerBuilder;
import org.springframework.security.config.annotation.web.builders.HttpSecurity;
import org.springframework.security.config.annotation.web.configuration.EnableWebSecurity;
import org.springframework.security.config.annotation.web.configuration.WebSecurityConfigurerAdapter;
import org.springframework.security.core.userdetails.UserDetailsService;
import org.springframework.security.crypto.bcrypt.BCryptPasswordEncoder;
```

```
import javax.annotation.Resource;
@EnableWebSecurity
public class SecurityDemo6 extends WebSecurityConfigurerAdapter {
  @Resource(name = "userDetailsServiceImpl")
  private UserDetailsService userDetailsService;
  @Override
  protected void configure(AuthenticationManagerBuilder auth) throws Exception {
    BCryptPasswordEncoder encoder = new BCryptPasswordEncoder();
    auth.userDetailsService(userDetailsService)
        .passwordEncoder(encoder);
  }
  @Override
  protected void configure(HttpSecurity http) throws Exception {
    // 启用 HTTP 请求权限控制
    http.authorizeRequests()
        // 限定 "/user/**" 请求必须有 user 或 admin 权限才能访问
        .antMatchers("/user/**").hasAnyRole("user", "admin")
        // 限定 "/admin/**" 请求必须有 admin 权限才能访问
        .antMatchers("/admin/**").hasAnyAuthority("admin")
        // 其他路径允许任何权限访问
        .anyRequest().permitAll()
        // 没有配置权限的其他请求可以匿名访问
        .and().anonymous()
        // 启用 Spring Security 默认的登录界面
        .and().formLogin()
        // 启用 HTTP 基础认证
        .and().httpBasic();
  }
}
```

示例代码 5-10 中使用的配置如下。

① HttpSecurity 对象的 authorizeRequests()方法会返回 ExpressionInterceptUrlRegistry 对象，该对象可以注册权限匹配地址。先使用 antMatchers()方法配置需要设置权限验证的 URL 路径，方法可以使用 Ant 风格的 URL 配置。Ant 风格支持使用通配符来简化 URL 的编写，可使用的通配符有“?”匹配任何单字符；“*”匹配 0 个或者任意数量的字符；“**”匹配 0 个或者更多的目录。在示例代码中限定了“/user/**”以及“/admin/**”两个地址，需要进行权限认证才可访问。

② 接着使用 hasAnyRole()方法以及 hasAnyAuthority()方法配置可访问对应的请求的权限。两者的区别：hasAnyRole()方法会在权限字符串前默认加入前缀“ROLE_”。

③ anyRequest()方法代表其他的没有限定的请求；permitAll()方法则表示无条件允许访问路径。

④ and()方法表示一个连接词，方法后会重新加入新的权限验证规则。示例代码中其后配置了 anonymous()方法，该方法允许匿名访问没有配置过的请求。

⑤ formLogin()方法代表启用 Spring Security 默认的登录页面，httpBasic()方法表示启用 HTTP 的 Basic 请求，输入用户名和密码。

authorizeRequests()的其他常用方法见表 5-4。

表 5-4 authorizeRequests 请求控制相关方法

名称	详情说明
antMatchers(String...)	开启 Ant 风格的路径匹配
mvcMatchers(String...)	开启 MVC 风格的路径匹配(与 Ant 风格类似)
and()	连接词，分隔之前限定前提规则
anyRequest()	匹配任意请求
regexMatchers(String...)	开启正则表达式的路径匹配
hasRole(String)	将访问权限赋予一个角色(会自动加入前缀 "ROLE_")
hasAnyRole(String...)	将访问权限赋予多个角色(会自动加入前缀 "ROLE_")
hasAuthority(String)	如含有指定权限则允许访问(不加入前缀 "ROLE_")
hasAnyAuthority(String...)	如含有指定多个权限中的一个则允许访问(不加入前缀 "ROLE")
denyAll()	无条件不允许任何访问
permitAll()	无条件允许任何访问
access(String)	匹配给定的 SpEL 表达式计算结果是否为 true
authenticated()	匹配已经登录认证的用户
fullyAuthenticated()	如果是完整验证(并非 RememberMe)，则允许访问
hasIpAddress(String)	访问如果是给定的 IP 地址，则允许访问
rememberMe()	用户通过 RememberMe 功能验证，则允许访问

(2)启动服务测试

在测试前添加一方法来测试无须权限的访问，该方法响应的 URL 地址为 "/"，代码如下。

```
@Controller
public class LoginController{
  @ResponseBody
   @RequestMapping("/")
public String demo(){ return " 无权限限制页面 "; }
}
```

启动 Spring Boot 项目，通过浏览器访问“http: //localhost: 8082”地址，因权限设置中无限制请求，因此该地址不需要登录即可访问，效果如图 5-8 所示。

图 5-8　地址 localhost:8082/ 页面

在访问“http: //localhost: 8082/admin/visit”地址时，访问请求被 Spring Security 拦截跳转到登录页面，此时使用含有 admin 权限和 ROLE_admin 权限的 admin 用户登录可正常访问，效果如图 5-9 和图 5-10 所示。

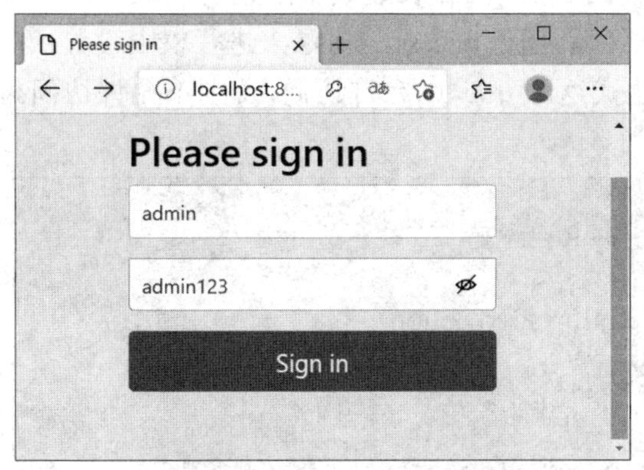

图 5-9　使用 admin 用户登录

图 5-10　地址 localhost:8082/admin/visit 页面

而使用只含有 ROLE_user 权限的 user 用户访问“http: //localhost: 8082/admin/visit”地址并登录时会出现如图 5-11 所示的禁止访问（Forbidden）页面，说明所配置的对不同权限用户的拦截设置已生效。

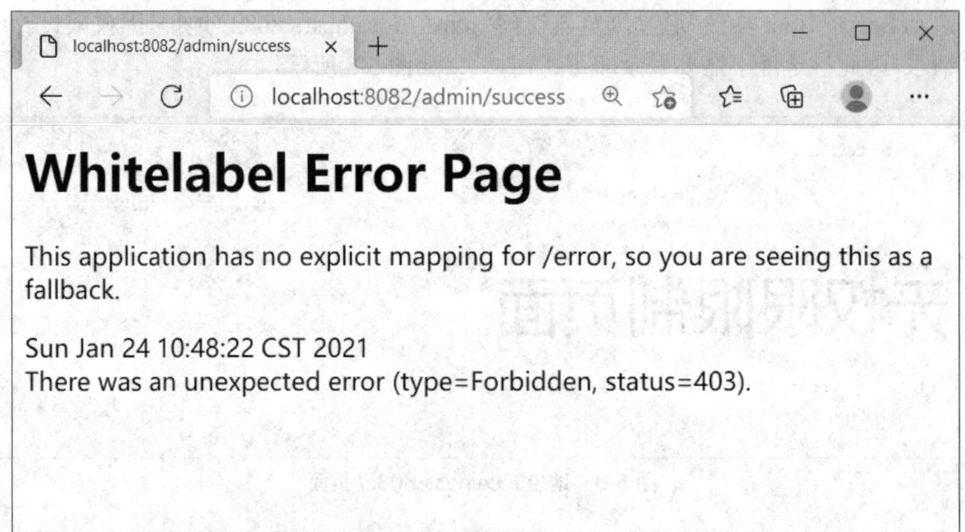

图 5-11 禁止访问页面

2. 使用 Spring 表达式配置访问权限

在 configure（HttpSecurity http）中还可以使用 Spring EL 表达式来配置访问权限，其方法为在调用 antMatchers（）方法配置完 URL 路径后，通过 access（）方法来进行权限设置，方法参数是一个 Spring EL 表达式。如果该表达式返回 true，则允许访问，否则不允许访问。方法示例如示例代码 5-11 所示。

示例代码 5-11：Spring 表达式配置访问权限

```
@Override
protected void configure(HttpSecurity http) throws Exception {
    // 方法启用请求验证
    http.authorizeRequests()
        // 限定 "/user/**" 请求必须有 Role_user 或 Role_admin 权限才能访问
        .antMatchers("/user/**").access("hasRole('user') or hasAnyRole('admin')")
        // 使用 Spring 表达式限定 "/admin/**" 请求必须有 admin 权限才能访问
        .antMatchers("/admin/**").access("hasAuthority('admin')")
        // 其他路径允许任何权限访问
        .anyRequest().permitAll()
        // 没有配置权限的其他请求可以匿名访问
        .and().anonymous()
        // 启用 Spring Security 默认的登录界面
        .and().formLogin()
        // 启用 HTTP 基础认证
        .and().httpBasic();
}
```

上述代码中使用 access（）方法配置了两个 Spring 表达式来实现访问权限限定。第一个表达式使用了 hasRole( )以及 hasAnyRole( )方法，两个方法之间用 or 连接符连接，通过它们的参数限定了角色“ROLE_user”或“ROLE_admin”拥有“/user/**”地址的访问权限；第二个表达式是使用 hasAuthority 方法，限定“ROLE_admin”角色拥有“/admin /**”地址的访问权限，并且要求访问是完整登录，不接受 RememberMe 这样的验证方式进行访问。

除代码中演示的这些正则式方法以外，Spring Security 还提供了其他的表达式方法，见表 5-5。

表 5-5　access( )方法可用的 Spring EL 表达式

名称	详情说明
authentication( )	用户认证对象
hasRole(String)	将访问权限赋予一个角色(会自动加入前缀 "ROLE_")
hasAnyRole(String...)	将访问权限赋予多个角色(会自动加入前缀 "ROLE_")
hasAuthority(String)	如含有指定权限则允许访问(不加入前缀 "ROLE_")
hasAnyAuthority(String...)	如含有指定多个权限中的一个则允许访问(不加入前缀 "ROLE")
hasIpAddress(String)	确定请求是否来自指定的 IP
permitAll( )	无条件允许任何访问
denyAll( )	拒绝任何访问
isAnonymous( )	是否是匿名访问
isAuthenticated( )	是否是已经登录认证的用户
isFullyAuthenticated( )	是否用户是完整验证，即非 RememberMe 认证才能通过认证
isRememberMe( )	是否是通过 RememberMe 认证的访问

技能点 4　自定义认证页面

1. 自定义用户登录页面

Spring Security 提供了默认登录页面来满足安全框架的登录验证需求，而在实际项目的开发中，需要开发者编写功能更丰富、用户界面(User Interface，UI)更美观的登录页面。Spring Security 可以通过 HttpSecurity 对象的 formLogin( )方法来获取登录表单配置器 FormLoginConfigurer，通过该配置器的方法来自定义配置登录页面。

FormLoginConfigurer 类中常用的方法见表 5-6。

表 5-6　FormLoginConfigurer 类常用方法

名称	详情说明
loginPage(String)	用户登录页面跳转路径，默认为 /login

续表

名称	详情说明
successForwardUrl（String）	用户登录成功后的重定向地址
successHandler（AuthenticationSuccessHandler）	用户登录成功后的处理
defaultSuccessUrl（String）	用户直接登录后默认跳转地址
failureForwardUrl（String）	用户登录失败后的重定向地址
failureUrl（String）	用户登录失败后的跳转地址，默认为 /login？error
failureHandler（AuthenticationFailureHandler）	用户登录失败后的错误处理
usernameParameter（String）	登录用户的用户名参数，默认为 username
passwordParameter（String）	登录用户的密码参数，默认为 password
loginProcessingUrl（String）	登录表单提交的路径，默认为 post 请求 /login
permitAll（）	无条件对请求进行放行

（1）编写登录过程 HTML 文件

自定义登录认证页面时，首先需编写登录页面的 HTML 网页，在项目的 resources/templates 目录下新建 loginPage.html 文件，如示例代码 5-12 所示。

示例代码 5-12：loginPage.html

```
<！ DOCTYPE html>
<html xmlns：th="http：//www.thymeleaf.org" lang="zh-CN">
<head>
    <title> 登录页面演示 </title>
    <meta charset="utf-8">
    <meta name="viewport" content="width=device-width，initial-scale=1">
    <link rel="stylesheet" type="text/css" th：href="@{assets/fonts/font-awesome-4.7.0/css/
font-awesome.min.css}">
    <link rel="stylesheet" type="text/css" th：href="@{assets/css/util.css}">
    <link rel="stylesheet" type="text/css" th：href="@{assets/css/main.css}">
</head>
<body>
    <div class="limiter">
        <div class="container-login100">
            <div class="wrap-login100">
                <div class="login100-form-title"
th：style=" 'background-image：url（/assets/images/bg-01.jpg）；' ">
                    <span class="login100-form-title-1"> 登 录 </span>
                </div>
```

```
            <form class="login100-form validate-form" th:action="@{/login}"
    method="post">
                <div class="wrap-input100 validate-input m-b-26"
    data-validate=" 用户名不能为空 ">
                    <span class="label-input100"> 用户名 </span>
                    <input class="input100" type="text" name="username"
    placeholder=" 请输入用户名 ">
                    <span class="focus-input100"></span>
                </div>

                <div class="wrap-input100 validate-input m-b-18"
    data-validate=" 密码不能为空 ">
                    <span class="label-input100"> 密码 </span>
                    <input class="input100" type="password" name="password"
    placeholder=" 请输入密码 ">
                    <span class="focus-input100"></span>
                </div>
                <div class="container-login100-form-btn">
                    <button class="login100-form-btn"> 登 录 </button>
                </div>
            </form>
        </div>
      </div>
    </div>
    <script th:src="@{assets/js/jquery-3.2.1.min.js}"></script>
    <script th:src="@{assets/js/main.js}"></script>
  </body>
  </html>
```

代码中通过 <form> 标签创建了一个登录表单，表单中用户名提交参数自定义为 username，用户密码提交参数自定义为 password，两个参数的 <input> 标签使用 JQuery 的 validate 验证功能进行非空验证，如提交时参数 input 的内容为空，则无法提交且提示“用户名不能为空”或“密码不能为空”。表单提交方法必须为 post，提交的 URL 定义为 /login。代码中还引入了 js 文件和 css 文件来渲染页面，这些静态文件都放置在 resources/static/ assets 目录下。

编写登录失败页面，在项目的 resources/templates/comm 目录下新建 loginFailure.html 文件，如示例代码 5-13 所示。

示例代码 5-13：loginFailure.html

```
<! DOCTYPE html>
<html xmlns:th="http://www.thymeleaf.org "lang="zh-CN">
<head>
  <meta charset="UTF-8">
  <title> 登录失败 </title>
  <meta name="viewport" content="width=device-width, initial-scale=1">
  <link rel="stylesheet" type="text/css" th: href="@{assets/fonts/font-awesome-4.7.0/css/
font-awesome.min.css}">
  <link rel="stylesheet" type="text/css" th:href="@{assets/css/util.css}">
  <link rel="stylesheet" type="text/css" th:href="@{assets/css/main.css}">
</head>
<body>
<div class="limiter">
  <div class="container-login100">
    <div class="wrap-login100">
      <div class="login100-form-title" th: style=" 'background-image: url(/assets/imag-
es/bg-01.jpg);'">
        <span class="login100-form-title-1"> 很抱歉,登录失败 </span>
      </div>
    </div>
  </div>
</div>
<script th:src="@{assets/js/jquery-3.2.1.min.js}"></script>
<script th:src="@{assets/js/main.js}"></script>
</body>
</html>
```

（2）编写登录 Controller

在 LoginController 类中添加访问登录页面的方法 login()，请求路径为“/login”，并跳转到 templates/comm 目录中的 loginPage.html。添加访问登录失败页面的方法 loginFailure()，请求路径为“/loginFailure”，并跳转到 templates/comm 目录中的 loginFailure.html，如示例代码 5-14 所示。

示例代码 5-14：登录用 Controller

```
@Controller
public class LoginController {
  @GetMapping(value = "/login")
  public String login(){
```

```
        return "comm/loginPage";
    }
    @RequestMapping("/loginFailure")
    public String loginFailure(){
        return "comm/loginFailure";
    }
}
```

（3）编写登录控制

下面通过重写 configure（HttpSecurity http）方法来实现自定义用户登录页面，如示例代码 5-15 所示。

示例代码 5-15：SecurityDemo8.java

```
@EnableWebSecurity
public class SecurityDemo8 extends WebSecurityConfigurerAdapter {
    @Resource(name = "userDetailsServiceImpl")
    private UserDetailsService userDetailsService;
    @Override
    protected void configure(AuthenticationManagerBuilder auth) throws Exception {
        BCryptPasswordEncoder encoder = new BCryptPasswordEncoder();
        auth.userDetailsService(userDetailsService)
                .passwordEncoder(encoder);
    }
    @Override
    protected void configure(HttpSecurity http) throws Exception {
        http.authorizeRequests()
                .antMatchers("/user/**","/").access("hasRole('user') or hasAnyRole('admin')")
                .antMatchers("/admin/**").hasAnyAuthority("admin")
                .anyRequest().permitAll()
                .and().anonymous()
                .and().httpBasic();
        // 启用登录设置
        http.formLogin()
                // 设置登录页面
                .loginPage("/login")
                // 获取从前端提交的用户名和密码
                .usernameParameter("username").passwordParameter("password")
                // 设置登录成功后跳转页面
```

```
                .defaultSuccessUrl("/user/hello")
                // 设置登录失败后跳转页面
                .failureForwardUrl("/loginFailure");
    }
}
```

示例代码 5-15 中使用的自定义登录配置如下。

① HttpSecurity 对象的 formLogin()方法用于获取登录表单配置器 FormLoginConfigurer 开启登录设置，使用其 loginPage()方法来配置自定义用户登录页面的跳转 URL 路径。

② usernameParameter()与 passwordParameter()方法用于接受从前端提交的用户名和密码，方法的参数值为前端表单中用户名、密码的 <input> 输入标签的 name 属性值。

③ defaultSuccessUrl()方法用于设置登录成功后跳转的 URL 地址，使用 failureForwardUrl()方法设置登录失败后跳转的 URL 地址。

(4)启动服务测试

启动 Spring Boot 项目，通过浏览器访问"http://localhost:8082/login"地址，会跳转到自定义登录页面，效果如图 5-12 所示。

图 5-12 自定义登录页面

登录成功后，系统会自动跳转并访问"http://localhost:8082/user/hello"地址，如输入错误的用户名或密码，，则会跳转到登录失败页面，效果如图 5-13 所示。

图 5-13　登录失败页面

2. 自定义用户登出

登录后还需要为用户提供登出功能，Spring Security 可以通过 HttpSecurity 对象的 logout()方法来获取登出配置器 LogoutConfigurer，进而通过该配置器的方法来自定义配置登出功能。

LogoutConfigurer 类的常用方法见表 5-7。

表 5-7　登出配置器常用方法

名称	详情说明
logoutUrl(String)	用户退出处理控制 URL，默认的请求为 post 类型，URL 地址为 /logout
logoutSuccessUrl(String)	用户退出成功后的重定向地址
logoutSuccessHandler(LogoutSuccessHandler)	用户退出成功后的处理器设置
deletcCookies(String…)	用户退出后删除指定 Cookie
invalidateHttpSession(boolean)	用户退出后是否立即清除 Session(默认为 true)
clearAuthentication(boolean)	用户退出后是否立即清除 Authentication 用户认证信息(默认为 true)

(1)编写含有登出功能的 HTML 文件

为实现自定义登出功能，首先需要在某个 HTML 页面中添加注销按钮，然后创建一个用以显示登录成功信息的 DemoPage.html 文件，如示例代码 5-16 所示。

示例代码 5-16：DemoPage.html

```
<! DOCTYPE html>
<html xmlns: th="http: //www.thymeleaf.org" lang="zh-CN">
```

```
    <head>
        <meta charset="UTF-8">
        <title>DemoPage</title>
        <meta name="viewport" content="width=device-width, initial-scale=1">
        <link rel="stylesheet" type="text/css" th: href="@{assets/fonts/font-awesome-4.7.0/css/
font-awesome.min.css}">
        <link rel="stylesheet" type="text/css" th:href="@{assets/css/util.css}">
        <link rel="stylesheet" type="text/css" th:href="@{assets/css/main.css}">
    </head>
    <body>
    <div class="limiter">
        <div class="container-login100">
            <div class="wrap-login100">
                <div class="login100-form-title"
    th:style="'background-image: url(/assets/images/bg-01.jpg);'">
                    <span class="login100-form-title-1"> 登录成功 </span>
                </div>
                <form class="login100-form-1 validate-form" th:action="@{/logoutMethod}"
    method="post">
                    <div class="m-b-26">
                        <span class="txt2"> 当前登录用户:</span>
                        <span class="txt2" th:text="${username}"></span>
                    </div>
                    <button class="login100-form-btn"> 注销用户 </button>
                </form>
            </div>
        </div>
    </div>
    </body>
    </html>
```

在代码的 <form> 标签中创建了一个表单，表单中显示了当前登录用户，使用 thymeleaf 标签获取后端传来的用户名参数，用表单提交的方法注销用户。注销用户的 URL 地址定义为 /logoutMethod，方法是 post。

（2）获取登录用户信息

Spring Security 的登录用户信息被封装为 SecurityContextImpl 对象被保存在会话 Session 对象中，对应的 key 为“SPRING_SECURITY_CONTEXT”。 SecurityContextImpl 对象包含了登录用户的用户名、用户权限等信息。获取用户信息的方法如示例代码 5-17 所示。

示例代码 5-17：获取用户信息并渲染到 DemoPage.html

```
@RequestMapping("/index")
public String index(HttpSession session, HttpServletRequest request){
  // 将 Session 中的元素保存为枚举类
  Enumeration<String> names = session.getAttributeNames();
  // 遍历枚举类
  while(names.hasMoreElements()){
    // 获取 Session 中元素名称
    String element = names.nextElement();
    // 获取 Session 中保存的 SecurityContextImpl 对象
    SecurityContextImpl attribute = (SecurityContextImpl)session.getAttribute(element);
    // 获取用户相关信息
    Authentication authentication = attribute.getAuthentication();
    // 将用户信息转换为可操作的 UserDetails 对象
    UserDetails userDetails = (UserDetails) authentication.getPrincipal();
    // 将用户名通过 HttpServletRequest 传至前端页面中
    request.setAttribute("username",userDetails.getUsername());
  }
  return "comm/DemoPage";
}
```

在代码中获取当前 HttpSession 对象，并使用方法遍历其中内容获取 Session 中保存的用户信息封装 Object 对象。该 Object 对象本质为一个 SecurityContextImpl 实例，所以将其强制转换为 SecurityContextImpl 对象，并获取其中的 Authentication 认证信息对象。使用认证信息对象的 getPrincipal()方法获取其中保存的 UserDetails 用户信息类，可通过操作该信息类获取用户的详细信息。最后，将获取的用户名通过 HttpServletRequest 对象传至前端进行渲染显示。

（3）编写登出控制

下面通过重写 configure(HttpSecurity http)方法来实现自定义用户登出功能，如示例代码 5-18 所示。

示例代码 5-18：SecurityDemo9.java

```
public class SecurityDemo9 extends WebSecurityConfigurerAdapter {
  @Resource(name = "userDetailsServiceImpl")
  private UserDetailsService userDetailsService;
  @Override
  protected void configure(AuthenticationManagerBuilder auth) throws Exception {
  BCryptPasswordEncoder encoder = new BCryptPasswordEncoder();
```

```
        auth.userDetailsService(userDetailsService)
                .passwordEncoder(encoder);
    }
    @Override
    protected void configure(HttpSecurity http) throws Exception {
        http.authorizeRequests()
            .antMatchers("/user/**","/").access("hasRole('user') or hasAnyRole('admin')")
                .antMatchers("/admin/**").hasAnyAuthority("admin")
                .anyRequest().permitAll()
                .and().anonymous()
                .and().httpBasic();
        http.formLogin()
                .loginPage("/login")
                .usernameParameter("username").passwordParameter("password")
                // 设置默认登录成功页面
                .defaultSuccessUrl("/index")
                .failureForwardUrl("/loginFailure");
        // 用户登出设置
        http.logout()
                // 设置登出方法的 URL
                .logoutUrl("/logoutMethod")
                // 设置登出后跳转的 URL 地址
                .logoutSuccessUrl("/login");
    }
}
```

该示例使用 HttpSecurity 对象的 logout()方法获取登出配置器 LogoutConfigurer 实例来开启登出设置，使用配置器的 logoutUrl()方法设置登出方法的 URL 地址为 /logoutMethod，该地址与 DemoPage.html 文件中的注销方法的 URL 须一致。使用 logoutSuccessUrl()方法设置登出后跳转的 URL 为 /login，即跳转回登录页面。

(4)启动服务测试

启动 Spring Boot 项目，通过浏览器访问“http://localhost:8082/login”地址，进行登录，登录成功后，页面跳转到 DemoPage.html 页面，效果如图 5-14 所示。

图 5-14　含有登出功能的页面

页面中显示了当前登录用户的用户名，以及“注销用户”的按钮。用鼠标左键点击“注销用户”按钮，页面即跳转回登录页面并且 Session 中不再保存该用户信息。

3. 自定义登录跳转

在登录成功或失败后，页面会重定向到设置的 URL，如果需要在这个过程中添加自定义功能，可通过 HttpSecurity 的方法来截取相关配置器进行功能配置。

（1）登录成功后处理

HttpSecurity 对象的 formLogin（）方法来获取登录表单配置器 FormLoginConfigurer，使用配置器的 successHandler（AuthenticationSuccessHandler）方法即可在身份验证成功后添加自定义功能，如示例代码 5-19 所示。

示例代码 5-19：登录成功后自定义处理

```
@Override
protected void configure(HttpSecurity http) throws Exception {
  http.authorizeRequests()
      .antMatchers("/user/**","/").access("hasRole('user') or hasAnyRole('admin')")
      .antMatchers("/admin/**").hasAnyAuthority("admin")
      .anyRequest().permitAll()
      .and().anonymous()
      .and().httpBasic();
  http.formLogin()
      .loginPage("/login")
```

```
        .usernameParameter("username").passwordParameter("password")
        .defaultSuccessUrl("/index")
        .successHandler(new AuthenticationSuccessHandler() {
      @Override
           public void onAuthenticationSuccess(HttpServletRequest httpServletRequest, HttpServletResponse httpServletResponse, Authentication authentication) throws IOException, ServletException {
          // 用于保存登录前的请求的缓存类
          RequestCache requestCache = new HttpSessionRequestCache();
          // 用于封装在身份验证过程中,实现重定向到跳转前 URL 所需要的功能的类
          SavedRequest savedRequest = requestCache
    .getRequest(httpServletRequest,httpServletResponse);
          // 判断 SavedRequest 实例对象是否为空,非空则证明存在登录前原始页面
          if(savedRequest != null){
            // 重定向到原始访问路径
            httpServletResponse.sendRedirect(savedRequest.getRedirectUrl());
          }else {
          // 直接登录的用户,根据用户角色分别重定向到后台首页和前台首页
            Collection<? extends GrantedAuthority> authorities = authentication.getAuthorities();
          boolean isAdmin = authorities.contains(
    new SimpleGrantedAuthority("ROLE_admin"));
          if(isAdmin){
             httpServletResponse.sendRedirect("/admin/visit");
          }else {
             httpServletResponse.sendRedirect("/user/hello");
          }
        }
      }
    })
    }
```

示例代码 5-19 中使用的配置如下。

① 使用 FormLoginConfigurer 配置器的 successHandler(AuthenticationSuccessHandler)方法配置登录成功后的处理,方法参数为 AuthenticationSuccessHandler 的自定义实现类。

② Spring Security 提供了 SavedRequest 类,该类封装了在身份验证过程中,实现重定向到跳转前 URL 所需要的功能。可以使用 Spring Security 的请求缓存类 RequestCache 的 getRequest()方法来实例化 SavedRequest 类,方法中需传入 HttpServlet 的请求和相应对象。

③ 通过判断实例化后 SavedRequest 是否为空,可以确定登录前是否有 URL 跳转,如

非空，则证明有跳转。此时，可使用 HttpServletResponse 对象实现重定向。

④ 如 SavedRequest 为空，则证明无跳转，此时可通过判断用户权限来实现不同权限登录跳转不同页面的功能。使用 Authentication 对象的 getAuthorities()方法来获取当前用户的权限集合，并通过 contains()方法来判断集合中是否有"ROLE_admin"权限，根据判断结果实现向"/admin/visit"与"/user/hello"的不同跳转。

（2）其他登录处理

除了登录成功后可进行自定义处理外，还可以使用配置器 FormLoginConfigurer 的 failureHandler(AuthenticationFailureHandler)方法来编写登录失败后自定义处理，以及使用 logoutSuccessHandler(LogoutSuccessHandler)方法编写登出成功后自定义处理，如示例代码 5-20 所示。

登录失败处理一般会用来实现记录登录失败时的 IP、时间、SessionId 等特征；发送登录失败提醒给用户，以保证账户安全；同时记录到日志中，或者发送到远端日志监控平台，分析是否是攻击行为等功能。

示例代码 5-20：其他登录处理

```
@Override
protected void configure(HttpSecurity http) throws Exception {
  http.authorizeRequests()
    .antMatchers("/user/**", "/").access("hasRole('user') or hasAnyRole('admin')")
    .antMatchers("/admin/**").hasAnyAuthority("admin")
    .anyRequest().permitAll()
    .and().anonymous()
    .and().httpBasic();
  http.formLogin()
    .loginPage("/login")
    .usernameParameter("username").passwordParameter("password")
    .defaultSuccessUrl("/index")
    .failureHandler(new AuthenticationFailureHandler() {
     @Override
      public void onAuthenticationFailure(HttpServletRequest httpServletRequest,
HttpServletResponse httpServletResponse, AuthenticationException e) throws IOException,
ServletException {
      // 编写登录失败后自定义方法
    }
  });
  http.logout()
    .logoutUrl("/logoutMethod")
    .logoutSuccessHandler(new LogoutSuccessHandler() {
     @Override
```

```
        public void onLogoutSuccess(HttpServletRequest httpServletRequest, HttpServletResponse httpServletResponse, Authentication authentication) throws IOException, ServletException {
            // 编写登出成功后自定义方法
        }
    });
}
```

(3)权限访问异常处理

当用户访问无权限页面时，会被拦截并转发到禁止访问页面(403 Forbidden)，使用 HttpSecurity 的 exceptionHandling()方法可获得 ExceptionHandlingConfigurer 异常配置处理器，并调用其 accessDeniedHandler()方法用以配置访问拒绝处理程序，方法参数为访问拒绝处理程序接口 AccessDeniedHandler 的实现类，在该实现类中可以实现自定义配置权限访问异常，如示例代码 5-21 所示。

示例代码 5-21：权限访问异常处理

```
@Override
protected void configure(HttpSecurity http) throws Exception {
  http.authorizeRequests()
      .antMatchers("/user/**", "/").access("hasRole('user') or hasAnyRole('admin')")
      .antMatchers("/admin/**").hasAnyAuthority("admin")
      .anyRequest().permitAll()
      .and().anonymous()
      .and().httpBasic();
  http.formLogin()
      .loginPage("/login")
      .usernameParameter("username").passwordParameter("password")
      .defaultSuccessUrl("/index")
      .failureForwardUrl("/loginFailure");
  http.exceptionHandling().accessDeniedHandler(new AccessDeniedHandler() {
  @Override
      public void handle(HttpServletRequest httpServletRequest, HttpServletResponse httpServletResponse, AccessDeniedException e) throws IOException, ServletException {
      // 如果是权限访问异常，则拦截到指定错误页面
      RequestDispatcher dispatcher = httpServletRequest.getRequestDispatcher("/error_403");
      // 转发 HttpServletRequest 与 HttpServletResponse
      dispatcher.forward(httpServletRequest, httpServletResponse);
      }
```

```
    });
}
```

技能点 5　记住我功能

开启身份验证后，用户每次重新访问网站，都需要执行一次登录，在一些安全性要求不高的应用中，这种情况会给用户带来很大的困扰。为了方便用户在下一次使用时直接登录，可以开启记住我（remember-me）功能，该功能会在用户登录一段时间内，通过确认用户标记来实现自动登录，免去了重复输入用户名和密码的过程。

Spring Security 可以通过 HttpSecurity 对象的 rememberMe（）方法来获取记住我配置器 RememberMeConfigurer，通过该配置器的方法来自定义记住我功能。该配置器的主要方法见表 5-8。

表 5-8　RememberMeConfigurer 常用方法

名称	详情说明
rememberMeParameter（String）	指示在登录时记住用户的 HTTP 参数
key（String）	记住我认证生成的 Token 令牌标识
tokenValiditySeconds（int）	记住我 Token 令牌有效期，单位为 s（秒）
tokenRepository（PersistentTokenRepository）	指定要使用的 PersistentTokenRepository，用来配置持久化的 Token
alwaysRemember（boolean）	是否应该始终创建记住我的 Cookie，默认为 false
clearAuthentication（boolean）	是否设置 Cookie 为安全的，如果设置为 true，则必须通过 HTTPS 进行连接请求

Spring Security 的记住我功能有两种常用实现方式，一种是基于简单加密的方式将 Token 保存在用户的 Cookie 中；另一种是通过将 Token 持久化保存在服务器数据库中，每次登录时验证用户的 Cookie 中存储的 Token 与数据库中存储的 Token 是否一致。

1. 基于简单加密的 Token 使用方式

简单加密的 Token 会把用户名、密码、Token 过期时间以及 Token 令牌标识通过信息摘要与（Message Digest 5，MD5）算法进行加密，然后会生成一个 Cookie，将加密后的字符串保存在 Cookie 中，该 Cookie 会在登录验证成功后发送到用户客户端。用户再次登录时，会通过用户名、密码以及 Token 令牌标识验证该 Token 的有效性，所以需要保证在此期间这三项数据没发生变化，验证成功后即可自动登录。

该种方式设置简单，但存在一定的安全隐患，如果用户记住我的 Token 被第三方盗用，那么他可以在该 Token 过期之前使用记住我功能随意登录用户账户，所以该方法只适用于对安全性要求不高的情况。

（1）前端添加记住我标记

启用记住我功能时，首先需要在登录页面 loginPage.html 中添加记住我标记，添加的部分如示例代码 5-22 中加粗部分所示。即在代码的 <form> 表单中，新增一个单选框（checkbox）类型的 <input> 标签，标签的 name 属性设置为 remember-me。

示例代码 5-22：loginPage.html

```
<form class="login100-form validate-form" th:action="@{/login}" method="post">
<div class="wrap-input100 validate-input m-b-26" data-validate=" 用户名不能为空 ">
  <span class="label-input100"> 用户名 </span>
  <input class="input100" type="text" name="username" placeholder=" 请输入用户名 ">
  <span class="focus-input100"></span>
</div>
 <div class="wrap-input100 validate-input m-b-18" data-validate=" 密码不能为空 ">
  <span class="label-input100"> 密码 </span>
   <input class="input100" type="password" name="password" placeholder=" 请输入密
码 ">
  <span class="focus-input100"></span>
</div>
 <div class="flex-sb-m w-full p-b-30">
  <div class="contact100-form-checkbox">
   <input class="input-checkbox100" id="ckb1" type="checkbox"
name="remember-me">
   <label class="label-checkbox100" for="ckb1"> 记住我 </label>
</div>
</div>
<div class="container-login100-form-btn">
  <button class="login100-form-btn"> 登 录 </button>
</div>
</form>
```

（2）配置记住我方法

接下来重写 configure（HttpSecurity http）方法，在验证登录设置后添加记住我功能配置，添加部分如示例代码 5-23 中加粗部分所示。

示例代码 5-23：SecurityDemo10.java

```
@EnableWebSecurity
public class SecurityDemo10 extends WebSecurityConfigurerAdapter {
  @Resource(name = "userDetailsServiceImpl")
```

```
    private UserDetailsService userDetailsService;
    @Override
    protected void configure(AuthenticationManagerBuilder auth) throws Exception {
        BCryptPasswordEncoder encoder = new BCryptPasswordEncoder();
        auth.userDetailsService(userDetailsService)
                .passwordEncoder(encoder);
    }
    @Override
    protected void configure(HttpSecurity http) throws Exception {
        http.authorizeRequests()
                .antMatchers("/user/**","/").access("hasRole('user') or hasAnyRole('admin')")
                .antMatchers("/admin/**").hasAnyAuthority("admin")
                .anyRequest().permitAll()
                .and().anonymous()
                .and().httpBasic();
        http.formLogin()
                .loginPage("/login")
                .usernameParameter("username").passwordParameter("password")
                .defaultSuccessUrl("/index")
                .failureForwardUrl("/loginFailure");
        http.logout()
                .logoutUrl("/logoutMethod")
                .logoutSuccessUrl("/login");
        // 开启记住我功能
        http.rememberMe()
                // 从前端获取是否开启记住我功能的参数
                .rememberMeParameter("remember-me")
                // 设置记住我功能生成 Token 的存留时间
                .tokenValiditySeconds(120);
    }
}
```

在该示例中，使用 HttpSecurity 对象的 rememberMe()方法获取 RememberMeConfigurer 配置器来配置记住我功能；使用配置器的 rememberMeParameter()方法来获取前端提交的记住我标记，方法的参数值为前端表单中 checkbox 类型的 <input> 标签的 name 属性值；使用配置器的 tokenValiditySeconds()方法来配置记住我功能生成的 Token 的有效时间，单位为秒(s)。

(3)启动服务测试

启动 Spring Boot 项目，通过浏览器访问“http://localhost:8082/login”地址登录页面，效

果如图 5-15 所示。

图 5-15　带有记住我功能的登录页面

登录成功后，在记住我 Token 有效期内再次访问页面时不需要重新认证登录，可在浏览器的 Cookie 管理器中找到记住我的 Cookie，如图 5-16 所示。

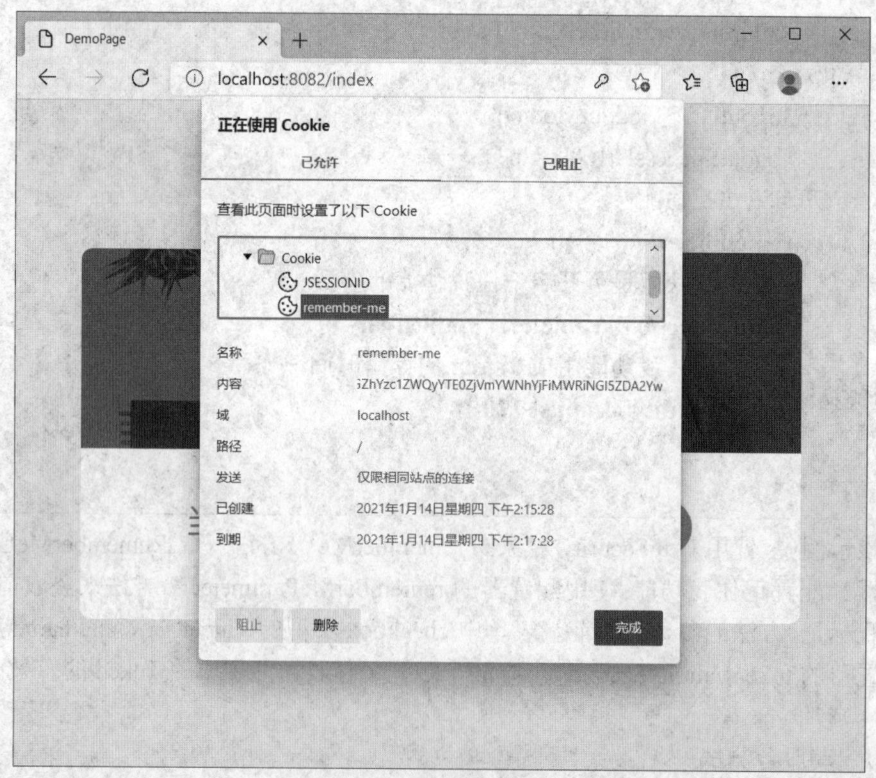

图 5-16　记住我的 Cookie

2. 基于数据持久化的 Token 使用方式

基于数据持久化的 Token 与简单加密的 Token 在用户客户端实现 remember-me 的思路一样，都是将生成的 Token 标记存入用户浏览器的 Cookie 中，等到用户下次访问时，在服务端验证 Token 的有效性。不同于简单加密的是，持久化 Token 登录成功后，会进行以下步骤以保证安全性。

第一步：将用户名、生成的 Token、生成的随机标记存入数据库中，同时将其组合保存在 Cookie 中发送给用户客户端。

第二步：下一次持有 Cookie 的用户登录时，会验证用户的 Cookie 数据与数据库中保存的 Cookie 数据是否一致，如一致，则生成一个新 Token 替换数据库以及用户的 Cookie 中的 Token，并将更新后的 Cookie 发送给用户。

以上逻辑可以保证当有人盗用用户的 Cookie 时可通过对比被立即发现，当服务端发现 Token 对比不一致时，会删除数据库中 Cookie 的 Token 记录，这样就保证下次登录时只能使用密码登录，使盗用的 Cookie 无效。

（1）在数据库中创建 Token 表

Spring Security 的 JdbcTokenRepositoryImpl 类提供了持久化 Token 的便捷操作，使用该类时，需提前注入数据源。查看该类的源代码，可以看到其包含了一个 CREATE_TABLE_SQL 属性，代码如下。

```
public class JdbcTokenRepositoryImpl extends JdbcDaoSupport implements PersistentTokenRepository {
    // 创建持久化 Token 数据库的 SQL 语句
    public static final String CREATE_TABLE_SQL = "create table persistent_logins (
    username varchar(64) not null,
    series varchar(64) primary key,
    token varchar(64) not null,
    last_used timestamp not null)";
    ......
}
```

该属性的值为数据库的建表语句，使用该语句在数据库中创建持久化 Token 所需要的表，表中字段 username 为该条持久化 Token 的所属用户， series 为存储时随机生成的标记，token 为用户访问时更新的 Token，last_used 为用户最近一次的登录日期。

（2）编写持久化记住我方法

建表完成后，在 Security 配置类中重写 configure（HttpSecurity http）方法，来实现持久化记住我功能的配置，如示例代码 5-24 所示。

示例代码 5-24：SecurityDemo11.java

```
@EnableWebSecurity
@Configuration
public class SecurityDemo11 extends WebSecurityConfigurerAdapter {
```

```
@Resource(name = "userDetailsServiceImpl")
private UserDetailsService userDetailsService;
@Autowired
private DataSource dataSource;
@Override
protected void configure(AuthenticationManagerBuilder auth) throws Exception {
   BCryptPasswordEncoder encoder = new BCryptPasswordEncoder();
   auth.userDetailsService(userDetailsService)
         .passwordEncoder(encoder);
}
@Override
protected void configure(HttpSecurity http) throws Exception {
   http.authorizeRequests()
         .antMatchers("/user/**", "/").access("hasRole('user') or hasAnyRole('admin')")
         .antMatchers("/admin/**").hasAnyAuthority("admin")
         .anyRequest().permitAll()
         .and().anonymous()
         .and().httpBasic();
   http.formLogin()
         .loginPage("/login")
         .usernameParameter("username").passwordParameter("password")
         .defaultSuccessUrl("/index")
         .failureForwardUrl("/loginFailure");
   http.logout()
         .logoutUrl("/logoutMethod")
         .logoutSuccessUrl("/login");
   http.rememberMe()
         .rememberMeParameter("remember-me")
         .tokenValiditySeconds(120)
         // 持久化 Token
         .tokenRepository(tokenRepository());
}
// 实例化 JdbcTokenRepositoryImpl 对象
@Bean
public JdbcTokenRepositoryImpl tokenRepository(){
```

```
        JdbcTokenRepositoryImpl jdbcTokenRepository = new JdbcTokenRepositoryImpl();
        // 为 JdbcTokenRepository 对象添加数据源
        jdbcTokenRepository.setDataSource(dataSource);
        return jdbcTokenRepository;
    }
}
```

在该示例中，首先需要使用方法实例化 JdbcTokenRepositoryImpl 对象，在实例化方法中为其添加 DataSource 数据源，并将其装配为 Bean 注入到测试类中。使用 RememberMeConfigurer 配置器的 tokenRepository 配置该 Bean 即可实现持久化 Token。

（3）启动服务测试

启动 Spring Boot 项目，通过浏览器访问“http：//localhost：8082/login”地址，登录后可查看数据库中的 persistent_logins 表，如图 5-17 所示。

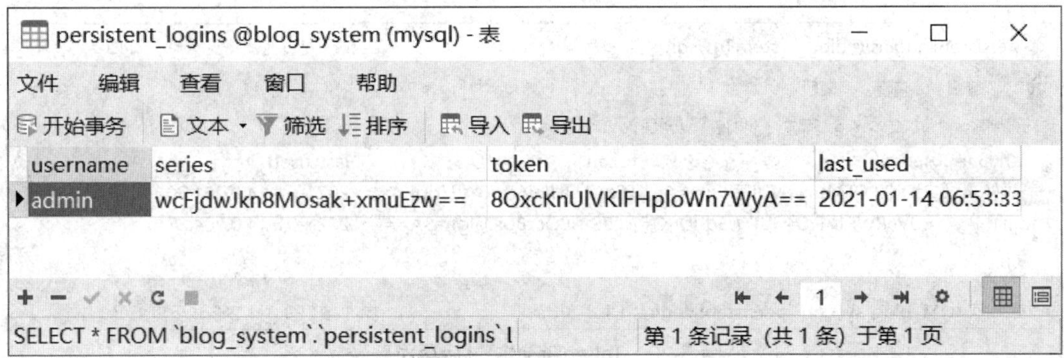

图 5-17　数据库中存储的 Token

可以看到，表中存储了本地登录时的用户名、上一次登录时间、生成的标记（series）以及 Token。关闭会话后，访问“http：//localhost：8082/index”页面，可以看到数据库中保存的上一次登录时间以及 Token 都发生了变化，如图 5-18 所示。

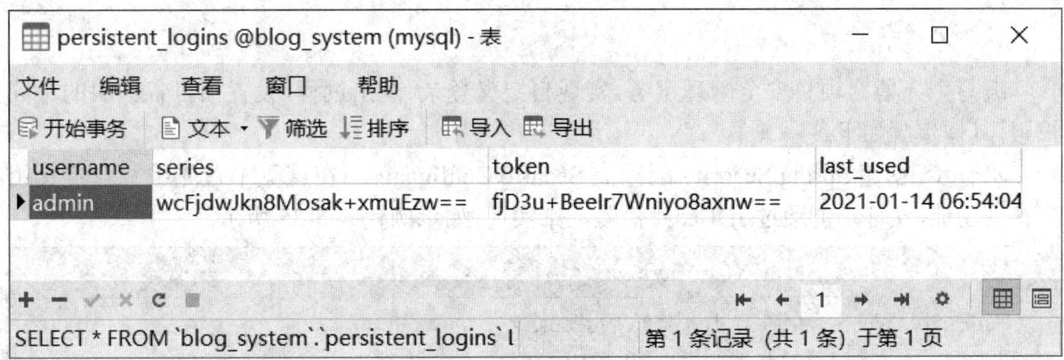

图 5-18　再次访问后的数据库中存储的 Token

用鼠标点击 Index 页面中的“注销”按钮使用户登出，再次观察数据库，可以发现数据库中 admin 用户的 Token 数据被清空，如图 5-19 所示。

图 5-19　用户主动注销后的数据库中存储的 Token

如果用户在 Token 有效期内未手动登出，那么本次有效期内使用的 Token 数据并不会主动消失，下次登录时如再使用记住我功能，会在数据库中新增一条 Token 数据，如图 5-20 所示。

图 5-20　Token 有效期过后再次登录

编写个人博客项目安全管理服务，实现自定义登录、登出过程，设置项目各模块的不同权限，具体步骤如下。

第一步，创建 Spring Security 配置类 SecurityConfig.java。在该类中首先编写自定义用户认证功能，认证功能通过 JDBC 身份验证来实现，如示例代码 5-25 所示。

示例代码 5-25：SecurityConfig.java

```
@EnableWebSecurity // 开启 MVC security 安全支持
public class SecurityConfig extends WebSecurityConfigurerAdapter {
 // 注入 DataSource 数据源
@Autowired
```

```
    private DataSource dataSource;
    /**
     * 重写configure(AuthenticationManagerBuilder auth)方法，进行自定义用户认证
     */
    @Override
    protected void configure(AuthenticationManagerBuilder auth) throws Exception {
        // 实例化密码编码器
        BCryptPasswordEncoder encoder = new BCryptPasswordEncoder();
        // 通过用户名查询用户详情SQL语句
        String userSQL ="select username,password,valid from t_user where username=? ";
        // 通过用户名查询用户权限SQL语句
    String authoritySQL ="select u.username,a.authority from t_user u,t_authority a," +
        "t_user_authority ua where ua.id=u.id and ua.authority_id=a.id and u.username =? ";
        // 使用JDBC进行身份认证
    auth.jdbcAuthentication().passwordEncoder(encoder)
            .dataSource(dataSource)
            .usersByUsernameQuery(userSQL)
            .authoritiesByUsernameQuery(authoritySQL);
    }
}
```

第二步，在SecurityConfig类中使用HttpSecurity对象添加自定义访问控制功能、自定义用户登录功能、自定义用户登出控制和自定义无权限访问提示功能，并开启记住我功能，如示例代码5-26所示。

示例代码5-26：SecurityConfig.java

```
    /**
    * 重写configure(HttpSecurity http)方法，实现权限管理以及自定义登录过程
    */
    @Override
    protected void configure(HttpSecurity http) throws Exception {
      // 自定义用户访问控制
      http.authorizeRequests()
          .antMatchers("/", "/page/**", "/article/**", "/login", "/recoverPassword", "/recover-
Method").permitAll()
          .antMatchers("/back/**","/assets/**","/user/**","/article_img/**").permitAll()
          .antMatchers("/admin/**").hasRole("admin")
          .anyRequest().authenticated();
          // 自定义用户登录控制
```

```
http.formLogin()
    .loginPage("/login")
    .usernameParameter("username").passwordParameter("password")
    .successHandler(new AuthenticationSuccessHandler() {
  @Override
        public void onAuthenticationSuccess(HttpServletRequest httpServletRequest, HttpServletResponse httpServletResponse, Authentication authentication) throws IOException, ServletException {
      // 从前端获取 URL 参数
      String url = httpServletRequest.getParameter("url");
      // 获取被拦截的原始访问路径
      RequestCache requestCache = new HttpSessionRequestCache();
          SavedRequest savedRequest = requestCache.getRequest(httpServletRequest, httpServletResponse);
      if(savedRequest ! =null){
        // 如果存在原始拦截路径,登录成功后重定向到原始访问路径
        httpServletResponse.sendRedirect(savedRequest.getRedirectUrl());
      }else if(url ! = null && ! url.equals(" ")){
        // 跳转到之前所在页面
        URL fullURL = new URL(url);
        httpServletResponse.sendRedirect(fullURL.getPath());
      }else{
        // 直接登录的用户,根据用户角色分别重定向到后台首页和前台首页
         Collection<? extends GrantedAuthority> authorities = authentication.getAuthorities();
          boolean isAdmin = authorities.contains(new SimpleGrantedAuthority("ROLE_admin"));
        if(isAdmin){ httpServletResponse.sendRedirect("/admin"); }
        else{ httpServletResponse.sendRedirect("/"); }
        }
      }
  })
    // 用户登录失败处理
        .failureHandler(new AuthenticationFailureHandler() {
          @Override
```

```
            public void onAuthenticationFailure(HttpServletRequest httpServletRequest,
HttpServletResponse httpServletResponse, AuthenticationException e) throws IOException,
ServletException {
                // 登录失败后，取出原始页面 URL 并追加在重定向路径上
                String url = httpServletRequest.getParameter("url");
                httpServletResponse.sendRedirect("/login? error&url="+url);
            }
        });
    // 设置用户登录后 Cookie 有效期，默认值
    http.rememberMe().alwaysRemember(true).tokenValiditySeconds("1800");
    // 自定义用户登出控制
    http.logout().logoutUrl("/logout").logoutSuccessUrl("/");
    // 针对访问无权限页面出现的 403 Fobidden 页面进行定制处理
    http.exceptionHandling().accessDeniedHandler(new AccessDeniedHandler() {
        @Override
        public void handle(HttpServletRequest httpServletRequest, HttpServletResponse
httpServletResponse, AccessDeniedException e) throws IOException, ServletException {
            // 如果是权限访问异常，则拦截到指定错误页面
            RequestDispatcher dispatcher = httpServletRequest
    .getRequestDispatcher("/errorPage/comm/error_403");
            dispatcher.forward(httpServletRequest, httpServletResponse);
        }
    });
}
```

第三步，编写用户自定义登录页面 login.html，在代码中编写登录表单，表单提交的 URL 为"/login"，如示例代码 5-27 所示。

示例代码 5-27：login.html

```
<! DOCTYPE html>
<html lang="en" xmlns:th="http://www.thymeleaf.org">
<head>
    <meta charset="utf-8">
    <meta http-equiv="Content-Type" content="text/html;">
    <meta name="viewport"
        content="width=device-width, initial-scale=1, maximum-scale=1, user-scalable=no">
    <title> 登录博客后台 </title>
    <link rel="shortcut icon" th:href="@{/user/img/pote.jpg}"/>
```

```
        <script th: src="@{/assets/js/jquery.min.js}"></script>
        <script th: src="@{/assets/js/amazeui.min.js}"></script>
        <link rel="stylesheet" th: href="@{/assets/css/amazeui.min.css}"/>
        <link rel="stylesheet" th: href="@{/assets/css/app.css}"/>
    </head>
    <body>
    <div class="log">
        <div class="am-g">
            <div class="am-u-lg-3 am-u-md-6 am-u-sm-8 am-u-sm-centered log-content">
                <h1 class="log-title am-animation-slide-top" style="color: black; " th:text="#{log-
in.welcomeTitle}"> 欢迎登录博客 </h1>
                <br>
                <div th: if="${param.error}" style="color: red" th: text="#{login.error}"> 用户名
或密码错误！ </div>
                <form class="am-form" id="loginForm" th: action="@{/login}" method="post">
                    <div>
                        <input type="hidden" name="url" th: value="${url}">
                    </div>
                    <div class="am-input-group am-radius am-animation-slide-left">
                        <span class="am-input-group-label log-icon am-radius">
                            <i class="am-icon-user am-icon-sm am-icon-fw"></i>
                        </span>
                            <input type="text" class="am-radius" th: placeholder="#{login.user-
name}"name="username"/>
                    </div>
                    <br>
                    <div class="am-input-group am-animation-slide-left log-animation-delay">
                        <span class="am-input-group-label log-icon am-radius">
                            <i class="am-icon-lock am-icon-sm am-icon-fw"></i>
                        </span>
                        <input type="password" class="am-form-field am-radius log-input" th: place-
holder="#{login.password}" name="password"/>
                    </div>
                    <div class="am-input-group am-animation-slide-left log-animation-delay">
                     <a th: href="@{/recoverPassword}" style="color: black"> 忘记密码？ </a>
```

```
            </div>
            <div style="padding-top: 10px;">
              <input type="submit" th:value="#{login.sub}"
                   class="am-btn am-btn-primary am-btn-block am-btn-lg am-radius am-an-
imation-slide-bottom log-animation-delay" />
            </div>
          </form>
        </div>
      </div>
      <footer class="log-footer">
          <p style="margin: 30px; color: #2E2D3C"><time class="comment-time" th:tex-
t="${#dates.format(new java.util.Date().getTime(), 'yyyy')}"></time> &copy; <a style="-
color: #0e90d2" rel="nofollow">Pote</a></p>
      </footer>
    </div>
    </body>
    </html>
```

第四步，编写无权限访问提示页面 error_403.html，如示例代码 5-28 所示。

示例代码 5-28：error_403.html

```
    <! DOCTYPE html>
    <html lang="en" xmlns:th="http://www.thymeleaf.org">
    <head>
      <meta charset="utf-8"/>
      <title>403(禁止访问)- My Blog</title>
          <meta name="viewport" content="width=device-width, initial-scale=1.0, maxi-
mum-scale=1.0, user-scalable=no"/>
      <meta http-equiv="X-UA-Compatible" content="IE=edge"/>
      <link rel="shortcut icon" th:href="@{/server/images/favicon.png}"/>
      <link th:href="@{/back/css/bootstrap.min.css}" rel="stylesheet" type="text/css"/>
      <link th:href="@{/back/css/style.min.css}" rel="stylesheet" type="text/css"/>
    </head>
    <body>
    <div class="wrapper-page">
      <div class="ex-page-content text-center">
        <h1>403! </h1>
        <h2> 抱歉，没有权限访问这个页面！ </h2>
        <br></br>
```

```
            <a class="btn btn-purple waves-effect waves-light" href="/"><i class="fa fa-angle-
left"></i> 返回首页 </a>
        </div>
    </div>
    </body>
    </html>
```

第五步，在 paging.html 页面的下拉菜单中添加用户“注销”按钮，如示例代码 5-29 所示。

示例代码 5-29：paging.html

```
<ul class="nav navbar-nav navbar-right pull-right">
    <li class="dropdown">
        <a th: href="@{index.html}" class="dropdown-toggle profile" data-toggle="dropdown"
  aria-expanded="true"><img th: src="@{/assets/img/me.jpg}" alt="user-img"
                              class="img-circle"/> </a>
        <ul class="dropdown-menu">
            <form name="logoutform" th: action="@{/logout}" method="post"></form>
            <li><a th: href="@{${commons.site_url()}}" target="_blank">
                <i class="fa fa-eye" aria-hidden="true"></i> 查看网站 </a></li>
            <li><a href="javascript: document.logoutform.submit();">
                <i class="fa fa-sign-out"></i> 用户注销 </a></li>
        </ul>
    </li>
</ul>
```

第六步，编写登录控制 Controller 类，在类中编写登录方法 login() 以及无权限访问跳转方法 AccessExecptionHandler()，如示例代码 5-30 所示。

示例代码 5-30 LoginController.java

```
@Controller
public class LoginController {
    // 向登录页面跳转，同时封装原始页面地址
    @GetMapping(value = "/login")
    public String login(HttpServletRequest request, Map map) {
        // 分别获取请求头和参数 URL 中的原始访问路径
        String referer = request.getHeader("Referer");
        String url = request.getParameter("url");
        System.out.println("referer="+referer);
        System.out.println("url="+url);
```

```
        // 如果参数“url”中已经封装了原始页面路径，直接返回该路径
        if(url! =null && ! url.equals("")){
            map.put("url",url);
            // 如果请求头本身包含登录，将重定向 URL 设为空，让后台通过用户角色选
择跳转
        }else if(referer! =null && referer.contains("/login")){
            map.put("url","");
        }else {
            // 否则的话，就记住请求头中的原始访问路径
            map.put("url", referer);
        }
        return "comm/login";
    }
    // 对 Security 拦截的无权限访问异常处理路径映射
    @GetMapping(value ="/errorPage/{page}/{code}")
    public String AccessExecptionHandler(@PathVariable("page") String page, @Path-
Variable("code") String code){
        return page+"/"+code;
    }
}
```

第七步，启动服务器，访问“http://localhost:8080/admin”地址，可以看到页面被拦截并跳转到登录页面，用户需要登录才能访问，如图 5-21 所示。

图 5-21　个人博客登录页面

使用含有“ROLE_admin”的账户登录后，即可访问“http：//localhost：8080/admin”地址，在页面上角的下拉菜单里可以使用“用户注销”按钮退出登录，如图 5-22 所示。

图 5-22 用户注销按钮

本任务通过对 Spring Security 的讲解，使读者掌握了使用 Spring Security 实现用户登录认证、应用权限管理、自定义认证页面以及记住我等功能。实现了个人博客项目的自定义登录、登出、设置项目各模块权限等功能。

一、选择题

1. Spring Security 提供了多种自定义认证方式，包括（　　）。（多选）

A. JDBC Authentication　　B. LDAP Authentication

C. AuthenticationProvider　　D. UserDetailsService

2. 下列关于使用 JDBC 身份认证方式创建用户 / 权限表以及初始化数据的说法，错误的是(　　)。

A. 用户表中的用户名 username 必须唯一

B. 创建用户表时,必须额外定义一个 tinyint 类型的字段

C. 初始化用户表数据时,插入的用户密码必须是对应编码器编码后的密码

D. 初始化角色表数据时,角色值必须带有“ROLE_”前缀

3. 下列 authorizeRequests 请求控制的相关方法中,用于赋予用户权限的有(　　)。(多选)

A. hasRoles(String...)　　B. hasAuthority(String)

C. hasAnyAuthority(String...)　　D. hasRole(String)

4. 下列关于 configure(HttpSecurity http)方法中参数 HttpSecurity 类的说法,正确的是(　　)。(多选)

A. authorizeRequests()方法开启基于 HttpServletRequest 请求访问的限制

B. formLogin()方法开启基于表单的用户登录

C. rememberMe()方法开启记住我功能

D. csrf()方法配置 CSRF 跨站请求伪造防护功能

5. 下列关于自定义用户登录中的相关说法,错误的是(　　)。

A. loginPage(String loginPage)指定用户登录页面跳转路径,默认为 GET 请求的 /login

B. failureUrl(String authenticationFailureUrl)指定用户登录失败后的跳转地址,默认为 login? failure

C. loginProcessingUrl(String loginProcessingUrl)指定登录表单提交的路径,默认为 POST 请求的 login

D. 项目加入 Security 后,可以不对 static 文件夹下的静态资源文件进行统一放行处理

二、简答题

请简述记住我功能中简单加密 Token 和持久化 Token 的区别。

项目六　个人博客项目消息队列

通过学习 RabbitMQ 的相关知识，掌握 RabbitMQ 的基本概念和主要组件，掌握 Spring AMQP 的主要组件，掌握使用 Spring AMQP 创建并声明队列、交换器、绑定的方法，掌握使用 Spring AMQP 生产消息和消费消息的方法。能够使用 Spring AMQP 编写个人博客项目消息队列服务，实现使用 RabbitMQ 发送密码重置邮件到用户邮箱的功能。

- 掌握 RabbitMQ 的基本概念和主要组件。
- 掌握 Spring AMQP 的主要组件。
- 掌握 Spring AMQP 创建并声明队列、交换器、绑定的方法。
- 掌握 Spring AMQP 生产消息和消费消息的方法。

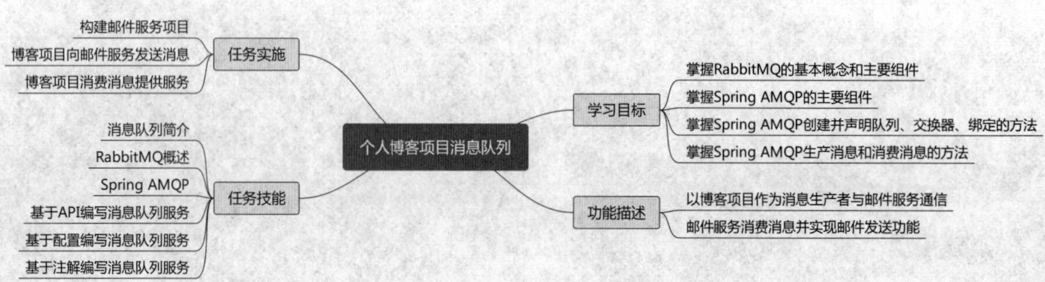

【情景导入】

在构建项目的过程中，随着项目体量的不断扩大，功能越发复杂，使用的用户数量也不断增多，如果只使用单项目应用会给开发带来很大困难，也会影响用户的使用体验。这时就涉及项目功能的拆分，有些拆分后的项目需要进行服务间的异步通信，此时就需要用到消息队列了。

【功能描述】

- 以博客项目作为消息生产者与邮件服务通信。
- 邮件服务消费消息并实现邮件发送功能。

技能点 1　消息队列简介

消息队列（message queue）是一种跨进程、异步的通信机制，用于上下游消息传递，由消息系统来确保消息的可靠传递。消息队列中的消息指的是在两个应用之间传递的数据，这个数据有很多种表达形式，可能是单纯的文本字符串、JSON 等，也可能包含嵌入式对象。

消息队列是在消息的传输过程中负责保存消息的容器。在消息队列中，一般有生产者和消费者两个角色分工。生产者负责发送数据到消息队列，消费者负责从消息队列中取出数据并对数据进行处理。消息队列一般用于应用解耦、异步处理、流量削锋、数据分发、错峰流控、日志收集、分布式事务管理等。

消息队列中间件是消息队列机制的重要组件，在它的帮助下能使消息队列实现更多的高级功能，它一般有两种传递模式：点对点（point-to-point，P2P）模式和发布 / 订阅（pub/sub）模式。

点对点模式是基于队列的，生产者发送消息到队列，消费者从队列中接收并处理消息，队列的模式使消息可以进行异步传输。

发布 / 订阅模式会创建一个内容节点来发布和订阅消息，此内容节点称为主题（topic），它为消息传递的中介，生产者将消息发布到某个主题，消费者则从主题中订阅消息。主题使消息的生产者与消息的消费者互相保持独立，并不需要特定的队列即可进行传输，发布 / 订阅模式常用于消息的一对多广播。

目前，开源的消息中间件有很多，主流的有 RabbitMQ、Kafka、ActiveMQ、RocketMQ 等，本书中主要学习 RabbitMQ。RabbitMQ 是一个由 Erlang 语言开发的高级消息队列协议（Advanced Message Queuing Protocol，AMQP）开源实现，它是应用层协议的一个开放标准，为面向消息的中间件设计。消息中间件主要用于组件之间的解耦，消息的发送者无须知道消息使用者的存在，反之亦然。AMQP 的主要特征是面向消息、队列、路由（包括点对点和发布 / 订阅）、可靠性高、安全。RabbitMQ 是一个开源的 AMQP 实现，其服务器端用 Erlang 语言编写，支持多种客户端，如 Python、Ruby、.NET、Java、JMS、C、PHP、ActionScript、XMPP、STOMP 等，且支持 AJAX。RabbitMQ 用于在分布式系统中存储和转发消息，它具有易用性高、扩展性高、可用性高等优点。

技能点 2 RabbitMQ 概述

1. RabbitMQ 安装

在 Windows 系统中使用 RabbitMQ 进行开发时，需要安装 Erlang 语言，可在 Erlang 官网 https://www.erlang.org/downloads 下载 Windows 版的安装包，如图 6-1 所示。

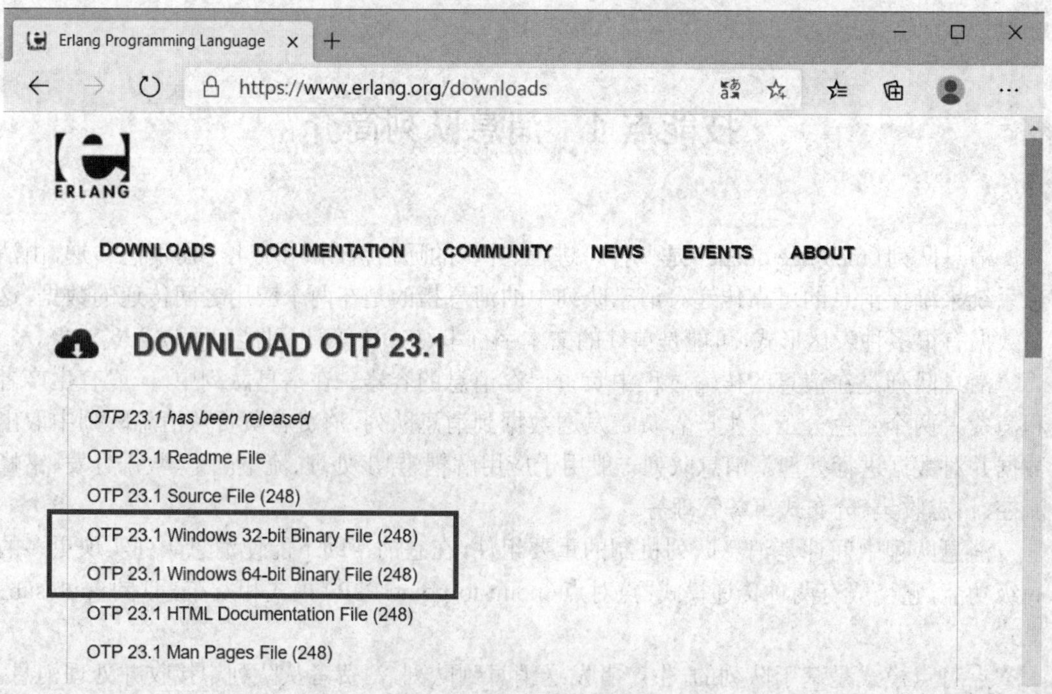

图 6-1 Erlang 下载页面

下载并安装完成后，在系统环境变量中配置 Erlang 地址，如图 6-2 所示。

图 6-2　配置 Erlang 环境变量

在 Windows 系统的命令提示符中输入“erl -version”验证配置是否成功，如图 6-3 所示。

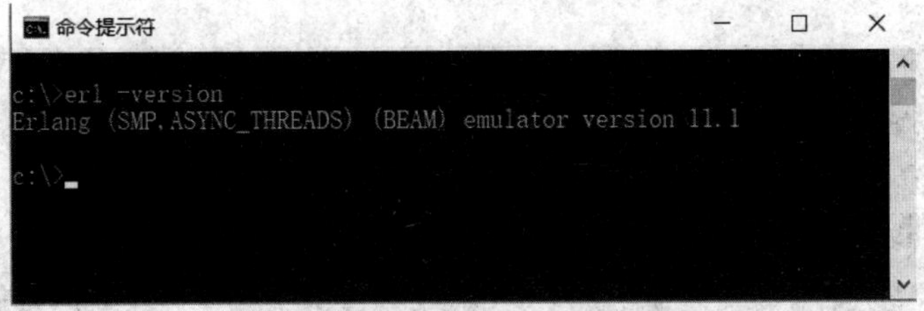

图 6-3　验证 Erlang 配置是否成功

接下来，安装 RabbitMQ 服务端，在 RabbitMQ 官网 https://www.rabbitmq.com/install-windows.html，下载其 Windows 版的安装包，如图 6-4 所示。

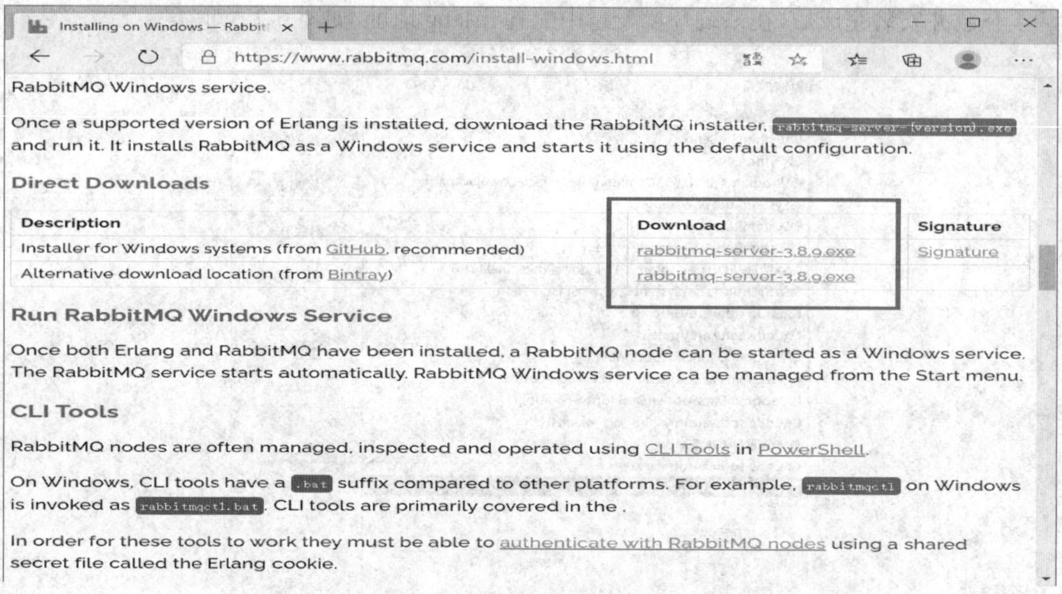

图 6-4　RabbitMQ 下载地址

下载并安装完成后，打开命令提示符，然后在安装目录的 sbin 文件夹下输入 rabbitmq-plugins enable rabbitmq_management 命令，安装管理 RabbitMQ 服务端页面的插件，如图 6-5 所示。

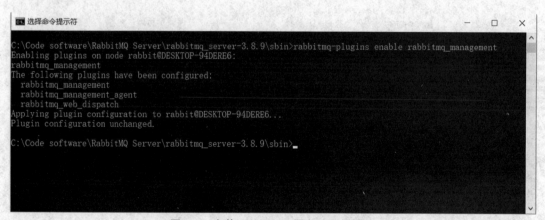

图 6-5　安装 RabbitMQ 服务端页面插件

安装完成后，使用 rabbitmq-server.bat 脚本启动 RabbitMQ 服务。启动完成后打开浏览器输入“http: //localhost: 15672”地址，访问 RabbitMQ 的管理界面，账号密码默认为 guest 和 guest，如图 6-6 所示。

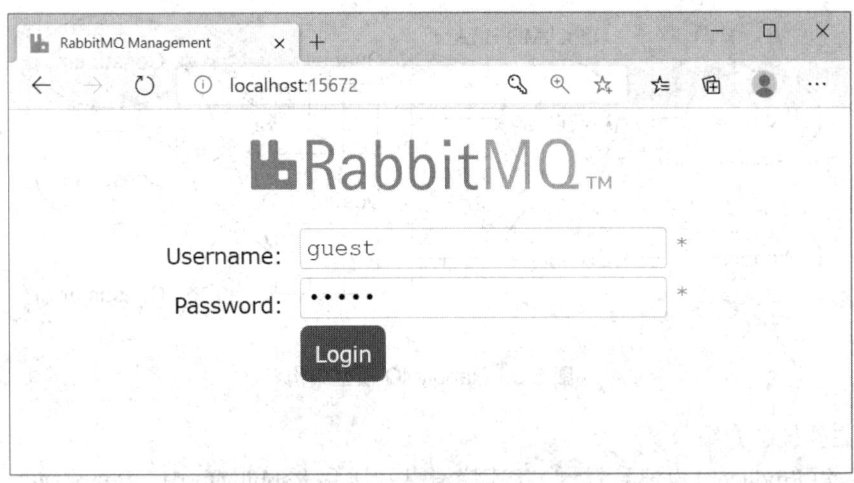

图 6-6　RabbitMQ 管理界面

登录后，管理界面如图 6-7 所示。至此 RabbitMQ 服务端安装完成。

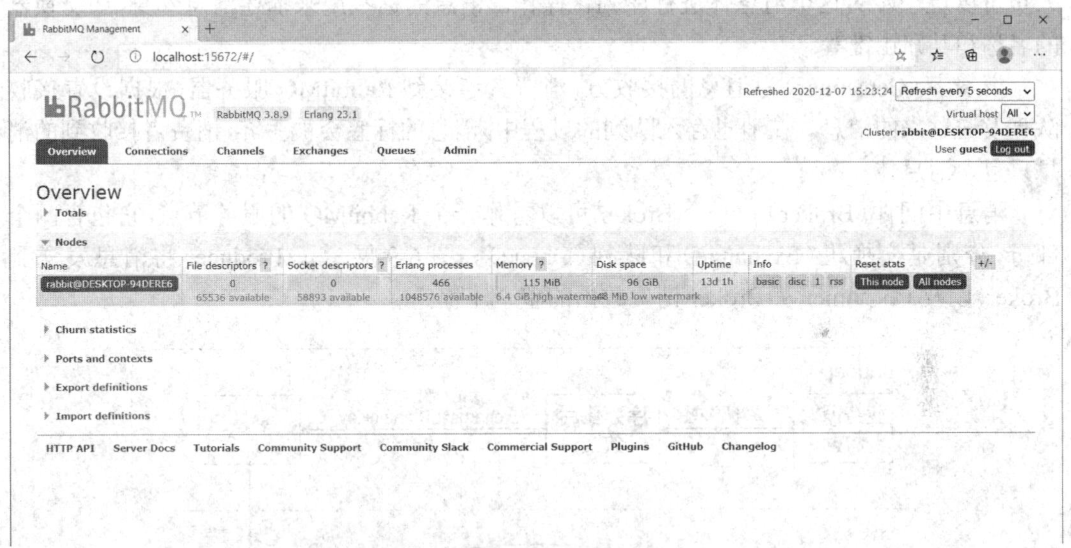

图 6-7　RabbitMQ 管理界面

2. RabbitMQ 的主要组件

RabbitMQ 是一个生产者与消费者模型，主要负责接受、存储和转发消息，它的整体模型架构如图 6-8 所示。

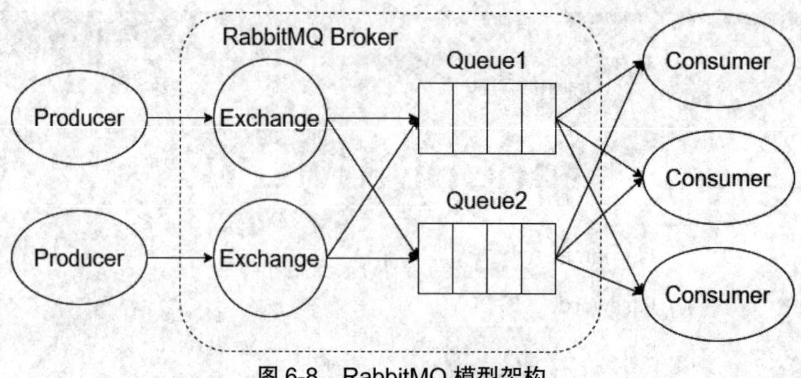

图 6-8 RabbitMQ 模型架构

(1)主要概念介绍

生产者(Producer)为消息发送方,其将消息发送到 RabbitMQ 中,RabbitMQ 会根据相关设置和消息内容将消息发送给消费者(Consumer)。消息一般包括两部分:消息体和消息标签。在实际项目中,消息体一般是带有项目业务逻辑的结构数据,如 JSON 字符串。开发人员可以进一步对这个消息体进行序列化操作。消息的标签用来描述消息本身,如交换器的名称和其路由键等。

消费者(Consumer)为消息的接收方,消费者连接到 RabbitMQ 服务器,根据设置接收队列发送过来的消息。在消息存入队列的过程中,消息的标签会被丢弃,消费者接收到的消息只包含消息体。

消息中间件(Broker)。一个 Broker 可以看作一个 RabbitMQ 的服务节点,它包含两个部分,分别是交换机(Exchange)和队列(Queue)。图 6-9 展示了 Producer 将消息发送到 Broker,以及 Consumer 从 Broker 消费数据的整个流程。

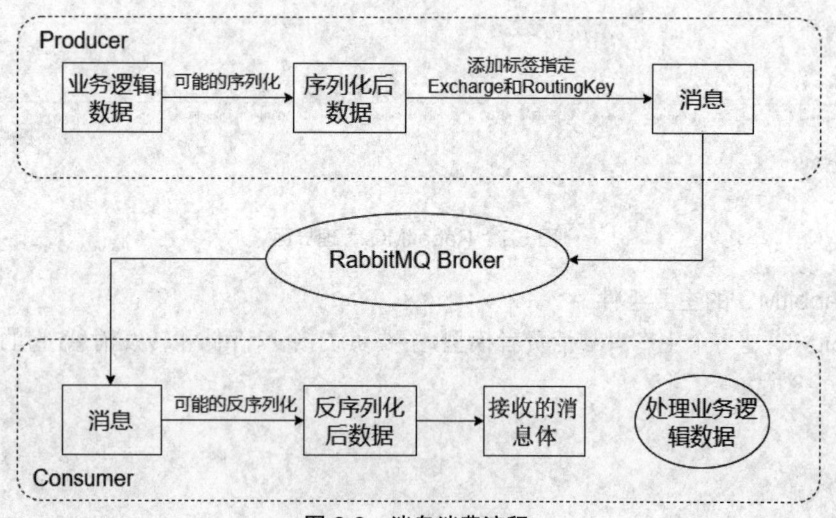

图 6-9 消息消费流程

生产者将业务逻辑数据与消息标签封装成消息(如需序列化则先进行序列化),发送到 Broker 中。消费者订阅并接收消息,经过可能的解包处理得到原始的数据,之后再进行业务逻辑数据的处理。

① Queue 队列。Queue 是 RabbitMQ 的内部对象，RabbitMQ 中的消息只存储在队列中，多个消费者可以订阅同一个队列，这时队列中的消息会被轮询发送给不同消费者进行处理，而不是每个消费者都收到所有消息。轮询发送的规则是队列会依序询问每一个消费者是否空闲，如消费者空闲即发送消息，发送结束或当前消费者忙碌时，队列会询问下一个消费者，周而复始。

② Exchange 交换机。交换机按一定的规则将生产者发送的消息转发到一个或多个队列中，若转发失败，消息会返回给生产者或被丢弃。

③ Routing Key 路由键。生产者将消息发送到交换机的过程中，一般会附带一个路由键，该路由键根据交换机类型和交换机的绑定键（Binding Key）来指定消息发送到哪个交换机中。

④ Binding Key 绑定键。RabbitMQ 中通过绑定键将交换机和队列关联在一起，实现消息的正确转发，如图 6-10 所示。

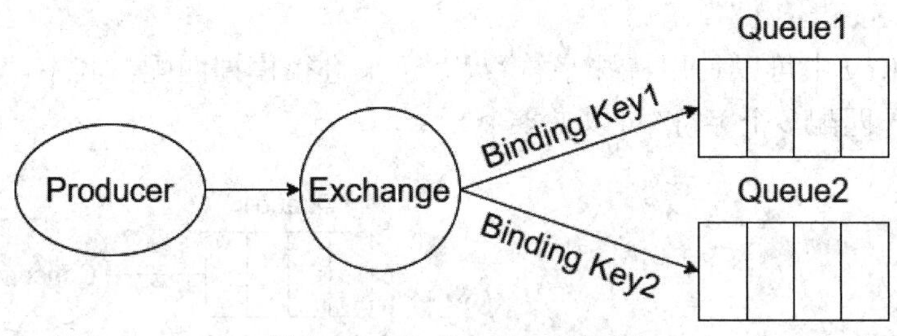

图 6-10　Binding Key 示意图

生产者将消息发送给交换机后，交换机会根据消息附带的路由键对消息进行分发，当消息的路由键（Routing Key）与队列的绑定键（Binding Key）相匹配时，消息就会被分发到对应的队列中。在交换机绑定多个队列的情况下，这些队列可以使用相同的 Binding Key 来进行绑定。交换机根据自身不同的类型来决定 Binding Key 是否起作用。RabbitMQ 中，交换机根据不同的转发规则有四种类型，其中 Fanout 类型交换机会无视 BindingKey，直接将消息路由到所有绑定到该交换机的队列中。

（2）交换机分类

RabbitMQ 的四种交换机类型为 Fanout、Direct、Topic 以及 Headers，它们的区别如下。

Fanout 类型交换机不需要配置 Routing Key，队列与交换机的 Binding Key 在该类型下也不起作用，交换机会将消息发送到每一个与交换机绑定的队列中，这些队列都可以对相同的消息进行接收，并将其发送给队列关联的消费者进行消费。该类型适用于不同消费者对相同业务消息进行处理的场合，如用户注册成功后，系统需要同时发送邮件和短信进行通知，那么邮件服务消费者和短信服务消费者需要共同消费注册成功这一业务消息。

Direct 类型交换机会把发送到交换机的消息发送到与 BindingKey 和 RoutingKey 相匹配的队列中。该类型适用于对不同消息进行分类处理的场合，如日志处理。其中，对不同级别的日志信息配置不同的路由键，进行分类发送处理，如图 6-11 所示。

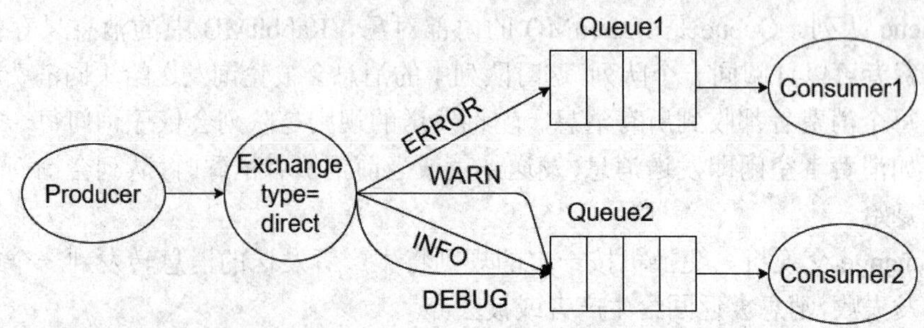

图 6-11　Direct 类型交换机示意图

Topic 类型交换机是在 Direct 类型上进行功能扩展，规则依然是把发送到交换机的消息发送到与 BindingKey 和 RoutingKey 相匹配的队列中，但其 BindingKey 和 RoutingKey 可进行模糊匹配，匹配的规则如下。

① BindingKey 和 RoutingKey 的值可以由不同字符部分组成，每部分之间由“.”号分隔，如“server.queue1.task1”。

② BindingKey 中可以使用特殊字符“#”“*”来进行模糊匹配。其中，“*”用于匹配单个字符，“#”用于匹配多个字符（可以是零个）。

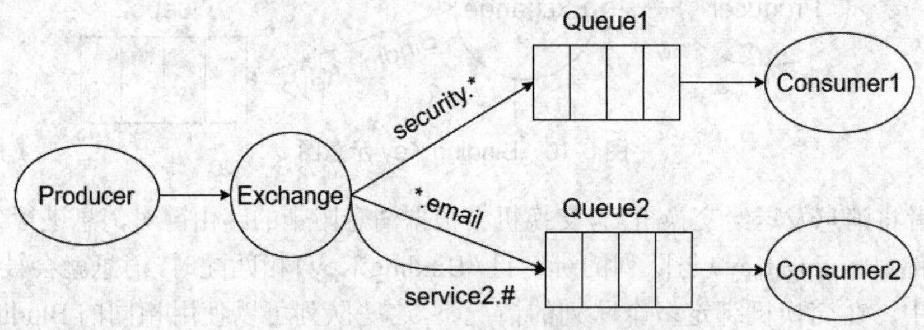

图 6-12　Topic 类型交换机示意图

Header 类型交换机不依赖于路由键的匹配规则来路由消息，而是根据消息内容中的 headers 属性进行匹配。在绑定队列和交换机时，制定一组键值对，当发送消息到交换机时，RabbitMQ 会获取到该消息的 headers，并对比其中的键值对是否完全匹配队列和交换机绑定时指定的键值对，如果完全匹配，则将消息发送到该队列。

（3）RabbitMQ 运行流程

在 RabbitMQ 的应用流程中，生产者与消费者都需要与 RabbitMQ Broker 建立连接（connection），连接是以传输控制协议（Transmission Control Protocol，TCP）建立的。建立成功后，客户端可以依托连接创建符合高级消息队列协议（Advanced Message Queuing Protocol，AMQP）的信道（channel），每个信道都会被标记一个唯一 ID。信道是建立在连接的上的虚拟连接，RabbitMQ 处理 AMQP 指令是通过信道完成的，信道与连接的关系模型如图 6-13 所示。

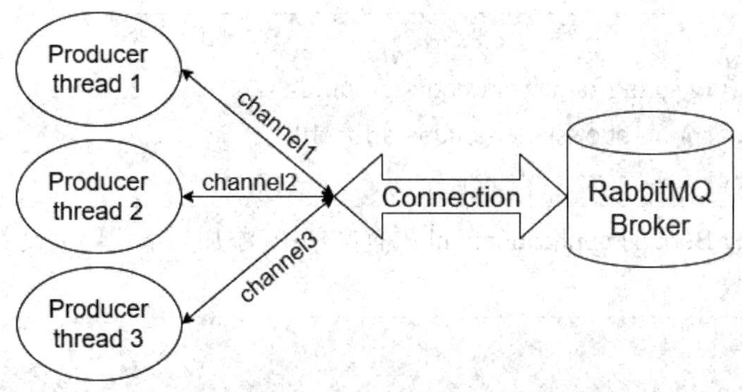

图 6-13　Channel 与 Connection 的关系模型

生产者通过信道发送消息的流程如下。

①生产者声明一个交换机，并设置交换机名称、类型、是否持久化等属性。

②生产者声明一个队列，并设置队列名称、是否排他、是否持久化、是否自动删除等属性。

③生产者将交换机和队列绑定起来，并设置它们之间的路由键。

④生产者发送消息至 RabbitMQ Broker，其中包含目标交换机、路由键等信息。

⑤交换机根据接收到的路由键查找相匹配的队列。如果找到，则将生产者发送过来的消息存入相应的队列中；如果没有找到，则根据生产者配置的属性选择丢弃或回退给生产者。

消费者消费信息的流程如下。

①消费者向 RabbitMQ Broker 请求消费相应队列中的消息，可能会设置相应的回调函数，以及做一些准备工作。

②等待 RabbitMQ Broker 回应并发送相应队列中的消息，消费者接收消息。

③消费者确认接收到消息。

④ RabbitMQ Broker 从队列中删除相应已经被确认接收的消息。

技能点 3　Spring AMQP

1.Spring AMQP 简介

Spring 提供了 Spring AMQP 项目来作为整合 RabbitMQ 的解决方案，它支持多种整合模式，包括基于 API 整合、基于配置类整合以及基于注解整合。使用 Spring AMQP 开发消息队列应用主要围绕着在三个部分来进行：使用 RabbitAdmin 类实例自动声明队列（queue）、交换机（exchange）和绑定（binding）；使用 RabbitTemplate 类实例来发送和接收消息；使用监听容器（listener container）异步消费和处理消息。

使用 Spring AMQP 时，首先需在 Maven 配置文件中引入依赖包 spring-boot-starter-amqp，代码如下。

```
<dependency>
  <groupId>org.springframework.boot</groupId>
  <artifactId>spring-boot-starter-amqp</artifactId>
</dependency>
```

接着在 Spring Boot 的 application.yml 配置文件中，添加如下代码来设置连接 RabbitMQ 服务器的必要参数。

```
spring:
 // 配置 RabbitMQ 连接
 rabbitmq:
  host: 127.0.0.1
  port: 5672
  username: guest
  password: guest
```

其中，host 为要连接的 RabbitMQ 的 IP 地址，port 为其运行端口号，username 与 password 分别为连接用户名和密码。

2.Spring AMQP 主要组件

Spring AMQP 提供了许多封装组件来帮助用户快速地实现消息队列应用的编写，开发中常用的组件有以下几种。

① Queue：队列的封装类，它的构造方法如下。

```
/**
* 用于构建实例的四个参数分别为
* name：队列名称
* durable：队列是否持久化
* exclusive：队列是否排他。设置为 true 则为排他，排他队列仅对声明它的连接可见，并在连接断开时自动删除
* autoDelete：当所有消费客户端连接断开后，是否自动删除队列
* 参数中只有 name 为必须参数，其他参数构建时可省略。省略时 durable 的默认值为 true，exclusive 的默认值为 false，autoDelete 的默认值为 false
*/
Queue（String name，boolean durable，boolean exclusive，boolean autoDelete）
```

② Exchange：交换机的封装类。根据不同类型 Exchange 有四种实现：DirectExchange、FanoutExchange、HeadersExchange、TopicExchange。它们的构造方法如下。

```
/**
* 用于构建实例的三个参数分别为
* name：交换机名称
* durable：交换机是否持久化
* autoDelete：当所有绑定队列都不在使用时，是否自动删除交换机
* 参数中只有 name 为必须参数，其他参数构建时可省略。省略时 durable 的默认值为 true，autoDelete 的默认值为 false
*/
DirectExchange(String name, boolean durable, boolean autoDelete)
TopicExchange(String name, boolean durable, boolean autoDelete)
FanoutExchange(String name, boolean durable, boolean autoDelete)
HeadersExchange(String name, boolean durable, boolean autoDelete)
```

③ Binding：绑定类。用于实现队列和交换机绑定的封装类，它的构造方法如下。

```
/**
* 用于构建实例的五个参数分别为
* destination：绑定的目标
* destinationType：绑定目标类型，枚举类，有 QUEUE 与 EXCHANGE 两种值
* exchange：绑定的交换机
* routingKey：绑定的路由键
* arguments：用于声明绑定的参数，可设置为 null
*/
Binding(String destination, Binding.DestinationType destinationType, String exchange,
        String routingKey, @Nullable Map<String, Object> arguments)
```

不同类型的交换机需要不同的绑定键来实现构造，代码如下。

```
// 使用固定的路由键将队列绑定到 Direct 类型的交换机
Binding("queue", Binding.DestinationType.QUEUE, "DirectExchange", "server.sms", null);
// 使用通配符键将队列绑定到 Topic 类型的交换机
Binding("queue", Binding.DestinationType.QUEUE, TopicExchange, "server.*", null);
// 将队列直接绑定到 Fanout 类型的交换机
Binding("queue", Binding.DestinationType.QUEUE, FanoutExchange , " ", null);
```

除了构造函数，还可以通过 BindingBuilder 类使用流式 API 的风格来实例化 Binding 类，代码如下。

```
Binding b = BindingBuilder.bind(Queue).to(Exchange).with("routingKey");
```

④ RabbitAdmin：用于简化 RabbitMQ 操作的类。该类可以创建、绑定、删除队列与交换机。在 Spring Boot 配置文件中配置好 RabbitMQ 连接后，Rabbit Admin 类可直接通过注解

注入使用，该类的主要方法名称和说明见表 6-1。

表 6-1 RabbitAdmin 的常用方法

名称	详情说明
declareExchange（Exchange）	声明一个新交换机
declareBinding（Binding）	声明一个新队列与交换机的绑定
declareQueue（Queue）	声明一个新队列
deleteExchange（String）	根据交换机名称删除一个交换机
deleteQueue（String）	根据名称删除队列
deleteQueue（String，boolean，boolean）	根据名称删除队列，可以设置是否验证队列正在使用以及是否验证队列中有未消费消息
purgeQueue（String）	清楚队列中的内容
removeBinding（Binding）	删除一个队列与交换机的绑定

⑤ RabbitTemplate：用于简化 RabbitMQ 发送和接收消息步骤的类。RabbitTemplate 类提供了丰富的消息发送方法，在 Spring Boot 配置文件中配置好 RabbitMQ 连接后，可直接通过注解注入使用，常用的发送方法为 convertAndSend，方法说明如下。

```
/**
 * 方法将 Java 对象转换为 AMQP 消息对象，然后根据参数设置发送到特定的交换机
 * 方法的参数说明如下，根据不同情况可选择含有不同数量参数的重载方法
 * @param exchange：消息发送的目标交换机名称
 * @param routingKey：发送时绑定的消息路由键
 * @param message：发送的消息主体
 * @param messagePostProcessor：amqp 提供的消息处理接口，用于在消息发送前对消息做进一步处理
 * @param correlationData：用于保存与发布者相关的数据的对象，该参数可为 null
 */
public void convertAndSend（String exchange，String routingKey，Object message，
MessagePostProcessor messagePostProcessor，@Nullable CorrelationData correlationData）；
```

技能点 4 基于 API 编写消息队列服务

基于 API 的方式实现消息队列服务，主要是使用技能点 3 中介绍的 Spring AMQP 组件来实现消息的发送和消费，实现过程主要分为三部分：创建并声明服务组件、生产者发送消

息到队列、消费者从队列消费消息。

1. 创建并声明服务组件

创建一个测试类，在类中使用 RabbitAdmin 组件来创建并声明 Fanout 类型的交换机和队列，并将他们绑定，如示例代码 6-1 所示。

示例代码 6-1：RabbitMQDemo1.java

```
import org.junit.Test;
import org.junit.runner.RunWith;
import org.springframework.amqp.core.Binding;
import org.springframework.amqp.core.BindingBuilder;
import org.springframework.amqp.core.FanoutExchange;
import org.springframework.amqp.core.Queue;
import org.springframework.amqp.rabbit.core.RabbitAdmin;
import org.springframework.beans.factory.annotation.Autowired;
import org.springframework.boot.test.context.SpringBootTest;
import org.springframework.test.context.junit4.SpringRunner;
@SpringBootTest
@RunWith(SpringRunner.class)
public class RabbitMQDemo1 {
  @Autowired
  private RabbitAdmin rabbitAdmin;
  @Test
  public void amqpAdminDemo(){
    // 创建 Fanout 类型的交换机对象
    FanoutExchange fanoutExchange = new FanoutExchange("fanout_exchange", true,
false);
    // 声明交换机
    rabbitAdmin.declareExchange(fanoutExchange);
    // 创建两个队列对象
    Queue queueA = new Queue("fanout_queue_A",true,false,false);
    Queue queueB = new Queue("fanout_queue_B",true,false,false);
    // 声明队列
    rabbitAdmin.declareQueue(queueA);
    rabbitAdmin.declareQueue(queueB);
    // 创建队列与交换机绑定对象
      Binding bindingA = new Binding("fanout_queue_A", Binding.DestinationType.
QUEUE,
    "fanout_exchange"," ",null);
```

```
        Binding bindingB = BindingBuilder.bind(queueB).to(fanoutExchange);
        // 声明绑定
        rabbitAdmin.declareBinding(bindingA);
        rabbitAdmin.declareBinding(bindingB);
    }
}
```

在该测试类中，首先通过注解自动注入了 RabbitAdmin 实例；接着通过该实例声明了持久化、不自动删除、Fanout 类型交换机（fanout_exchange），并定义了 fanout_queue_A 与 fanout_queue_B 两个消息队列；最后分别用构造函数以及流式 API 的方法将两个队列与交换机绑定在一起。

执行测试类，通过 URL 打开 RabbitMQ 服务器可视化页面，在 Exchanges 标签页中可以看到名为 fanout_exchange 的交换机创建成功，交换机类型为 Fanout。其他，以“amq”开头的交换机都是系统自动生成的，可视化页面效果如图 6-14 所示。

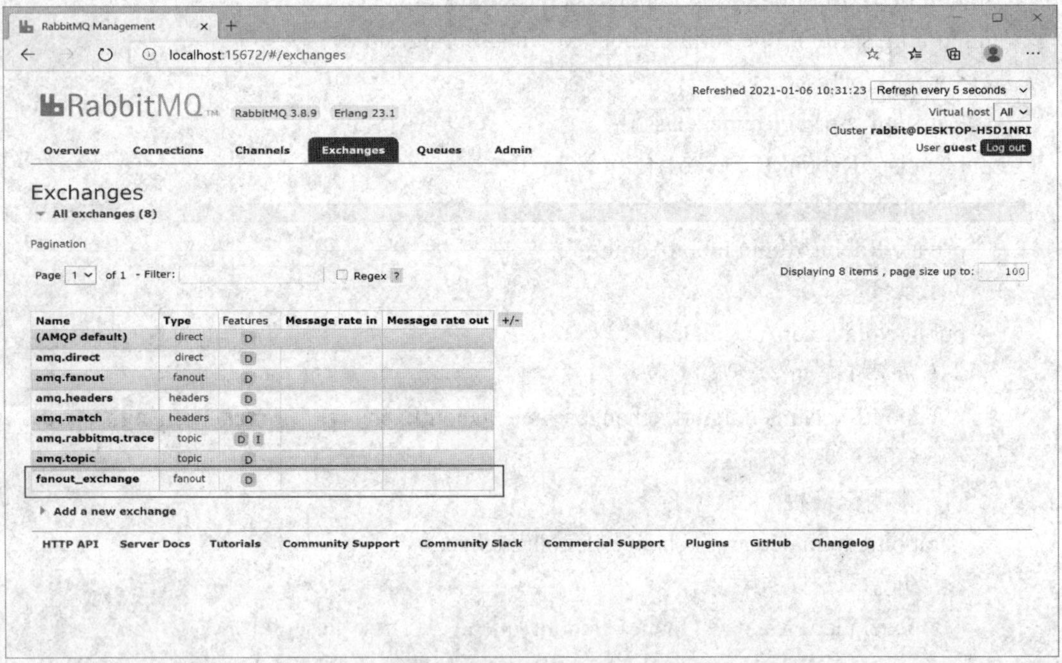

图 6-14　RabbitMQ 可视化管理界面

用鼠标点击 fanout_exchange 交换机，进入图 6-15 所示页面。在页面中可以看到交换机绑定的队列详情，该交换机绑定了 fanout_queue_A 与 fanout_queue_B 两个队列，由于交换机是 Fanout 类型的，所以路由键“Routing key”一栏为空。

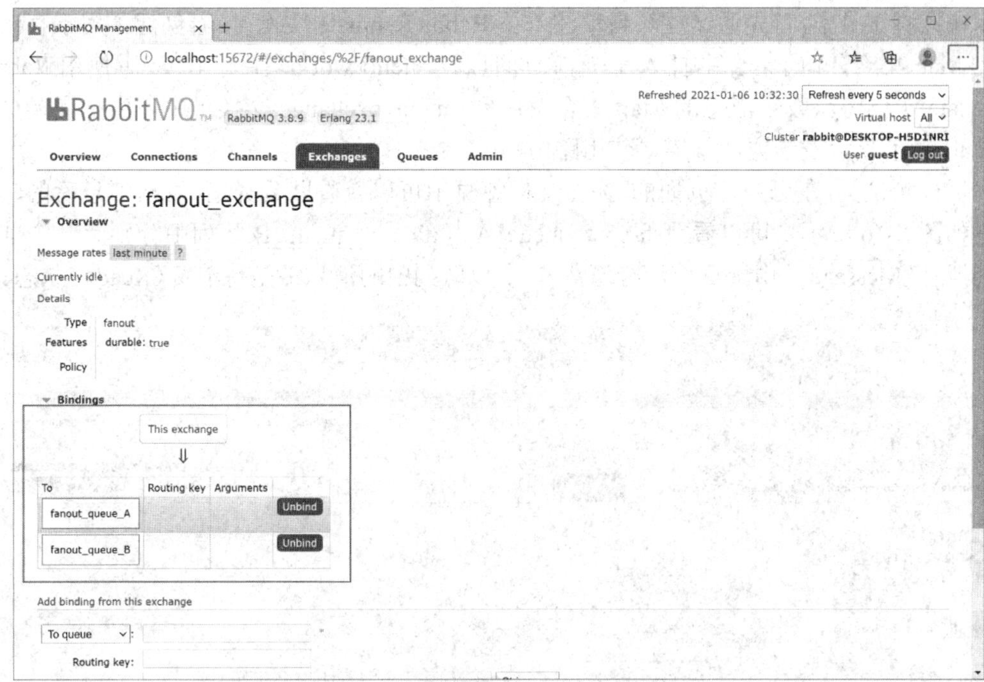

图 6-15 RabbitMQ 可视化管理界面的 Exchange 标签页

2. 生产者发送消息到队列

接下来，再创建一个测试类，类中使用 RabbitTemplate 组件来发送消息到交换机中，如示例代码 6-2 所示。

示例代码 6-2：RabbitMQDemo2.java

```
@SpringBootTest
@RunWith(SpringRunner.class)
public class RabbitMQDemo2 {
  @Autowired
  private RabbitTemplate rabbitTemplate;
  @Test
  public void publisherDemo(){
    // 创建消息信息 Map 集合，并向集合中赋值
    Map<String, String> messageMap = new HashMap();
    SimpleDateFormat sdf = new SimpleDateFormat("yyyy-MM-dd HH:mm:ss");
    messageMap.put("message", "hello world");
    messageMap.put("publishTime", sdf.format(new Date()).toString());
    // 使用 RabbitTemplate 对象发送消息
    rabbitTemplate.convertAndSend("fanout_exchange", " ", messageMap);
  }
}
```

在该测试类中，首先通过注解自动注入了 RabbitTemplate 实例；接着创建了用于存储消息的 Map 集合，并向集合中插入了当前时间以及“hello world”字符串；最后使用 RabbitTemplate 实例将承载了消息的 Map 集合发送到 fanout_exchange 交换机中。由于交换机是 Fanout 类型，所以发送方法不需要使用路由键。

执行测试类，在可视化页面的 Queues 标签页中可以查看队列详情，发送消息后的队列详情如图 6-16 所示。可以看到 fanout_queue_A 与 fanout_queue_B 队列中分别有一个准备中的消息（Message），由于目前没有消费者，所以队列中的消息都是待消费（Ready）状态。

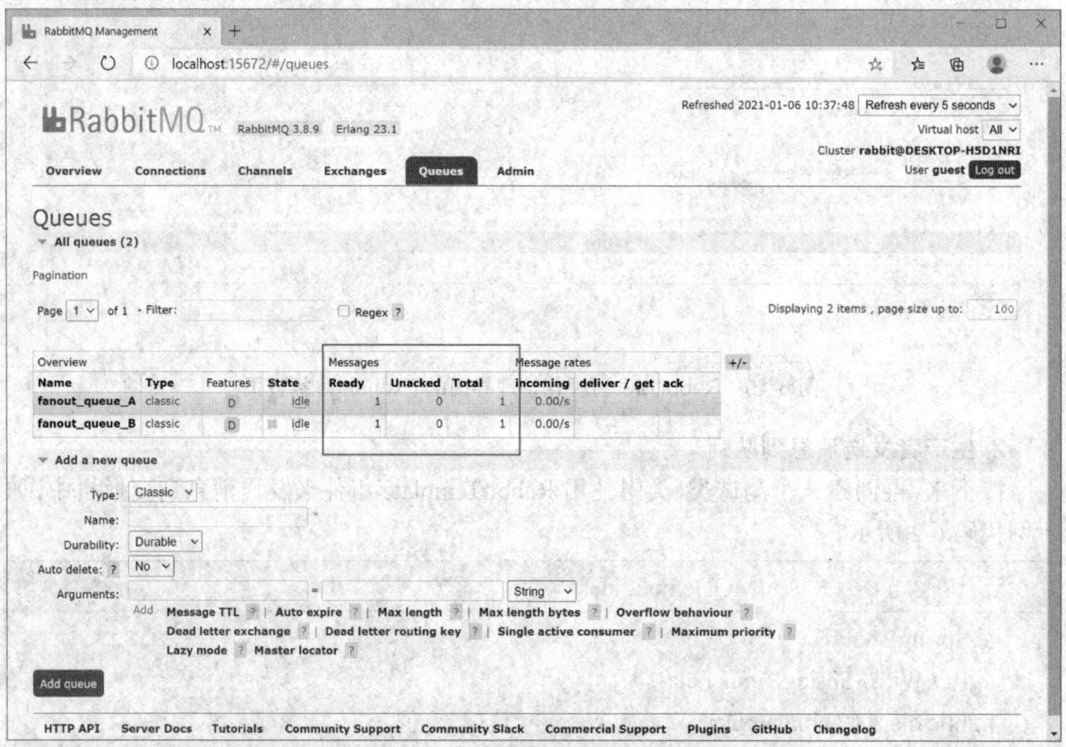

图 6-16 RabbitMQ 可视化管理界面的 Queues 标签页

3. 消费者从队列中消费消息

Spring AMQP 提供了 @RabbitListener 注解用来配置目标方法作为信息的消费者，通过设置注解的 queues 参数，可以使被注解的方法监听一个或多个队列，如示例代码 6-3 所示。

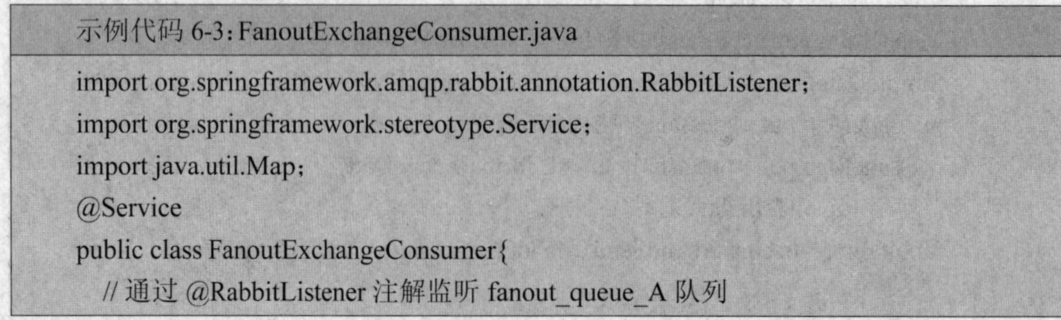

示例代码 6-3：FanoutExchangeConsumer.java

```
import org.springframework.amqp.rabbit.annotation.RabbitListener;
import org.springframework.stereotype.Service;
import java.util.Map;
@Service
public class FanoutExchangeConsumer{
    // 通过 @RabbitListener 注解监听 fanout_queue_A 队列
```

```
    @RabbitListener(queues = "fanout_queue_A")
    public void consumerA(Map map){
        System.out.println(" 队列 A 获取的信息为 "+map.toString());
    }
    // 通过 @RabbitListener 注解监听 fanout_queue_B 队列
    @RabbitListener(queues = "fanout_queue_B")
    public void consumerB(Map map){
        System.out.println(" 队列 B 获取的信息为 "+map.toString());
    }
}
```

代码中使用 consumerA()与 consumerB()方法作为监听方法，使用 @RabbitListener 注解配置两者，使它们分别监听 fanout_queue_A 队列和 fanout_queue_B 队列。当某队列中存在待消费的消息时，对应的监听方法会立即接收并消费该消息。在监听方法中，可配置参数用以接收消息，参数与消息的类型需一致；也可以使用 Object 类或 Message 类来接收，使用这两类进行接收时，获得的对象除了消息本体外，还有含有消息 Properties 属性的 MessageProperties 类。

启动消费者服务，fanout_queue_A 队列和 fanout_queue_B 队列中的消息都消费成功，控制台上输出的效果如图 6-17 所示。

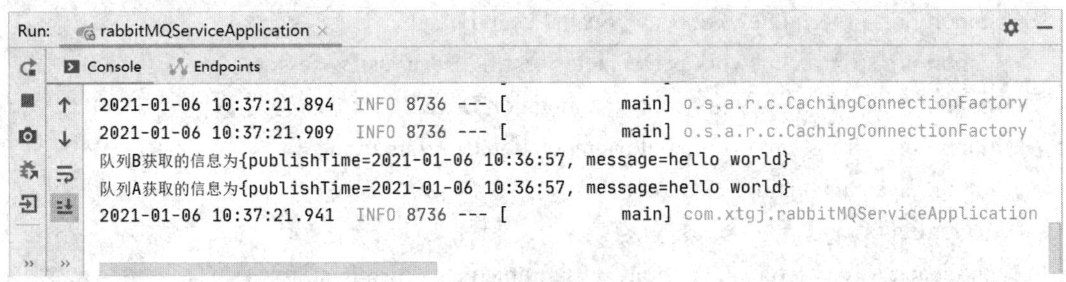

图 6-17　控制台内容

技能点 5　基于配置编写消息队列服务

基于配置的方式实现消息队列服务，主要是将 Spring AMQP 消息发送组件实例化成 Bean，通过 Spring 框架管理并注入来实现消息的发送。

1. 创建配置类

创建一个配置类，该类继承后置处理器 BeanPostProcessor 接口，该接口的作用是在项目的 Bean 对象实例化和依赖注入完成后，在显示调用初始化方法的前后添加自定义逻辑。接口的代码如下。

```
public interface BeanPostProcessor {
// 实例化和依赖注入完毕，在调用显示的初始化之前执行
@Nullable
    default Object postProcessBeforeInitialization(Object bean, String beanName) throws
BeansException {
        return bean;
    }
// 实例化、依赖注入、初始化完毕时执行
@Nullable
    default Object postProcessAfterInitialization(Object bean, String beanName) throws
BeansException {
        return bean;
    }
}
```

配置类的示例代码如 6-4 所示。

示例代码 6-4：TopicExchangeRabbitConfig.java

```
import org.springframework.amqp.core.*;
import org.springframework.amqp.rabbit.core.RabbitAdmin;
import org.springframework.beans.BeansException;
import org.springframework.beans.factory.config.BeanPostProcessor;
import org.springframework.context.annotation.Bean;
import org.springframework.context.annotation.Configuration;
import javax.annotation.Resource;
@Configuration
public class TopicExchangeRabbitConfig implements BeanPostProcessor{
    @Resource
    private RabbitAdmin rabbitAdmin;
    @Bean
    public TopicExchange rabbitmqDemoTopicExchange() {
        // 配置 TopicExchange 交换机
        return new TopicExchange("topic_exchange", true, false);
}
// 创建三个队列
    @Bean
    public Queue topicExchangeQueueA() {
        return new Queue("topic_queue_A", true, false, false);
    }
```

```
    @Bean
    public Queue topicExchangeQueueB() {
      return new Queue("topic_queue_B", true, false, false);
    }
    @Bean
    public Queue topicExchangeQueueC() {
      return new Queue("topic_queue_C", true, false, false);
  }
  // 将三个队列使用不同的 routing Key 绑定到 topic_exchange 交换机
    @Bean
    public Binding bindTopicA() {
      return BindingBuilder.bind(topicExchangeQueueA())
            .to(rabbitmqDemoTopicExchange())
            .with("*.a.rabbit");
    }
    @Bean
    public Binding bindTopicB() {
      return BindingBuilder.bind(topicExchangeQueueB())
            .to(rabbitmqDemoTopicExchange())
            .with("*.b.rabbit");
    }
    @Bean
    public Binding bindTopicC() {
      return BindingBuilder.bind(topicExchangeQueueC())
            .to(rabbitmqDemoTopicExchange())
            .with("#.rabbit");
  }
  // 实例化、依赖注入、初始化完毕时执行
    @Override
  public Object postProcessAfterInitialization(Object bean, String beanName) throws
BeansException {
    // 声明 topic_exchange 交换机与三个队列
      rabbitAdmin.declareExchange(rabbitmqDemoTopicExchange());
      rabbitAdmin.declareQueue(topicExchangeQueueA());
      rabbitAdmin.declareQueue(topicExchangeQueueB());
```

```
        rabbitAdmin.declareQueue(topicExchangeQueueC());
        return null;
    }
}
```

配置类中使用 TopicExchange 类实例化了一个 TopicExchange 交换机对象，交换机名称为 topic_exchange。使用 Queue 类实例化了队列对象 topic_queue_A、topic_queue_B 以及 topic_queue_C。使用 Binding 对象实例化了三个绑定对象，并将三个队列与 topic_exchange 交换机通过不同的 Routing Key 绑定起来。最后在重写的 postProcessAfterInitialization()方法中使用 rabbitAdmin 实例，声明以上 Spring AMQP 对象。

2. 生产者发送消息到队列

接下来再创建一个测试类，类中使用 RabbitTemplate 组件来发送消息到 topic_exchange 交换机中，发送时的 Routing Key 为 demo.a.rabbit，如示例代码 6-5 所示。

示例代码 6-5：RabbitMQDemo3.java

```
@SpringBootTest
@RunWith(SpringRunner.class)
public class RabbitMQDemo3 {
    @Autowired
    private RabbitTemplate rabbitTemplate;
    @Test
    public void publisherDemo(){
        // 创建消息信息 Map 集合，并向集合中赋值
        Map<String, String> messageMap = new HashMap();
        SimpleDateFormat sdf = new SimpleDateFormat("yyyy-MM-dd HH:mm:ss");
        messageMap.put("message", "topic_exchange_message");
        messageMap.put("publishTime", sdf.format(new Date()).toString());
        // 使用 RabbitTemplate 对象发送消息
        rabbitTemplate.convertAndSend("topic_exchange", "demo.a.rabbit", messageMap);
    }
}
```

执行测试类后，在可视化页面的 Queues 标签页中查看队列详情，如图 6-18 所示。发送的 Routing Key 匹配 topic_queue_A 与 topic_queue_C 队列，所以在这两个队列中各有一条待消费消息。

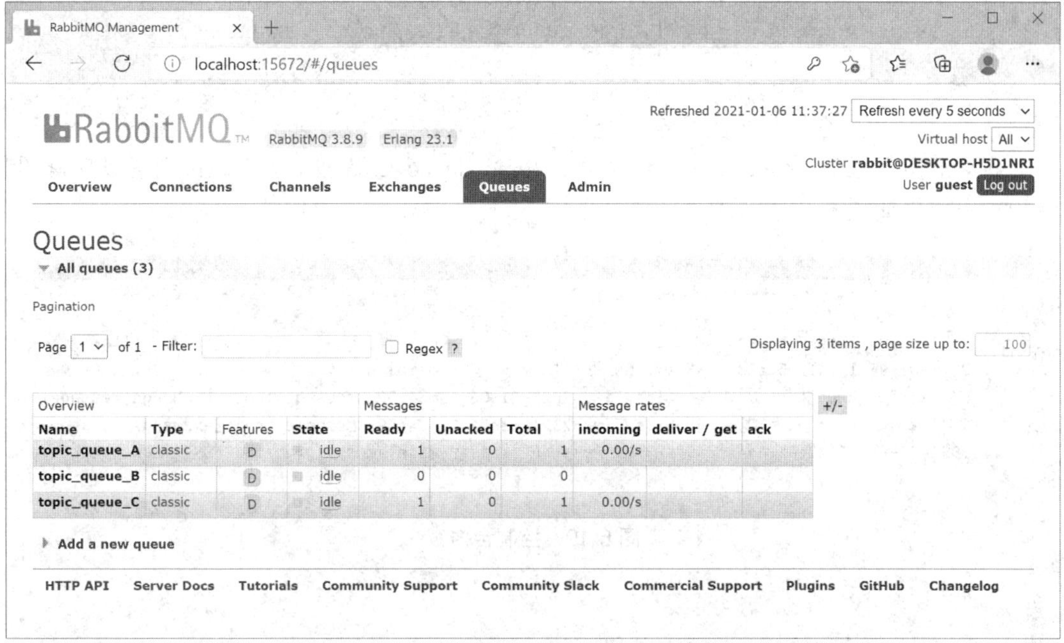

图 6-18　RabbitMQ 可视化管理界面的 Queues 标签页

3. 消费者从队列中消费消息

创建消费者类，使用 @RabbitListener 注解用来配置三个消费者分别监听三个队列，如示例代码 6-6 所示。

示例代码 6-6：TopicExchangeConsumer.java

```
import org.springframework.amqp.rabbit.annotation.*；
import org.springframework.stereotype.Service；
import java.util.Map；
@Service
public class TopicExchangeConsumer {
  @RabbitListener（queues = （"topic_queue_A"））
  public void process（Map<String，Object> map）{
    System.out.println（" 队列 [topic_queue_A] 收到消息：" + map.toString（））；
  }
  @RabbitListener（queues = （"topic_queue_B"））
  public void process2（Map<String，Object> map）{
    System.out.println（" 队列 [topic_queue_B] 收到消息：" + map.toString（））；
  }
  @RabbitListener（queues = （"topic_queue_C"））
  public void process3（Map<String，Object> map）{
```

```
        System.out.println(" 队列 [topic_queue_C] 收到消息:" + map.toString());
    }
}
```

启动消费者服务，topic_queue_A 队列和 topic_queue_C 队列中的消息都消费成功，控制台的输出效果如图 6-19 所示。

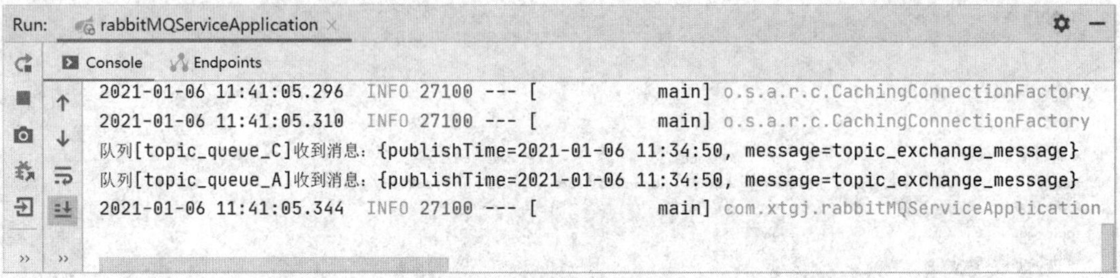

```
Run:   rabbitMQServiceApplication
Console   Endpoints
2021-01-06 11:41:05.296  INFO 27100 --- [           main] o.s.a.r.c.CachingConnectionFactory
2021-01-06 11:41:05.310  INFO 27100 --- [           main] o.s.a.r.c.CachingConnectionFactory
队列[topic_queue_C]收到消息: {publishTime=2021-01-06 11:34:50, message=topic_exchange_message}
队列[topic_queue_A]收到消息: {publishTime=2021-01-06 11:34:50, message=topic_exchange_message}
2021-01-06 11:41:05.344  INFO 27100 --- [           main] com.xtgj.rabbitMQServiceApplication
```

图 6-19 控制台内容

技能点 6 基于注解编写消息队列服务

基于注解的方式实现消息队列服务，主要是使用 @RabbitListener 注解来实现的。在之前已经讲过使用该注解通过 queues 属性创建消费者监听方法的过程，该方式是在 RabbitMQ 服务器中存在被监听的队列时使用的，如在不存在队列的情况下启动只含有 queues 属性的监听服务，程序会报 DeclarationException 异常。

@RabbitListener 注解还可以主动创建交换机、队列，并将两者绑定和声明，代码如下。

```
@RabbitListener(bindings = @QueueBinding(
        value = @Queue(value = "queueName"),
        exchange = @Exchange(value = "exchangeName", type = ExchangeTypes.TOPIC),
        key ="routing key"))
public void consumerMethod()
```

@RabbitListener 注解的 bindings 属性可以用 @QueueBinding 注解来声明配置交换器与队列的绑定。@QueueBinding 注解的三个属性如下。

value：使用 @Queue 注解设置，用于声明队列。value 属性为队列名称，该注解还可以配置队列的其他属性，如 durable 属性（设置是否持久化队列），autoDelete 属性（设置是否自动删除队列）等。

exchange：使用 @Exchange 注解设置，用于声明交换机。type 属性为交换机类型，该注解同样有其他用于设置交换机的属性。

key：交换机与队列绑定的 Routing Key。

1. 创建消费者类

创建一个消费者类，在类中使用 @RabbitListener 注解配置两个消费者分别创建并监听两个队列，如示例代码 6-7 所示。

示例代码 6-7：DirectExchangeConsumer.java

```
import org.springframework.amqp.rabbit.annotation.Exchange;
import org.springframework.amqp.rabbit.annotation.Queue;
import org.springframework.amqp.rabbit.annotation.QueueBinding;
import org.springframework.amqp.rabbit.annotation.RabbitListener;
import org.springframework.stereotype.Service;
import java.util.Map;
@Service
public class DirectExchangeConsumer {
    // 通过 @RabbitListener 注解创建队列 direct_queue_A 以及交换机 direct_exchange 并将其绑定，创建后开始监听该队列
    @RabbitListener(bindings = @QueueBinding(value = @Queue("direct_queue_A"),
            exchange = @Exchange(value = "direct_exchange",type ="direct"),
            key = "direct_key_A"))
    public void directProducer1(Map map){
        System.out.println(" 队列 A 获取的信息为 "+map.toString());
    }
    // 通过 @RabbitListener 注解创建队列 direct_queue_B 以及交换机 direct_exchange 并将其绑定，创建后开始监听该队列
    @RabbitListener(bindings = @QueueBinding(value = @Queue("direct_queue_B"),
            exchange = @Exchange(value = "direct_exchange",type = "direct"),
            key = "direct_key_B"))
    public void directProducer2(Map map){
        System.out.println(" 队列 B 获取的信息为 "+map.toString());
    }
}
```

执行该消费者类，在可视化页面的 Exchange 标签页中打开 direct_exchange 交换机，查看交换机与队列的绑定详情，如图 6-20 所示。可以看到，与交换机绑定的两个队列以及它们各自的 Routing Key。

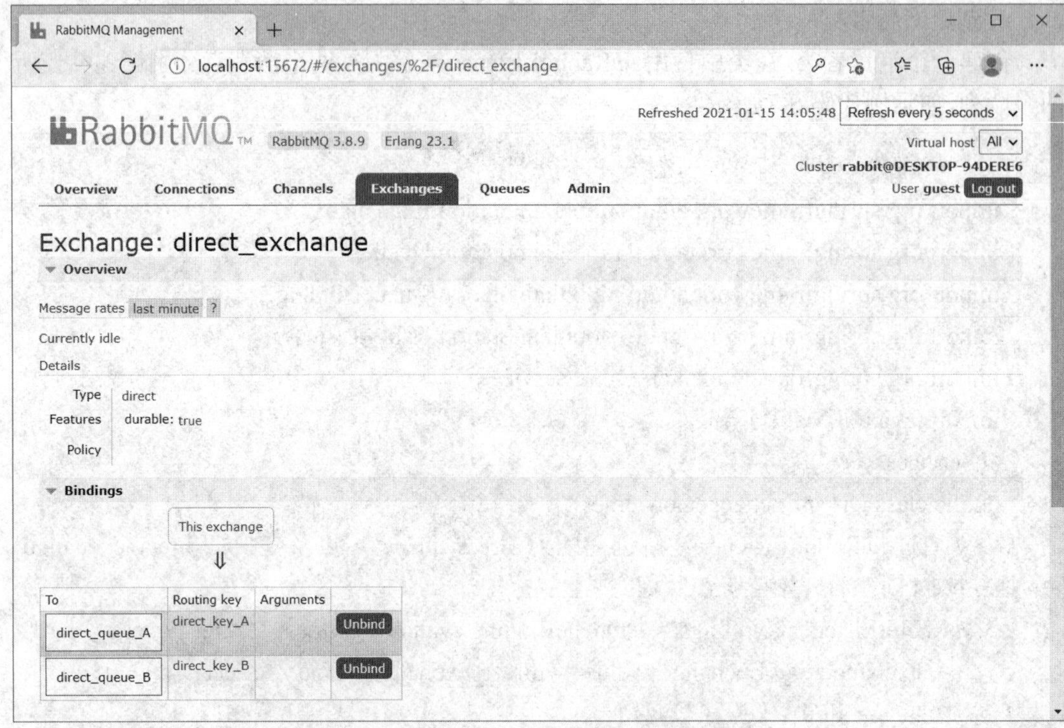

图 6-20　RabbitMQ 可视化管理界面的 Exchange 标签页

2. 生产者发送消息到队列

创建测试类，类中使用 RabbitTemplate 组件来发送消息到 direct_exchange 交换机中，发送时的 routing Key 为 direct_key_B，如示例代码 6-8 所示。

示例代码 6-8：RabbitMQDemo4.java

```
@SpringBootTest
@RunWith（SpringRunner.class）
public class RabbitMQDemo4 {
   @Autowired
   private RabbitTemplate rabbitTemplate;
   @Test
   public void publisherDirect（）{
      SimpleDateFormat sdf = new SimpleDateFormat（“yyyy-MM-dd HH：mm：ss”）；
      String sendTime = sdf.format（new Date（））；
      Map<String，Object> map = new HashMap<String，Object>（）；
      map.put（“sendTime”，sendTime）；
      map.put（“msg”，”hello direct”）；
      rabbitTemplate.convertAndSend（“direct_exchange”，”direct_key_B”，map）；
   }
}
```

执行测试类后，消息发送成功，可以在消费者服务的控制台上看到消息被发送到队列 direct_queue_B 中并被消费者消费成功，如图 6-21 所示。

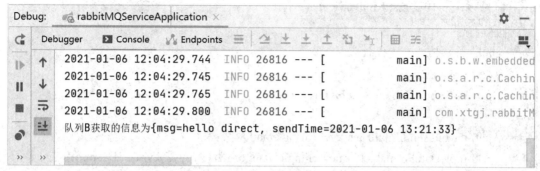

```
Debug:  rabbitMQServiceApplication ×
Debugger   Console   Endpoints
2021-01-06 12:04:29.744  INFO 26816 --- [           main] o.s.b.w.embedded
2021-01-06 12:04:29.745  INFO 26816 --- [           main] o.s.a.r.c.Cachin
2021-01-06 12:04:29.765  INFO 26816 --- [           main] o.s.a.r.c.Cachin
2021-01-06 12:04:29.800  INFO 26816 --- [           main] com.xtgj.rabbitM
队列B获取的信息为{msg=hello direct, sendTime=2021-01-06 13:21:33}
```

图 6-21　控制台内容

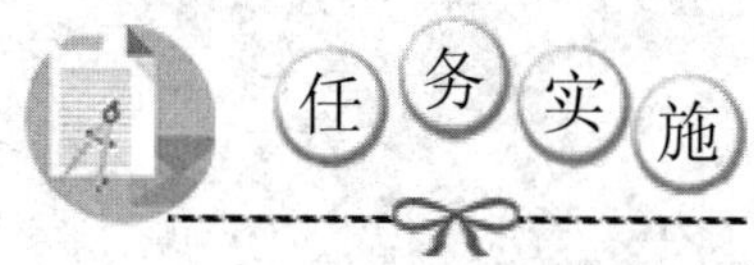

编写个人博客项目消息队列服务，使用 RabbitMQ 实现用户忘记密码时发送密码重置邮件到用户邮箱的功能，具体步骤如下。

第一步，新建 Spring Boot 项目来编写邮件服务，创建 EmailService 邮件服务消费类，类中通过 @ RabbitListener 注解声明 direct 类型的 email_service 交换机，并与队列绑定，绑定时的 Routing Key 为 recoverPassword，如示例代码 6-9 所示。

示例代码 6-9：EmailService.java

```
@Service
public class EmailService {
  @RabbitListener(bindings = @QueueBinding(value = @Queue("email_queue_A"),
       exchange = @Exchange(value = "email_service",type ="direct"),
       key = "recoverPassword"))
public void recoverPassword(Map map) throws Throwable {
  // 从消息中获取用户地址 emailAddress
  String emailAddress = (String) map.get("emailAddress");
  // 使用第三方邮件服务 API，向用户邮箱地址发送邮件
    SendMail.sendEEmail(emailAddress);
}
}
```

消费者方法从消费的消息中获取用户地址，使用第三方邮件服务 API 实现找回密码邮件发送功能。

第二步，在个人博客项目中编写邮件服务生产者 EmailMQService 类，类中的生产者方法使用 RabbitTemplate 组件来发送消息到 email_service 交换机中，如示例代码 6-10 所示。

示例代码 6-10：EmailMQService.java

```
@Service
public class EmailMQService {
  @Resource
  private RabbitTemplate rabbitTemplate;

  public void sendEmailMessage(String email) {
    SimpleDateFormat sdf = new SimpleDateFormat("yyyy-MM-dd HH:mm:ss");
    String sendTime = sdf.format(new Date());
    // 将用户邮箱地址以及消息发送时间保存在 Map 集合中
    Map<String, Object> message = new HashMap<>();
    message.put("sendTime", sendTime);
    message.put("emailAddress", email);
    // 发送消息，Routing Key 为 recoverPassword
    rabbitTemplate.convertAndSend("email_service","recoverPassword",message);
  }
}
```

第三步，在登录页面 login.html 的 form 表单中添加忘记密码链接，如示例代码 6-11 所示。

示例代码 6-11：login.html

```
<form class="am-form" id="loginForm" th:action="@{/login}" method="post">
  <div>
    <input type="hidden" name="url" th:value="${url}">
  </div>
  <div class="am-input-group am-radius am-animation-slide-left">
    <input type="text" class="am-radius"
th:placeholder="#{login.username}" name="username"/>
    <span class="am-input-group-label log-icon am-radius">
      <i class="am-icon-user am-icon-sm am-icon-fw"></i>
    </span>
  </div>
  <br>
  <div class="am-input-group am-animation-slide-left log-animation-delay">
    <input type="password" class="am-form-field am-radius log-input"
th:placeholder="#{login.password}" name="password"/>
```

```
        <span class="am-input-group-label log-icon am-radius">
          <i class="am-icon-lock am-icon-sm am-icon-fw"></i>
        </span>
      </div>
      <div class="am-input-group am-animation-slide-left log-animation-delay">
        <a th:href="@{/recoverPassword}" style="color: black"> 忘记密码？ </a>
      </div>
      <div style="padding-top: 10px;">
          <input type="submit" th: value="#{login.sub}" class="am-btn am-btn-primary am-
btn-block am-btn-lg am-radius am-animation-slide-bottom log-animation-delay"/>
      </div>
    </form>
```

第四步，编写通过输入邮箱找回密码的页面 recoverPassword.html。在页面中，通过表单提交用户邮箱，用户邮箱输入框 <input> 的 name 属性为 email，提交 URL 为 /recoverMethod，如示例代码 6-12 所示。

示例代码 6-12：recoverPassword.html

```
<! DOCTYPE html>
<html xmlns: th="http://www.thymeleaf.org">
<head>
  <meta charset="utf-8">
  <meta http-equiv="X-UA-Compatible" content="IE=edge">
  <meta name="viewport"
      content="width=device-width, initial-scale=1, maximum-scale=1, user-scalable=no">
  <title> 找回密码 </title>
  <meta http-equiv="Cache-Control" content="no-siteapp"/>
  <link rel="shortcut icon" th: href="@{/user/img/bloglogo.jpg}"/>
  <script th: src="@{/assets/js/jquery.min.js}"></script>
  <script th: src="@{/assets/js/amazeui.min.js}"></script>
  <link rel="stylesheet" th: href="@{/assets/css/amazeui.min.css}"/>
  <link rel="stylesheet" th: href="@{/assets/css/app.css}"/>
</head>
<body>
<div class="log">
  <div class="am-g">
    <div class="am-u-lg-3 am-u-md-6 am-u-sm-8 am-u-sm-centered log-content">
      <h1 class="log-title am-animation-slide-top" style="color: black; " > 找回密码 </h1>
```

```
            <br>
            <form class="am-form" id="loginForm" th:action="@{/recoverMethod}"
    method="post">
                <div class="am-input-group am-animation-slide-left log-animation-delay"> 请
输入要重置密码的邮箱 </div>
              <div class="am-input-group am-radius am-animation-slide-left">
                <input type="text" class="am-radius" th:placeholder=" 用户邮箱 "
    name="email"/>
                <span class="am-input-group-label log-icon am-radius">
                    <i class="am-icon-user am-icon-sm am-icon-fw"></i>
                </span>
              </div>
              <div style="padding-top: 10px; ">

            <input type="submit" th: value=" 下一步 " class="am-btn am-btn-primary am-btn-
block am-btn-lg am-radius am-animation-slide-bottom log-animation-delay"/>
              </div>
            </form>
          </div>
        </div>
        <footer class="log-footer">
            <p style="margin: 30px; color: #2E2D3C"><time class="comment-time" th: tex-
t="${#dates.format(new java.util.Date().getTime(), 'yyyy')}"></time> &copy; Powered
By <a style="color: #0e90d2" rel="nofollow">CrazyStone</a></p>
        </footer>
    </div>
    </body>
    </html>
```

第五步，编写发送成功页面 sendSuccess.html，如示例代码 6-13 所示。

示例代码 6-13：sendSuccess.html

```
    <body>
    <div class="log">
        <div class="am-g">
            <div class="am-u-lg-3 am-u-md-6 am-u-sm-8 am-u-sm-centered log-content">
                <h1 class="log-title am-animation-slide-top" style="color: black; " > 密码找回链
接已发送至邮箱 </h1>
```

```
        </div>
    </div>
    <footer class="log-footer">
        <p style="margin：30px；color：#2E2D3C"><time class="comment-time" th：tex-
t="${#dates.format(new java.util.Date().getTime()，'yyyy')}"></time> &copy；Powered
By <a style="color：#0e90d2" rel="nofollow">CrazyStone</a></p>
    </footer>
</div>
</body>
```

第六步，编写找回密码 Controller()方法。recoverPassword()方法用于跳转到找回密码页面；recoverMethod()方法用于接受前端传过来的用户 email 值，执行找回密码功能，并跳转到发送成功页面，如示例代码 6-14 所示。

示例代码 6-14：找回密码 Controller()方法

```
@Autowired
EmailMQService emailMQService；
@GetMapping(value = "/recoverPassword")
public String recoverPassword(){
  // 跳转到找回密码页面
return "comm/recoverPassword"；
}
@PostMapping(value ="/recoverMethod")
public String recoverMethod(String email){
  // 执行找回密码功能，发送消息到交换机
emailMQService.sendEmailMessage(email)；
return "comm/sendSuccess"；
}
```

第七步，启动个人博客项目服务以及邮件服务，通过“http：//localhost：8080/login”地址跳转到登录页面，如图 6-22 所示。

图 6-22 博客登录页面

用鼠标点击“忘记密码”进入密码找回页面，在页面中输入测试邮箱的地址，如图 6-23 所示。之后，点击“下一步”进行测试。

图 6-23 找回密码页面

打开测试邮箱可以看到密码找回邮件，效果如图 6-24 所示。

图 6-24　收到的密码找回邮件

本任务通过对 RabbitMQ 和 Spring AMQP 的相关知识的讲解，使读者掌握 RabbitMQ 与 Spring AMQP 的基本概念和主要组件，学会了使用 Spring AMQP 创建并声明队列、交换机、绑定，并使用 Spring AMQP 生产消息和消费消息的方法，实现使用 RabbitMQ 发送密码重置邮件到用户邮箱的功能。

1. 开发中，使用到消息服务的需求场景主要包括(　　)。(多选)

A. 异步处理　　B. 应用解耦

C. 流量削锋　　D. 分布式事务管理

2. 以下关于消息中间件的说法，错误的是(　　)。

A. RabbitMQ 是使用 Erlang 语言开发的开源消息队列系统，基于 AMQP 协议

B. Redis 服务可以作为消息中间件提供服务

C. RocketMQ 是 Apache 的顶级项目，具有高吞吐量、高可用等特点

D. ActiveMQ 是 Apache 出品的基于 JMS 协议的高性能中间件

3. RabbitMQ 中提供了哪几种交换机类型？（　　）。（多选）

A. Direct　　B. Fanout　　C. Topic　　D. Headers

4. RabbitMQ 提供的工作模式不包括（　　）。

A. 单点模式　　B. 发布订阅模式　　C. 路由模式　　D. Headers 模式

5. 以下关于基于注解方式定制 RabbitMQ 消息组件中的相关注解及说法，错误的是（　　）。

A. 需要使用 @EnableRabbit 开启基于注解的支持

B. @RabbitListener 标记在消息消费者方法上，会立即监听并消费消息队列中的消息

C. @RabbitListener 注解的 queues 属性可以定制消息队列

D. @QueueBinding 注解包括有 value、type、key 等属性

二、简答题

请简述 RabbitMQ 生产者发送消息以及消费者消费消息的运行流程。

项目七　初识 Spring Cloud

通过学习基本的应用架构，了解微服务的基本概念，学习 Spring Cloud 的基本概念和特点，熟悉基本的服务发现机制，实现用户注册登录服务。

- 掌握传统单体应用框架和微服务架构。
- 掌握 Spring Cloud 基本概念。
- 掌握服务发现。
- 掌握负载均衡器。
- 掌握声明式客户端 Feign。

【情境导入】

随着网络的不断发展，对于软件应用框架的使用也是一种考验，传统单体应用框架是大部分企业的选择，但随着用户的增加、功能的改变，这种应用模式已经满足不了用户的需求，一种全新的思想需要被探索出来，微服务（microservice）就被提出了。

【功能描述】

- 数据库与基本结构设计。
- 搭建 user-provider 服务。

技能点 1　传统单体应用框架和微服务架构

1. 传统单体应用框架

平时所接触的项目，大部分为传统单体应用架构，采用模块功能化的设计理念，如在项目中分为登录模块、用户模块和管理模块等。在编写设计完成整体项目之后，会进行打包处理，并上传服务器进行部署，常见的为 war 包或者 jar 包。在用户数量不多、功能不复杂的情况下，这种模式易于开发调试和部署，可以满足用户的需求，如图 7-1 所示。

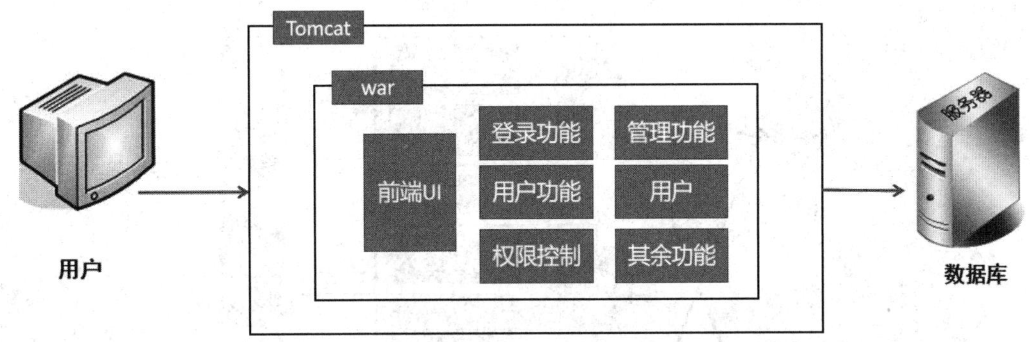

图 7-1　传统单体应用框架

但随着用户数量的增多，系统负载的能力又有限，这种开发模式显现一些弊端，出现的一些问题如下。

① 单体应用框架二次开发、维护难度大。随着应用的使用和技术的发展，在原有项目上进行更新迭代是不可避免的，但整体项目一旦变得复杂，对于后期的维护是巨大的挑战，任何单体应用框架都很难进行二次开发。

② 单体应用的整体性受限。项目部署之后，由于整体项目的运行均在一个进程中，当任意一个模块出现问题，整体项目可能也会因此崩溃，进而影响整个应用。

③ 技术更新难度大。在一般项目中，都会基于一种技术进行开发，此后的功能扩展都将基于该技术，若需要对该技术进行革新修改，那么整个项目都将进行重新开发，成本将会是巨大的。

④ 开发过程效率低：在开发过程中，每次调试都需要对整体项目进行运行，项目启动的速度影响着开发效率。

对于这种情况，可以使用系统水平扩展的方法进行扩展，增加服务器的数量，将打包好的应用放置在不同服务器中，应用 Apache、Nginx 等负载均衡器，实现水平扩展。这样确实可以缓解服务器的压力，但是也造成了服务器资源的浪费。例如，对于一些使用率比较低的模块功能，水平扩展之后和之前效率基本不变，但是依旧分配了资源，久而久之就会造成极大的浪费。

2. 微服务架构的基本概念

微服务的核心思想是服务拆分与解耦，降低复杂性。微服务主要是将原有项目的功能进行合理拆解，尽可能保证每个服务的功能单一，按照单一责任原则（single responsibility principle）明确角色。这些应用对外提供公共的 API，可以独立承担对外的职责，将各个服务做轻，从而做到灵活、可复用，亦可根据各个服务自身资源需求，单独布署，单独做横向扩展。基于这种思想所构建开发的软件服务应用被称为“微服务”，而围绕“微服务”思想所构建的一系列体系结构，被称之为“微服务架构”。

根据这种思想，可以将传统单体应用进行微服务化的拆分，使其每一个功能都是独立的服务，可访问自己的数据库，对应提供公共的 API，各个服务之间可以互相调用，如图 7-2 所示。

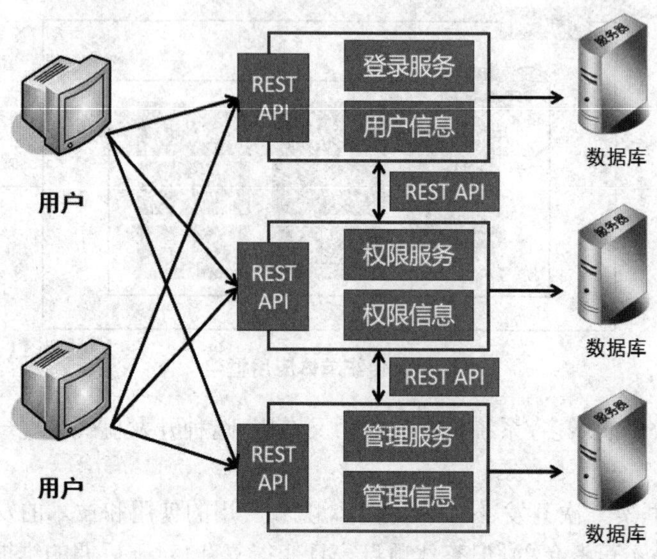

图 7-2 微服务架构示例

微服务架构相比于传统单体应用架构，具有很多优点，列举如下。

①易于开发和维护。一个服务只关注一个特定的业务功能，所以业务清晰、代码量少。开发和维护单个微服务比较简单。整个应用是通过若干个微服务构建而成的，所以整个应用被维持在一个可控的状态。

②单个服务启动快。相比于单体应用，单个服务项目的代码量少、启动快。

③局部修改易部署。在微服务架构中，对于个别服务的修改，只需重新部署该服务即可，其余服务无须暂停。

④可应用多种技术和第三方集成。在微服务架构中，可根据开发人员特点和功能需求，选择合理的技术栈。

⑤按功能需求对服务器收缩或扩展。在微服务架构中，对于使用频率很高的部分服务，可根据实际需求增加该服务下服务器的配置。

3. 微服务架构的基本组件

微服务需要很多组件共同应用来实现服务架构的功能，包括服务注册、服务网关、服务配置中心、服务框架等。其中，有一些常用的组件，可以共同协作来完成整体的微服务框架。具体功能如下。

①服务注册。服务注册者将自己调用地址注册到服务注册中心，调用者有需要的时候，每次都先去查询“服务注册”即可，免去人工维护微服务节点的信息同步问题。

②服务网关。它也被称为 API 网关，指提供给外部系统调用的是统一网关，主要为安全、动态路由和权限控制等。

③服务配置中心。微服务的配置中心用于统一管理所有微服务节点的配置信息，由于同一程序可能适用于多种环境，因此在微服务的实践中，有必要将程序与配置分开并集中管理配置。这些配置包括微服务节点信息配置、程序运行时配置、变量配置、数据源配置、日志配置、版本配置等。

④服务框架。它用来规范各个微服务节点之间的通信标准，包括微服务间通信采用的协议，数据传输的方式、格式等，用来保证各个节点之间的高效率协同。

⑤服务监控。在整体的微服务启动运行之后，为了能够实时掌握整体项目的状况，需要对各个服务节点进行监控，收集数据进行分析，形成预警。

⑥服务追踪。在应用微服务架构之后，请求会经过多个微服务节点的处理，这样就形成了一条调用链条。当出现问题时，需要进行问题追踪和故障定位，因此需要对调用链条进行记录，以方便查询。

⑦服务治理。需要准备一些策略和方案，来保障整个微服务架构在生产环境遇到极端情况下也能正常提供服务，如熔断、限流、隔离等。

⑧负载平衡。服务提供方使用多种形式提供服务，使用负载平衡的方式能够让服务调用找到最合适的节点。

⑨服务容错。它指服务的保护机制如熔断器，用于保证在用户使用的过程中，降低在远程资源访问失败时对用户的影响。

技能点 2　Spring Cloud 基本概念

Spring Cloud 是在 Spring Boot 的基础上构建的，它利用 Spring Boot 的开发便利性，巧妙地简化了分布式系统基础设施的开发，如服务发现注册、配置中心、消息总线、负载均衡、断路器、数据监控等，都可以用 Spring Boot 的开发风格做到一键启动和部署。Spring Cloud 的 logo 如图 7-3 所示。

图 7-3　Spring Cloud 的 logo

1. Spring Cloud 的基础功能

Spring Cloud 是一系列框架的有序集合，为开发人员进行微服务框架开发提供了一套细致的解决方法。利用 Spring Cloud 特有的一些功能，开发人员可轻松完成微服务功能。Spring Cloud 具有以下常用模块和基础功能。

①服务发现（Spring Cloud Eurek）。

②客户端负载均衡（Spring Cloud Ribbon）。

③服务容错保护（Spring Cloud Hystrix）。

④声明式服务调用（Spring Cloud Feign）。

⑤ API 网关服务(Spring Cloud Zuul)。

⑥ 分布式配置中心(Spring Cloud Config)。

2. Spring Cloud 的优点

Spring Cloud 是基于 Spring Boot 实现的微服务框架开发架构。其巧妙地利用 Spring Boot 便利性,可以说是再次将 Spring Boot 进行了封装。Spring Cloud 已成为当下流行的微服务框架,主要因为其具有如下的优点。

①开箱即用,快速启动:基于 Spring Boot 无须进行繁多的配置,使用约定代替配置,实现快速开发。

②易于扩展维护:每一个服务都是独立的,可在其基础上进行扩展维护。

③环境适用性强:可应用于多个场景,如 PC 端、云服务器以及各种容器等。

④具有良好的容错机制: Spring Cloud 提供了 Hystrix 组件,专门用于错误处理,保证各个模块在出错之后能够启动备用模块进行快速处理以及善后。

⑤组件功能强大: Spring Cloud 为微服务架构提供了非常完整的支持,如配置管理、服务发现、断路器、微服务网关等。

3. Spring Cloud 的版本兼容性

Spring Cloud 版本名称一般是由"版本号 + 小版本名称"组成的。这样能够使开发人员更加便捷地管理每一个 Spring Cloud 子项目,避免版本号混淆。常见的 Spring Cloud 版本标识名称见表 7-1。

表 7-1 Spring Cloud 版本标识名称及其具体含义

标识名称	具体含义
SNAPSHOT	快照版,表示开发版本,随时可能修改
M	M 是 milestone 的缩写,一般同时标注 PRE,表示预览版本
SR	Service Release,SR1 表示第一个正式版本,一般同时标注 GA(Generally Available),表示稳定版本

Spring Cloud 与 Spring Boot 的更新迭代非常频繁,导致在开发过程中对两者的版本选择非常困难。若版本选择出现问题,可能会因调试兼容性导致在开发中占用非常多的时间。二者具体版本对应关系见表 7-2。

表 7-2 Spring Cloud 版本与 Spring Boot 版本关系

Spring Cloud	Spring Boot
Hoxton	2.2.x,2.3.x(starting with SR5)
Greenwich	2.1.x
Finchley	2.0.x
Edgware	1.5.x
Dalston	1.5.x

在本书中，Spring Boot 选择 2.2.8.RELEASE 版本，对应地 Spring Cloud 使用 Hoxton. SR6 版本，之后的案例中都会使用上述版本，请读者注意。

技能点 3　服务发现

1. Eureka 简介

Eureka 是 Netflix 开发的服务发现框架，其本身是一个基于 REST 的服务（Eureka Server），主要用于 AWS 云的定位服务，以实现中间层服务器的负载平衡和故障转移。Eureka 附带了一个基于 Java 的客户端组件 Eureka Client，其使与服务的交互变得更加容易。

Eureka 包含两大组件：Eureka Server 提供注册服务；Eureka Client 是客户端组件，可作为服务消费者和服务提供者。在各个节点启动后，会在 Eureka Server 中进行注册，这样 Eureka Server 中的服务注册表将会存储所有可用服务节点的信息，服务节点的信息可以在界面中直观地看到。

Eureka 的服务发现机制如图 7-4 所示。

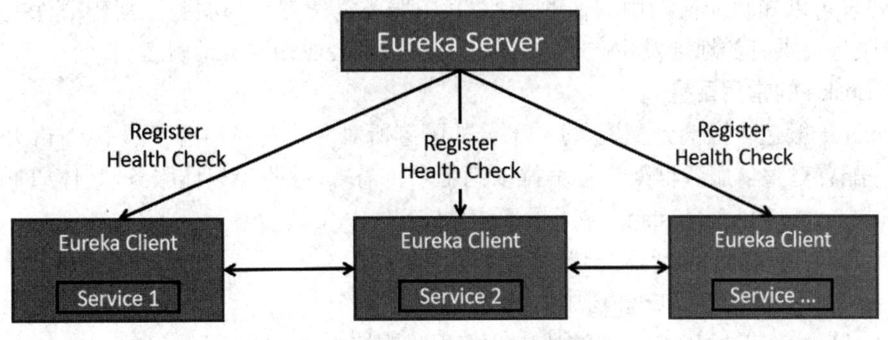

图 7-4　Eureka 的服务发现机制

从图 7-4 中可以看出，Eureka Client 所注册的服务都是通过 REST 进行调用的，当客户端通过注解等方式嵌入程序代码运行时，客户端发现组件就会向注册中心注册自身所提供的服务，并周期性地发送心跳来更新服务（默认时间为 30 s，若三次心跳都不能发现服务，那么 Eureka 就会将该服务节点从服务注册中移除）。此时，客户端发现组件也会对服务端所查询出的注册信息进行缓存，即使 Eureka Server 出现死机，客户端也可通过缓存信息调用服务节点进行服务。所以 Eureka 系统的高可用性、灵活性、可伸缩性都是依赖于心跳机制和缓存更新机制。

2. Eureka 角色关系

Eureka 的服务发现机制包含三个角色，分别是 Eureka Server 服务注册中心、Eureka Client 服务消费者和服务提供者 Eureka Provider，三者关系如图 7-5 所示。

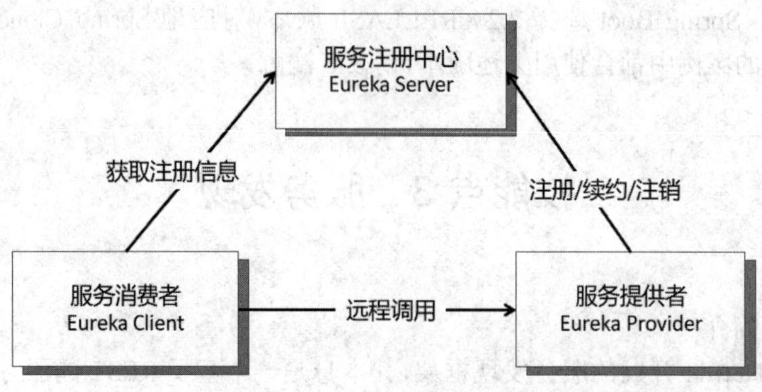

图 7-5 Eureka 的角色关系

①服务注册中心。在各个服务启动后，会在服务注册中心（Eureka Server）中进行注册，并存储所有节点信息，以供其他服务调用和查看。

②服务提供者。在 Eureka Client 启动之后，服务提供者将会通过 REST 请求将其注册在 Eureka Server 中，并使用心跳机制来维护更新，进行服务的续约，防止被 Eureka Server 踢出服务列表。

③服务消费者。其用于获取 Eureka Server 注册的服务列表，该列表每 30 s 更新一次。通过该列表可以知晓如何调用其他的服务，其中服务消费者会与服务注册中心保持心跳连接，一旦服务提供者的地址发生改变，服务注册中心就会通知服务消费者。

3. Eureka 的常用配置

Eureka 一般包含四部分配置，每一部分的配置所对应的前缀和内容都不一样，开发者可通过这些配置对整体服务框架进行编排以满足项目需求，并监控整体服务，具体如下。

① instance：Eureka Instance 实例配置信息。

② client：Eureka Client 客户端特性配置。

③ server：Eureka Server 注册中心特性配置。

④ dashboard：Eureka Server 注册中心仪表盘配置。

其中，有两个比较重要的机制，这两种机制需要配置对应的属性来实现。

（1）自我保护机制

自我保护机制主要用于客户端和 Eureka Server 之间存在网络分区场景时，Eureka Server 会判断是否存在大量续约失败的服务，从而确定是否开启自我保护。默认情况下，在 Eureka Server 运行期间，以心跳的接收比例为标准，若低于标准则进入保护模式，此时 Eureka Server 将会尝试保护其服务注册表中的信息，不再删除服务注册表中的数据；当网络故障恢复后，Eureka Server 节点会自动退出保护模式，开发人员可以在 Eureka Server 中进行配置，来关闭保护模式，以确保服务中心不可用的服务实例被及时清除。

在默认情况下，Eureka Server 的自我保护机制是开启的，可在配置文件中使用如下代码进行关闭。

```
eureka.server.enable-self-preservation=false
```

在关闭自我保护机制的基础上，可设置 eureka.server.eviction-interval-timer-in-ms=30000 属性，来定时移除已经失效的服务信息，代码如下。

```
eureka.server.eviction-interval-timer-in-ms=30000
```

（2）心跳机制

在应用启动后，服务节点即 Eureka Client 将会向 Eureka Server 发送心跳，默认周期为 30 s，如果 Eureka Server 在多个心跳周期内没有接收到某个节点的心跳，Eureka Server 将会从服务注册表中把这个服务节点移除（默认 90 s）。

应用 lease-renewal-interval-in-seconds 属性，可设置 Eureka Client 向 Eureka Server 发送周期性心跳的时间间隔为 25 s，代码如下。

```
eureka.instance.lease-renewal-interval-in-seconds=25
```

Eureka Server 在一定时间内没有接收到 Eureka Client 实例的心跳，就会从服务中心将其移除。应用 lease-expiration-duration-in-seconds 属性，可设置 Eureka Server 每隔 35 s 刷新服务列表，并将无效的实例移除，代码如下。

```
eureka.instance.lease-expiration-duration-in-seconds=35
```

（3）其他配置

除了自我保护机制和心跳机制，Eureka 还有很多配置。其中，通过对 Eureka Instance 实例信息配置，即可获得有关实例信息配置。这些配置均以 eureka.instance 为前缀，常见的 Eureka Instance 配置表 7-3。

表 7-3　Eureka Instance 实例信息配置

名称	详情说明
eureka.instance.appname	获取 spring.application.name 的值（应用名），如果取值为空，则取默认 unknown
eureka.instance.instance-id	注册到 Eureka 上的唯一实例 ID，不能与相同 appname 的其他实例重复
eureka.instance.hostname	配置主机名，不配置的时候将配置操作系统的主机名
eureka.instance.ip-address	实例的 IP 地址

Eureka Client 为客户端特性配置，用于配置服务注册的相关信息，均以 eureka.client 为前缀的格式进行配置，其常见的配置见表 7-4。

表 7-4　Eureka Client 客户端特性配置

名称	详情说明
eureka.client.enabled	启用 Eureka 客户端
eureka.client.registry-fetch-interval-seconds	从 Eureka 客户端获取注册信息的间隔时间，单位为秒
eureka.client.instance-info-replication-interval-seconds	更新实例信息的变化到 Eureka 客户端的间隔时间，单位为秒
eureka.client.fetch-registry	是否从 Eureka 客户端获取注册信息

Eureka Server 为注册中心端配置，用于配置注册中心端的相关信息，均以 eureka.server 为前缀的格式进行配置，其常见的配置见表 7-5。

表 7-5　Eureka Server 注册中心端配置

名称	详情说明
eureka.server.enable-self-preservation	开启自我保护
eureka.server.renewal-threshold-update-interval-ms	续约数阈值更新频率
eureka.server.renewal-percent-threshold	自我保护续约百分比阈值因子。如果实际续约数小于续约数阈值，则开启自我保护，默认值为 0.85 s
eureka.server.peer-eureka-nodes-update-interval-ms	Eureka Server 节点更新频率

4. Eureka 注册实例

【实例】应用 Eureka 实现服务发现，需要在项目中的服务端发现以及客户端发现组件，创建服务注册中心 eureka-server、服务消费者 eureka-consumer 和服务提供者 eureka-provider。创建之后的整体结构如图 7-6 所示。

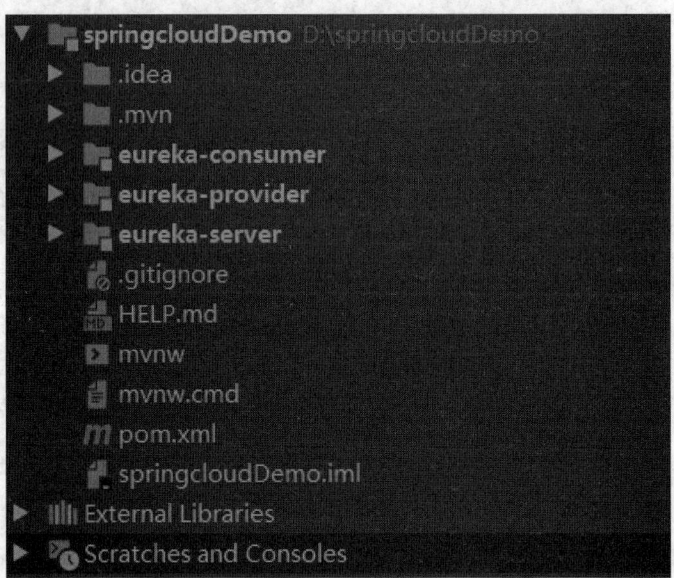

图 7-6　整体结构图

第一步，创建父类项目 springcloudDemo，使用 Initializr 方式新建项目，引入对应的依赖。Spring Boot 为 2.1.9.RELEASE 版本，并删除 src 文件夹。springcloudDemo 类为整体架构的父类，服务注册中心、消费者和提供者均为其下的模块（Module）。其 Pom 文件如示例代码 7-1 所示。

示例代码 7-1：springcloudDemo 项目的 Pom 文件

```
<? xml version="1.0" encoding="UTF-8"? >
```

```
<project xmlns="http://maven.apache.org/POM/4.0.0" xmlns:xsi="http://www.w3.org/2001/XMLSchema-instance"
        xsi:schemaLocation="http://maven.apache.org/POM/4.0.0 https://maven.apache.org/xsd/maven-4.0.0.xsd">
    <modelVersion>4.0.0</modelVersion>
    <parent>
      <groupId>org.springframework.boot</groupId>
      <artifactId>spring-boot-starter-parent</artifactId>
      <version>2.2.8.RELEASE</version>
      <relativePath/> <!-- lookup parent from repository -->
    </parent>
    <groupId>com.example</groupId>
    <artifactId>springcloud</artifactId>
    <version>0.0.1-SNAPSHOT</version>
    <name>springcloud</name>
    <description>Demo project for Spring Boot</description>

    <properties>
      <java.version>1.8</java.version>
    </properties>
    <build>
      <plugins>
        <plugin>
          <groupId>org.springframework.boot</groupId>
          <artifactId>spring-boot-maven-plugin</artifactId>
        </plugin>
      </plugins>
    </build>
  </project>
```

第二步，在父类项目中新建 Module 子模块，如图 7-7 所示。设置于模块的名称为 eureka-server（此处名称均可自行修改），该模块为一个基础的 Spring Boot 项目，对应 Pom 文件的主要依赖关系如示例代码 7-2 所示。

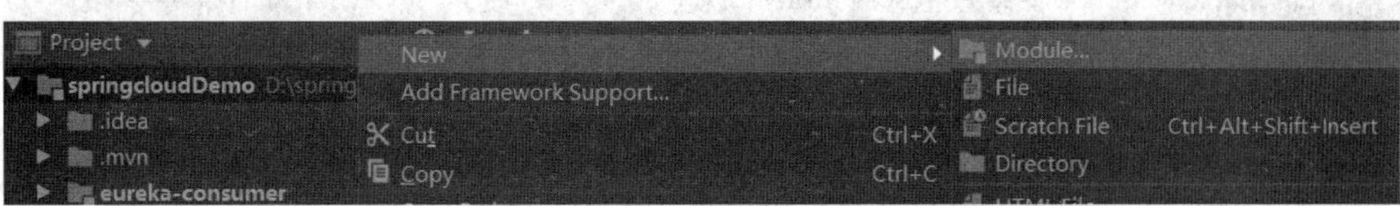

图 7-7　IDEA 创建 Module 子模块

示例代码 7-2：eureka-server 子模块的 Pom 文件

```
<dependencies>
    <dependency>
        <groupId>org.springframework.cloud</groupId>
        <artifactId>spring-cloud-starter-netflix-eureka-server</artifactId>
    </dependency>
    <dependency>
        <groupId>org.springframework.boot</groupId>
        <artifactId>spring-boot-starter-test</artifactId>
        <scope>test</scope>
    </dependency>
</dependencies>
```

第三步，编写配置文件，在配置文件中增加端口号、名称、注册 IP、URL 等。配置端口号为 8761，所有的服务实例都需要向此端口进行注册。本模块为注册模块，所以并不需要向自己注册和检索服务，将 register-with-eureka 和 fetch-registry 属性设置为 false。具体配置如示例代码 7-3 所示。

示例代码 7-3：eureka-server 子模块的配置文件

```
server.port=8761
spring.application.name=eureka-service
// eureka Service 配置
eureka.client.register-with-eureka=false
eureka.client.fetch-registry=false
// 拼接 service-url 为 "http：//${eureka.instance.hostname}：${server.port}/eureka/" 或直接编写 URL
eureka.client.service-url.defaultZone= http：//localhost：8761/eureka/
// 注册成 IP
eureka.instance.prefer-ip-address=true
// 关闭自我保护机制
// Eureka 考虑到生产环境中可能存在的网络分区故障，会导致微服务与 Eureka Server 之间无法正常通信。架构理念是同时保留所有微服务（正常的微服务和不正常的微服务都会保留），不盲目注销任何健康的微服务
eureka.server.enable-self-preservation=false
```

第四步，修改对应的启动类 EurekaServerApplication，在其类上编写 @EnableEurekaServer 注解，将该模块声明为一个 Eureka Server，如示例代码 7-4 所示。

示例代码 7-4：eureka-server 子模块的 EurekaServerApplication 启动类

```
@EnableEurekaServer
```

```
@SpringBootApplication
public class EurekaServerApplication {
    public static void main(String[] args) {
        SpringApplication.run(EurekaServerApplication.class, args);
    }
}
```

第五步，完成这些配置之后，运行该项目，并在浏览器访问“http://localhost:8761/”地址，即可看到 Eureka 的信息面板，效果如图 7-8 所示。

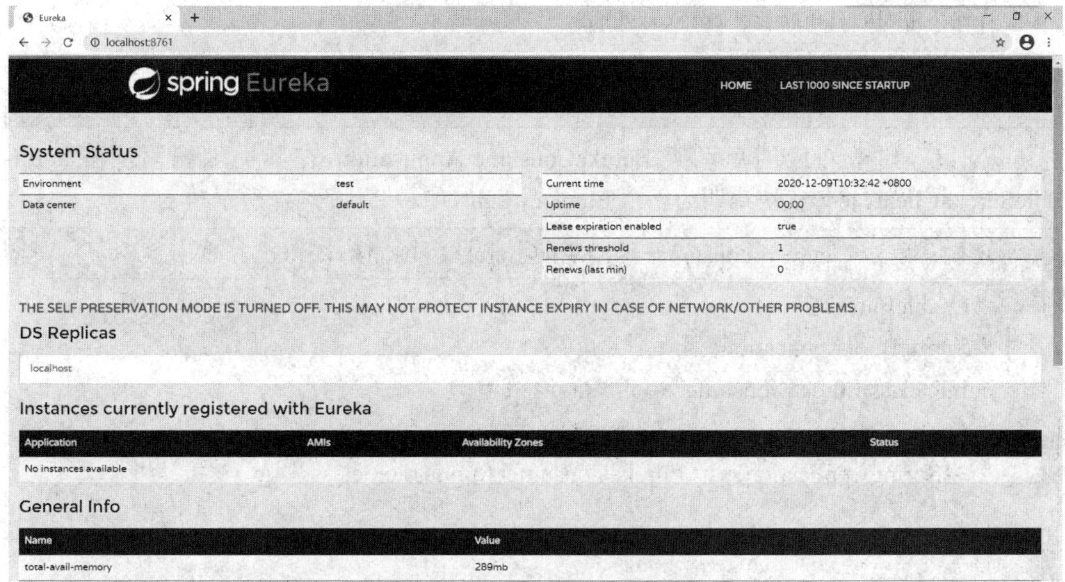

图 7-8　Eureka Server 信息面板

对应的信息面板成功显示，说明 Eureka Server 配置完成。但是，在 Instances currently registered with Eureka 信息栏中显示“No instances available”，表示还没有任何可用的实例。接下来就需要搭建客户端组件项目，完成注册。

【实例】编写 Eureka 服务消费者 eureka-consumer。

第一步，创建 Spring Boot 子模块 eureka-consumer，引入相关依赖，如示例代码 7-5 所示。

示例代码 7-5：eureka-consumer 子模块的 Pom 文件

```
<dependencies>
    <dependency>
        <groupId>org.springframework.boot</groupId>
        <artifactId>spring-boot-starter-web</artifactId>
    </dependency>
    <dependency>
```

```
        <groupId>org.springframework.cloud</groupId>
        <artifactId>spring-cloud-starter-netflix-eureka-client</artifactId>
    </dependency>
</dependencies>
```

第二步，编写 eureka-consumer 子模块的配置文件，设置端口号、名称、显示 IP 以及 URL，如示例代码 7-6 所示。

示例代码 7-6：eureka-consumer 子模块的配置文件

```
server.port=7002
spring.application.name=eureka-consumer
eureka.instance.prefer-ip-address=true
eureka.client.service-url.defaultZone=http://localhost:8761/eureka/
```

第三步，修改对应的启动类 EurekaConsumerApplication，在其类上编写注解 @EnableEurekaClient，将该模块声明为一个 Eureka Client，如示例代码 7-7 所示。

示例代码 7-7：eureka-consumer 子模块的 EurekaConsumerApplication 启动类

```
@EnableEurekaClient
@SpringBootApplication
public class EurekaConsumerApplication {
  public static void main(String[] args) {
    SpringApplication.run(EurekaConsumerApplication.class, args);
  }
}
```

第四步，完成这些配置之后，运行该项目，并在浏览器访问“http://localhost:8761/”地址，即可看到 Eureka 的信息面板，效果如图 7-9 所示。

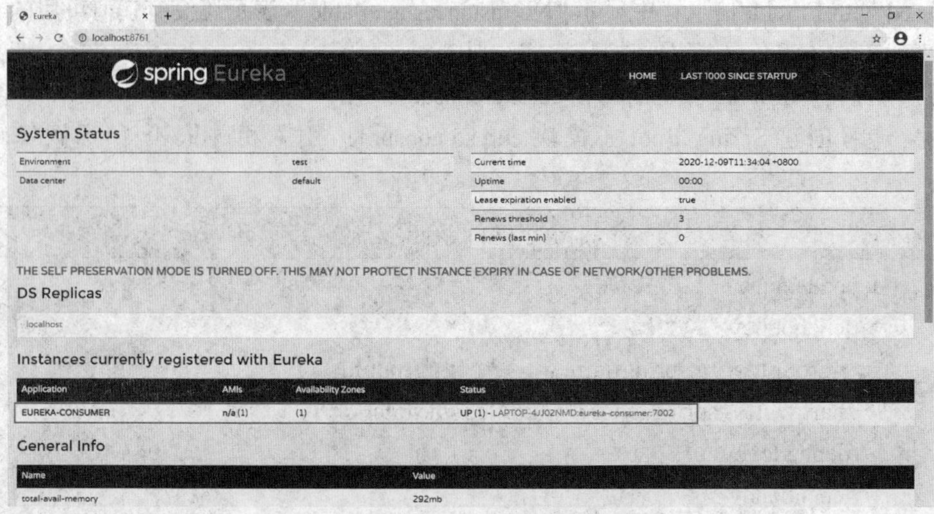

图 7-9 Eureka Server 显示服务消费者

在信息面板中看到对应的实例信息 EUREKA-CONSUMER，说明成功地将服务消费者注册到服务中心。

【案例】编写 Eureka 服务提供者 eureka-provider。

第一步，创建 Spring Boot 子模块 eureka-provider，引入相关依赖，如示例代码 7-8 所示。

示例代码 7-8：eureka-provider 子模块的 Pom 文件

```
<dependencies>
  <dependency>
    <groupId>org.springframework.boot</groupId>
    <artifactId>spring-boot-starter-web</artifactId>
  </dependency>
  <dependency>
    <groupId>org.springframework.cloud</groupId>
    <artifactId>spring-cloud-starter-netflix-eureka-client</artifactId>
  </dependency>
</dependencies>
```

第二步，编写 eureka-provider 子模块的配置文件，设置端口号、名称、主机名（hostname）以及 URL，如示例代码 7-9 所示。

示例代码 7-9：eureka-provider 子模块的配置文件

```
server.port=7006
spring.application.name=eureka-provider
eureka.instance.hostname=localhost
eureka.client.service-url.defaultZone=http://localhost:8761/eureka/
```

第三步，修改对应的启动类 EurekaProviderApplication，在其类上编写注解 @EnableEurekaClient，将该模块声明为一个 Eureka Client，如示例代码 7-10 所示。

示例代码 7-10：eureka-provider 子模块的 EurekaProviderApplication 启动类

```
@EnableEurekaClient
@SpringBootApplication
public class EurekaProviderApplication {
  public static void main(String[] args) {
    SpringApplication.run(EurekaProviderApplication.class, args);
  }
}
```

第四步，完成这些配置之后，依次运行 EurekaServer、eureka-consumer 和 eureka-provider，并在浏览器访问“http://localhost:8761/”地址，即可看到 Eureka 的信息面板，效果如图 7-10 所示。

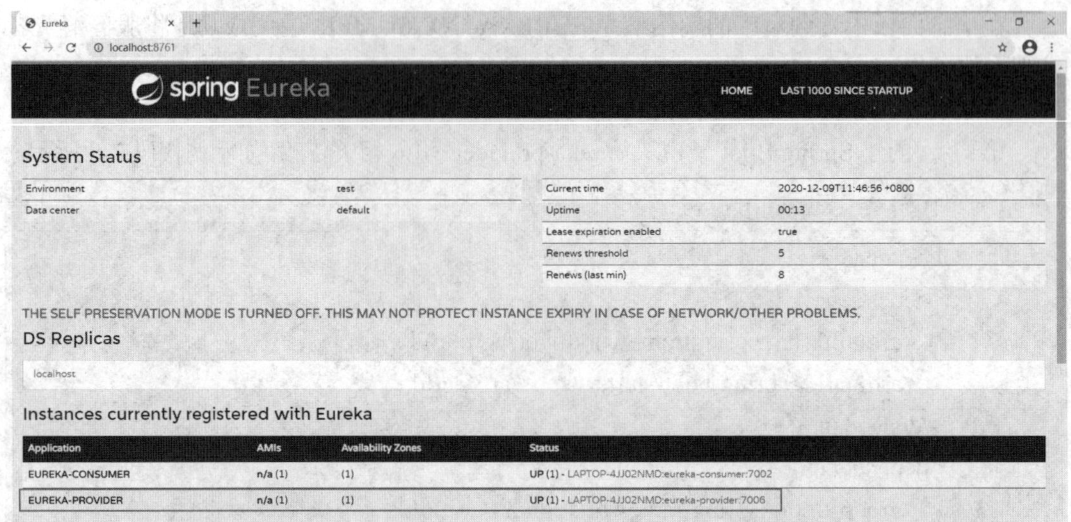

图 7-10 Eureka Server 显示服务消费者和提供者

在 Eureka 的信息面板中可以看到 EUREKA-PROVIDER 和 EUREKA-CONSUMER，说明成功将服务提供者和服务消费者注册到服务中心。

5. 服务之间的调用

服务之间的调用通过创建 RestTemplate 实例来实现，RestTemplate 是 Spring 提供用于访问 Rest 服务的客户端实例，它提供了很多便捷的方式来访问 HTTP 服务，能够有效地提高客户端的效率。

【案例】创建 RestTemplate 实例来实现服务间的调用。

第一步，编写 eureka-provider 控制器 MyProvider，编写 HelloEureka()方法，使用 @Value 获取端口号，并在页面上进行显示，如示例代码 7-11 所示。

示例代码 7-11：eureka-provider 子模块的 MyProvider 控制器

```
@RestController
public class MyProvider {
    @Value("${server.port}")// 获取端口号 7006
    String port;
    @RequestMapping("/hello/test")
    public String HelloEureka()
{
    System.out.println(" 被 eureka-consumer 所调用 ");
        return "Hello Eureka:"+port;
    }
}
```

第二步，编写 eureka-consumer 控制器 MyControllerTest，编写 String test()方法，使用 restTemplate.getForObject()方法，输入 URL 即所调用服务的“主机地址 + 端口号 + 方法地址”，如示例代码 7-12 所示。

示例代码 7-12：eureka-provider 子模块的 MyControllerTest 控制器

```
@RestController
public class MyControllerTest {
   @Autowired
   private RestTemplate restTemplate;
   @RequestMapping("/hello/test")
   public String test()
{
   System.out.println(" 调用 eureka-provider 方法 ");
      return restTemplate.getForObject("http://localhost:7006/hello/test",String.class);
   }
}
```

第三步，编写对应的 properties 配置文件，设置端口号、名称等信息，如示例代码 7-13 所示。

示例代码 7-13：eureka-provider 子模块的配置文件

```
server.port=7002
spring.application.name=eureka-consumer
eureka.client.service-url.defaultZone=http://localhost:8761/eureka/
eureka.client.register-with-eureka=false
```

第四步，运行 Eureka Server、eureka-provider 和 eureka-consumer，在浏览器访问“http://localhost:7002/hello/test”地址，可看到如图 7-11 所示的显示信息。

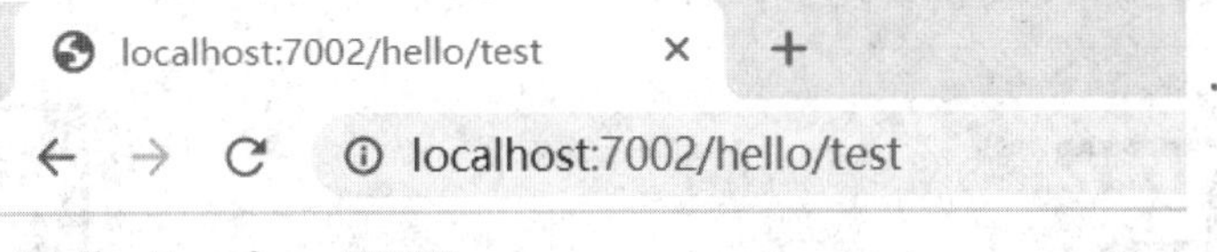

图 7-11　访问 hello/test 方法

在服务消费者 eureka-consumer 的控制台上看到的信息如图 7-12 所示。

```
2020-12-11 10:57:52.396  INFO 11704 --- [trap-executor-0]
2020-12-11 11:02:52.404  INFO 11704 --- [trap-executor-0]
调用eureka-provider方法
```

图 7-12　eureka-consumer 控制台调用提示词

在服务提供者 eureka-provider 的控制台上看到的信息如图 7-13 所示。

```
2020-12-11 11:03:33.237  INFO 8684 --- [trap-executor-0]
被eureka-consumer所调用
```

图 7-13　eureka-provider 控制台调用提示词

技能点 4 负载均衡器

由于现有网络的各个核心部分随着业务量的提高及访问量和数据流量的快速增长，服务器的处理能力和计算强度也相应地增大，使单一的服务器设备根本无法承担。在这种情况下，只增加硬件方面的投入，从长远来看并不能满足当前业务增长的需求，反而会浪费更多的资源。这就是负载均衡（load balance）概念的产生背景。

1. 负载均衡的概念

首先，来看没有使用负载平衡的 Web 项目，如图 7-14 所示。

图 7-14 无负载平衡的 Web 项目

由图 7-4 可知，用户通过网络与服务器连接，在访问数量和数据量骤增时，受限于服务器本身性能以及网络质量，用户所提交的信息和操作可能受到阻碍，若服务器正常运行但速度低下，会产生响应缓慢、无法连接的情况，这时的用户是体验是极差的，也是需要改变的。

随着这种情况的不断出现，负载均衡理念进入人们的视线。引入负载均衡，可将用户请求分摊到多个服务器上运行，共同完成任务。通过心跳检测来清除故障的服务器端节点，在用户访问服务器并发送请求时，负载均衡器会从服务节点列表中按照算法将请求发送至不同服务器进行处理，从而提高系统整体的可用性和稳定性，如图 7-15 所示。

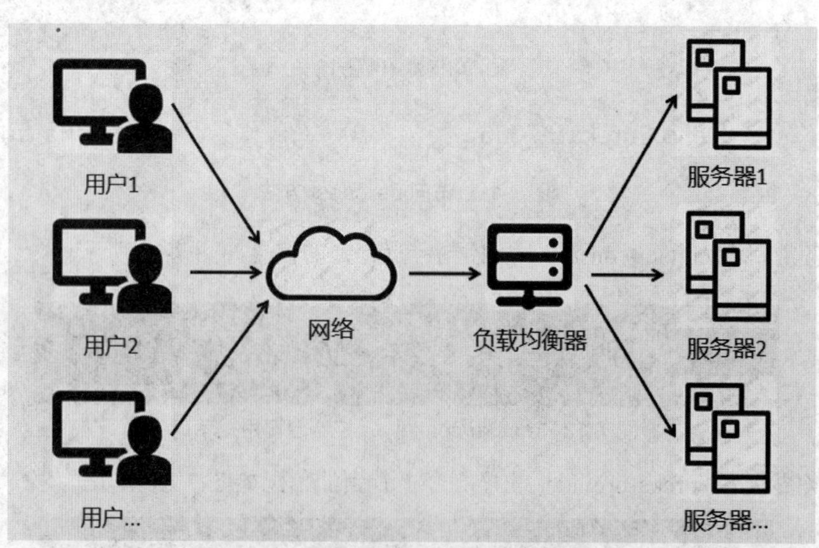

图 7-15 应用负载平衡的 Web 项目

2. 负载均衡实现方式分类

负载均衡应用很广泛，对于不同的项目需求，所应用的负载均衡技术也不同。基本分为软件、硬件、本地、全局和链路等实现方式。

（1）软件负载均衡技术

软件负载均衡技术应用在一些中小型的网站系统中，可以满足一般的均衡负载需求。软件负载均衡技术是在一个或多个交互的网络系统中的多台服务器上安装一个或多个相应的负载均衡软件来实现的一种负载均衡技术。软件可以很方便地安装在服务器上，并且实现一定的负载均衡功能。软件负载均衡技术配置简单、操作方便，最重要的是成本很低。

（2）硬件负载均衡技术

硬件负载均衡技术适用于流量高的大型网站系统。这种方式更加稳定，应用于有较大规模的企业网站、政府网站。硬件负载均衡技术是在多台服务器间安装相应的负载均衡设备，也就是负载均衡器，来达到负载均衡的目的，与软件负载均衡技术相比，其效果更加稳定、显著。

（3）本地负载均衡技术

本地负载均衡技术指对本地服务器群进行负载均衡处理。该技术通过对服务器进行性能优化，使流量能够平均分配在服务器群中的各个服务器上。本地负载均衡技术不需要购买昂贵的服务器或优化现有的网络结构。

（4）全局负载均衡技术

全局负载均衡技术适用于拥有多个服务器集群的大型网站系统。全局负载均衡技术是对分布在全国各个地区的多个服务器进行负载均衡处理，该技术通过获取访问用户的 IP 地址，根据 IP 自动转向最近的服务器，以达到负载均衡的目的。

（5）链路集合负载均衡技术

链路集合负载均衡技术将网络系统中的多条物理链路当作单一的聚合逻辑链路来使用，使网站系统中的数据流量由聚合逻辑链路中所有的物理链路共同承担。使用这种技术可以在不改变现有线路结构的基础上，提高网络数据的吞吐量。

3. Ribbon 简介

Ribbon 是基于 Netflix Ribbon 的由 Netflix 公司研发的一款基于 HTTP 和 TCP 的客户端负载均衡工具。Ribbon 虽然只是一个工具类框架，不需要像服务注册中心那样进行独立部署，但是它几乎存在于每一个 Spring Cloud 构建的微服务和基础设施中，因为微服务之间的调用都是通过 Ribbon 来实现的。Ribbon 提供了很多负载均衡算法，如轮询、随机算法等，同时也可自定义算法。

Ribbon 有两种使用方式：一种是结合使用 RestTemplate；另一种是结合使用 Feign，其中 Feign 已经集成了 Ribbon。

Ribbon 包括很多模块，这些模块集成了很多功能，见表 7-6。

表 7-6　Ribbon 模块功能

名称	详情说明
ribbon-loadbalancer	负载均衡模块

续表

名称	详情说明
Ribbon	负载均衡算法
ribbon-eureka	基于 Eureka 封装的模块，能够快速、方便地集成 Eureka
ribbon-transport	基于 Netty 实现多协议的支持，如 HTTP、TCP、UDP 等
ribbon-httpclient	基于 Apache HttpClient 封装的 REST 客户端，集成了负载均衡模块，可以直接在项目中使用，来调用接口
ribbon-example	Ribbon 使用代码示例
ribbon-core	客户端 API 的一些配置和其他 API 的定义

在 Eureka 的自动配置依赖模块 spring-cloud-starter-netflix-eureka 中，已经集成了 Ribbon，可以直接使用 Ribbon 来实现客户端的负载均衡。Ribbon 会获取 Eureka 中的服务信息列表，在调用服务实例的时候，进行负载均衡处理，其依赖关系如图 7-16 所示。

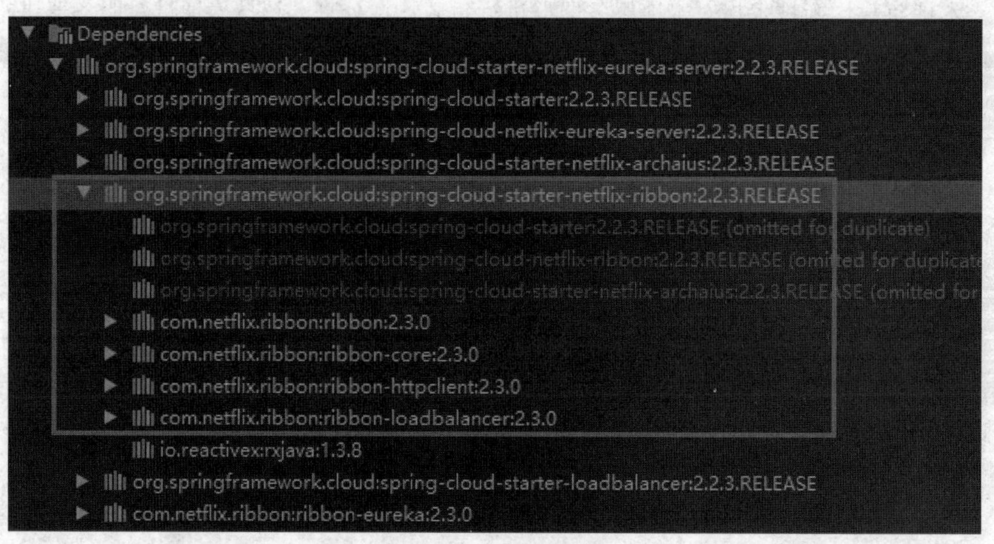

图 7-16 依赖关系图

4. Ribbon 应用实例

在使用 Spring Cloud Ribbon 进行客户端负载均衡的时候，可以给 RestTemplate Bean 添加 @LoadBalanced 注解，该 RestTemplate 在请求时拥有客户端负载均衡的能力。基本原理是当 RestTemplate 发起一个请求，请求就会被 LoadBalancerInterceptor 拦截，对应的 URL 会被拦截处理，最终的实际请求是由 loadBalancer 发出的。

【案例】应用 Ribbon 实现负载均衡，访问服务提供者实例，观察访问实例对应的端口号。

第一步，编写两个服务提供者 eureka-provider2 和 eureka-provider3，修改 eureka-provider。将三个服务提供者的 spring.application.name 属性，统一编写为 eureka-provider，service-url 统一指向 http://localhost:8761/eureka/。不同的是端口号，三个服务提供者的端口号

分别为 7007、7008 和 7006。以 eureka-provider2 为例，相关代码如示例代码 7-14 所示。

示例代码 7-14：eureka-provider2 服务提供者的配置文件

```
server.port=7007 //eureka-provider2 定义的端口号
spring.application.name=eureka-provider
eureka.client.service-url.defaultZone=http://localhost:8761/eureka/
```

第二步，编写名称为 MyProvider 的控制器，使用 @RestController 注解，编写对应的方法，获取当前访问的服务端口号，以查看负载均衡的效果。以 eureka-provider2 为例，相关代码如示例代码 7-15 所示。

示例代码 7-15：eureka-provider2 服务提供者的 MyProvider 控制器

```
@RestController
public class MyProvider {
  @Value("${server.port}")// 获取 properties 文件中定义的端口号
  String port;
  @RequestMapping("/getprovider/test")// 访问该方法的路径
  public String getpros()
{
    return " 现在访问的是 provider2，端口号为："+port;
  }
}
```

以 eureka-provider3 为例，相关代码如示例代码 7-16 所示。

示例代码 7-16：eureka-provider3 服务提供者的 MyProvider 控制器

```
@RestController
public class MyProvider {
  @Value("${server.port}")// 获取 properties 文件中定义的端口号
  String port;
  @RequestMapping("/getprovider/test")// 访问该方法的路径
  public String getpros()
{
    return " 现在访问的是 provider3，端口号为："+port;
  }
}
```

以 eureka-provider 为例，相关代码如示例代码 7-17 所示。

示例代码 7-17：eureka-provider 服务提供者的 MyProvider 控制器

```
@RestController
public class MyProvider {
  @Value("${server.port}")// 获取 properties 文件中定义的端口号
```

```
    String port;
    @RequestMapping("/getprovider/test")// 访问该方法的路径
    public String getpros()
  {
      return " 现在访问的是 provider,端口号为:"+port;
    }
  }
```

第三步,创建子模块 my-ribbons,实现客户端的负载均衡,其本质为服务消费者,需引入 eureka-client 的依赖,如示例代码 7-18 所示。

示例代码 7-18:my-ribbons 子模块的 Pom 文件

```
<dependencies>
  <dependency>
    <groupId>org.springframework.cloud</groupId>
    <artifactId>spring-cloud-starter-netflix-eureka-client</artifactId>
  </dependency>
  <dependency>
    <groupId>org.springframework.boot</groupId>
    <artifactId>spring-boot-starter-web</artifactId>
  </dependency>
  <dependency>
    <groupId>org.springframework.boot</groupId>
    <artifactId>spring-boot-starter-test</artifactId>
    <scope>test</scope>
  </dependency>
</dependencies>
```

第四步,创建 RestTemplate 实例来实现服务之间的调用。与服务之间的调用不同,在进行负载均衡时, URL 应修改为“实例名称 + 方法”,实例名称是对应服务提供者中 properties 文件中设置的 spring.applicaiton.name 属性值,如示例代码 7-19 所示。

示例代码 7-19:my-ribbons 子模块的 MyRibbonController 控制器

```
@RestController
public class MyRibbonController {
  @Autowired
  private RestTemplate restTemplate;

  @RequestMapping("/consumer/test")
```

```
    public String test()
    {
        return restTemplate.getForObject("http: //EUREKA-PROVIDER/getprovider/test",
String.class);
    }
}
```

第五步，编写 my-ribbons 的启动文件，在 restTemplate 方法上使用 @LoadBalanced 注解，开启负载均衡，如示例代码 7-20 所示。

示例代码 7-20：my-ribbons 子模块的 MyRibbonsApplication 启动类

```
@EnableEurekaClient
@SpringBootApplication
public class MyRibbonsApplication {
    public static void main(String[] args) {
        SpringApplication.run(MyRibbonsApplication.class, args);
    }
@Bean
// 在应用 RestTemplate 的基础上增加 @LoadBalanced 注解，开启负载均衡
@LoadBalanced
    public RestTemplate restTemplate()
    {
        return new RestTemplate();
    }
}
```

第六步，编写 my-ribbons 的配置文件，如示例代码 7-21 所示。

示例代码 7-21：my-ribbons 子模块的配置文件

```
server.port=7000   // 定义端口为 7000
spring.application.name=my-ribbons
eureka.instance.prefer-ip-address=true
eureka.client.service-url.defaultZone=http://localhost:8761/eureka/
eureka.client.register-with-eureka=false
```

第七步，完成配置之后，依次启动 EurekaServer、eureka-provider、eureka-provider2、eureka-provider3 和 my-ribbons。访问“http: //localhost: 8761/”地址并查看 Eureka 的信息面板，效果如图 7-17 所示。

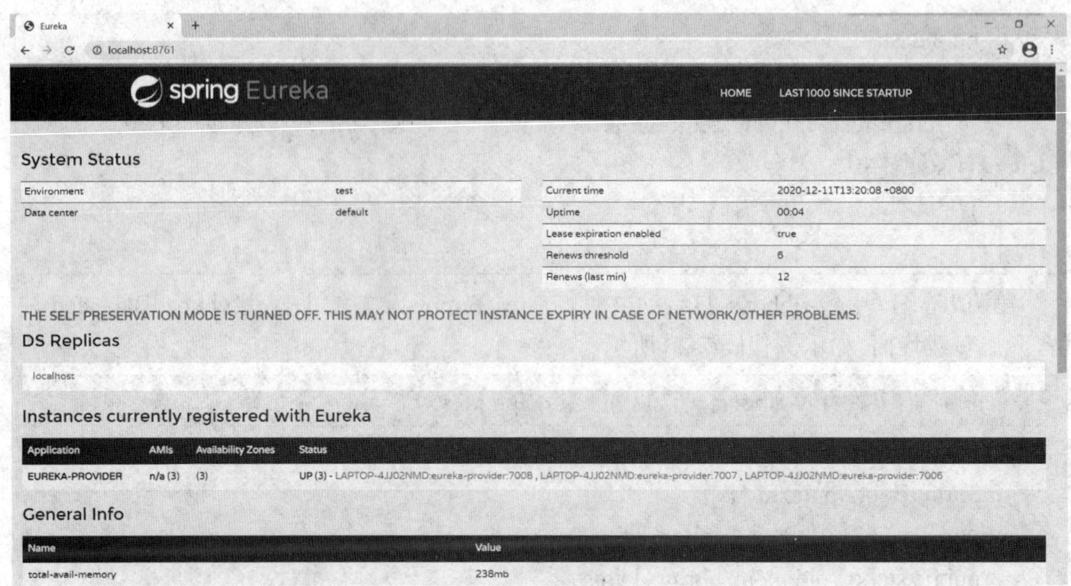

图 7-17 Eureka 的信息面板

由 Eureka 的信息面板可知，注册了名为 EUREKA-PROVIDER 的实例，共有三个实例，端口号分别是 7006、7007 和 7008。之后，通过 my-ribbons 服务消费者，访问服务提供者中的 getpros()方法。由于部署注册了三个实例，在访问时会通过 Ribbon 进行负载均衡，当通过浏览器访问“http://localhost:7000/consumer/test”地址时，效果如图 7-18 所示。

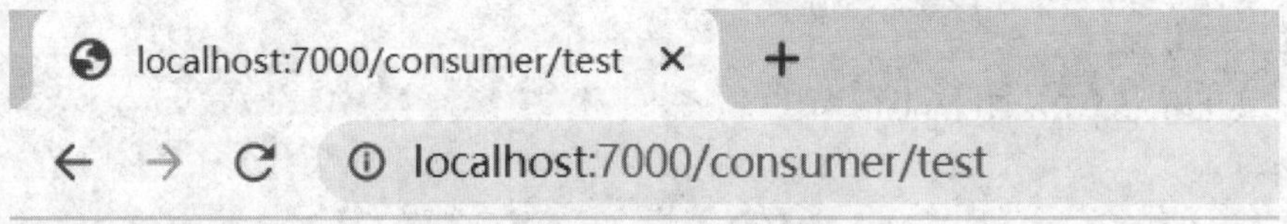

图 7-18 访问 provider3 方法的效果

此时访问的是 provider3，端口号为 7008。再次刷新页面，重新访问，效果如图 7-19 所示。

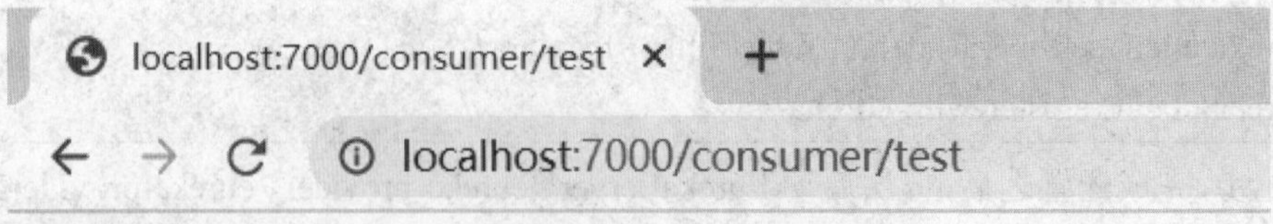

图 7-19 访问 provider2 方法的效果

此时可以看到地址栏并没有改变，但此时访问的为 provider2，端口号为 7007。再次刷新页面，重新访问，效果如图 7-20 所示。

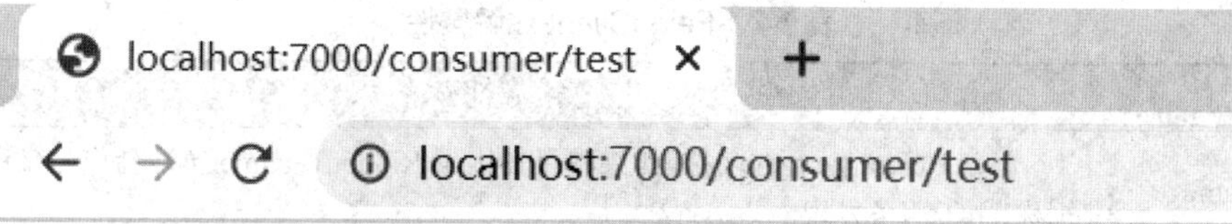

现在访问的是provider,端口号为：7006

图 7-20　访问 provider 方法的效果

经过第二次刷新并重新访问，访问到了 provider 服务，端口号为 7006。至此，所部署的三个实例均被访问到，表明成功使用 Ribbon 实现了客户端的负载均衡。

技能点 5　声明式客户端 Feign

之前使用 Ribbon 和 RestTemplate 实现了服务之间的调用，是利用了 RestTemplate 对 HTTP 请求的封装处理，形成了一套模版化的调用方法。但是在实际的开发中，往往一个接口会被多处调用，所有的参数需要在请求的 URL 中进行拼接，给代码编写增加了难度，效率低下。Feign 在此基础上做了进一步封装，来帮助我们自定义和实现依赖服务接口的定义。开发人员通过编写简单的接口和插入注解，就可以定义 HTTP 请求的参数、格式、地址等信息，从而简化了使用 Spring Cloud Ribbon 时，自动封装服务调用客户端的开发过程。

1. Feign 简介

Feign 是 Netflix 公司开发的声明式、模板化的 HTTP 客户端，可以帮助开发人员更快捷、优雅地调用 HTTP API。在 Spring Cloud 中，对 Feign 进行了增强，整合了 Ribbon 和 Eureka，Feign 也支持 Spring MVC 注解、符合 JAX-RS 标准的注解及可插拔式编码器和解码器。在技能点 4 中，利用 Ribbon 维护了服务列表信息，并且通过轮询实现了客户端的负载均衡。与 Ribbon 不同的是，Feign 只需要定义服务绑定接口且以声明式的方法，优雅而且简单地实现了服务调用。Feign 强大的功能让其使用更加方便。

在应用 Feign 时，需要在 Pom 文件中添加对应的依赖关系，之后便可以使用 Feign 进行开发，代码如下。

```
<dependency>
    <groupId>org.springframework.cloud</groupId>
    <artifactId>spring-cloud-starter-openfeign</artifactId>
</dependency>
```

通常，在开发中使用 @FeignClient 注解来完成开发，该注解的部分常用属性及其详情见表 7-7。

表 7-7　FeignClient 注解属性

名称	属性详请
name	指定 FeignClient 的名称，如果项目使用了 Ribbon，name 属性会作为微服务的名称，用于服务发现
url	url 一般用于调试，可以手动指定 @FeignClient 调用地址
configuration	设置 Feign 配置类
fallback	定义容错的处理类，当调用远程接口失败或超时时，会调用对应接口的容错逻辑
fallbackFactory	工厂类，用于生成 fallback 实例，通过这个属性可以实现每个接口通用的容错逻辑，减少代码冗余
path	定义当前 FeignClient 的统一前缀

2. Feign 应用实例

【案例】应用 Feign 实现服务间的调用和负载均衡。

第一步，创建新子模块 my-feign，添加 client、web 和 openfeign 等依赖关系，如示例代码 7-22 所示。

示例代码 7-22：my-feign 子模块的 Pom 文件

```
<dependencies>
  <dependency>
    <groupId>org.springframework.boot</groupId>
    <artifactId>spring-boot-starter-web</artifactId>
  </dependency>
  <dependency>
    <groupId>org.springframework.cloud</groupId>
    <artifactId>spring-cloud-starter-netflix-eureka-client</artifactId>
  </dependency>
  <dependency>
    <groupId>org.springframework.cloud</groupId>
    <artifactId>spring-cloud-starter-openfeign</artifactId>
  </dependency>
  <dependency>
    <groupId>org.springframework.boot</groupId>
    <artifactId>spring-boot-starter-test</artifactId>
    <scope>test</scope>
  </dependency>
</dependencies>
```

第二步，编写 my-feign 子模块和配置文件 application.properties，添加名称、端口号和 service-url 信息，如示例代码 7-23 所示。

示例代码 7-23：my-feign 子模块的配置文件

```
spring.application.name=my-feign
server.port=7010
eureka.client.service-url.defaultZone=http://localhost:8761/eureka/
```

第三步，编写 my-feign 子模块，创建 service 文件夹，在其中创建 FeignService 接口，编写对应的方法，在接口上使用 @Service 注解，自动注册到 Spring 容器中。应用 @FeignClient（name = "eureka-provider"）注解，name 属性指定 FeignService 接口所要调用的是 eureka-provider，此处的 name 属性必须与服务提供者（此处为之前所编写的 eureka-provider）中配置文件 application.properties 的属性 spring.application.name 保持一致，如示例代码 7-24 所示。

示例代码 7-24：my-feign 子模块的 FeignService 接口

```
@Service
@FeignClient(name ="eureka-provider")
public interface FeignService {
  @RequestMapping(value = "/getprovider/test", method = RequestMethod.GET)
  String getprovider();
}
```

第四步，编写 my-feign 子模块，创建 controller 文件夹，在其中创建 FeignController 类，编写对应的方法。类上标记 @RestController 将返回值转化为字符串形式，调用 feignService 中的 getprovider（）方法，如示例代码 7-25 所示。

示例代码 7-25：my-feign 子模块的 FeignController 控制器

```
@RestController
public class FeignController {
  @Autowired
  private FeignService feignService;

  @RequestMapping("/getprovider/test")
  public String getpros(){
    return feignService.getprovider();
  }
}
```

第五步，编写启动类 MyFeignApplication，在其类上编写注解 @EnableEurekaClient 声明为一个 Eureka client，并使用 @EnableFeignClients 注解，应用声明式客户端 Feign，如示例代码 7-26 所示。

示例代码 7-26：my-feign 子模块的 MyFeignApplication 启动类

```
@EnableEurekaClient
@EnableFeignClients
@SpringBootApplication
public class MyFeignApplication {
  public static void main(String[] args) {
    SpringApplication.run(MyFeignApplication.class, args);
  }
}
```

第六步，依次启动 EurekaServer、eureka-provider、eureka-provider2、eureka-provider3 和 my-feign，其中 eureka-provider、eureka-provider2、eureka-provider3 为之前所编写的服务提供者，无须更改其代码内容。启动成功之后，打开浏览器访问“http://localhost:7010/getprovider/test”地址，效果如图 7-21 所示。

图 7-21 Feign 整合负载均衡访问 provider 方法

此时，页面上显示的是 provider，端口号为 7006。再次刷新该页面，效果如图 7-22 所示。

图 7-22 Feign 整合负载均衡访问 provider3 方法

经过刷新之后，所访问的是 provider3 服务提供者的方法，端口号为 7008，说明 Feign 实现了负载均衡。该实现方法不需要额外进行设置，由于整合了 Ribbon，只需定义服务绑定接口且以声明式的方法，就可以优雅而且简单地实现负载均衡。此时，再次刷新该页面，效果如图 7-23 所示。

图 7-23 Feign 整合负载均衡访问 provider2 方法

此时，所调用的是 provider2 服务提供者的方法，其端口号为 7007。至此，完成了应用 Feign 实现服务间的调用以及负载均衡。

运用 Spring Cloud 微服务知识，设计服务提供者（user-provider），实现用户登录注册数据交互功能。登录页面效果如图 7-24 所示。

图 7-24　用户登录页面效果图

1. 数据库与基本结构设计

数据库设计，创建表 user，对应的字段名见表 7-8。

表 7-8　数据库表 user 字段属性

字段名	类型	长度	说明
id	int	0	自增主键
name	varchar	255	登录用户名
password	varchar	255	登录密码
iphone	varchar	255	手机号码

传统登录模块功能一般包括登录页面显示、获取用户登录信息、信息校验、返回结果和呈现页面。在微服务架构下，重构模块功能，将这个模块进行拆分，分为负责显示页面获取

用户信息参数、处理相关请求的服务消费者（user-consumer）和该服务远程调用的服务提供者（user-provider）。服务提供者负责定义用户登录注册的接口 Dao，封装相关的接口并负责与数据库进行数据交互，包括注册新用户、对已有用户进行登录校验等。微服务登录模块结构如图 7-25 所示。

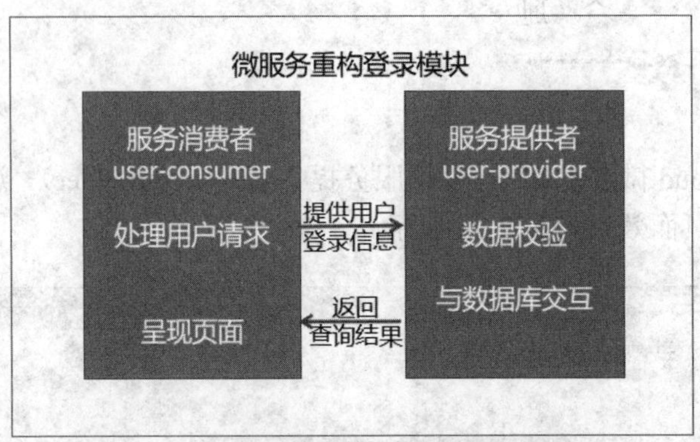

图 7-25 微服务重构登录模块

注册中心沿用之前所使用的 eureka-server，无须更改。创建项目 user-consumer 和 user-provider。项目结构如图 7-26 所示。

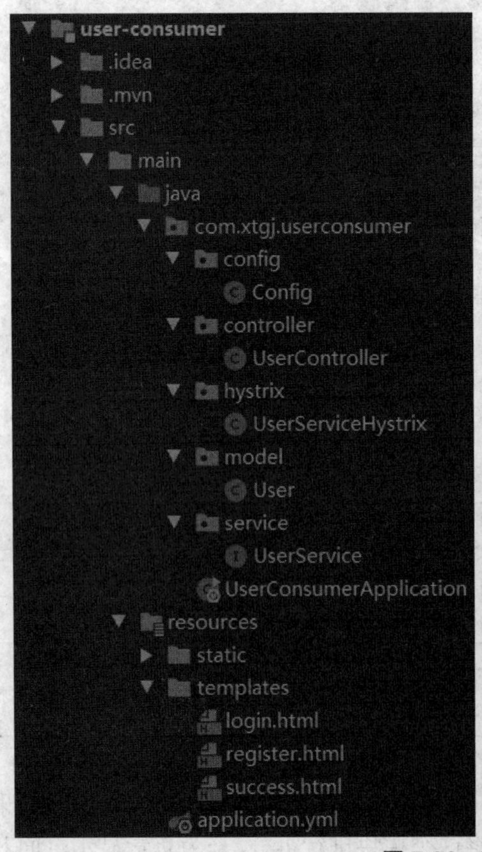

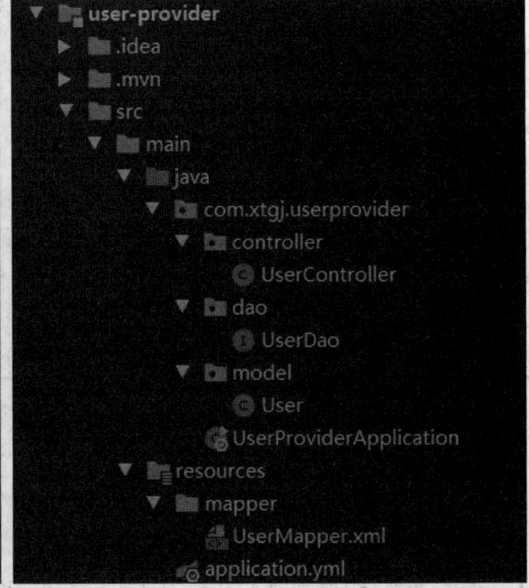

图 7-26 项目结构图

2. 搭建 user-provider 服务

第一步，搭建 user-provider 作为用户模块服务提供者。通过 Spring Initializr 创建名称为 user-provider 的 Spring Boot 项目，引入 EurekaClient、Web、Mybatis、MySQL 的依赖，如示例代码 7-27 所示。

示例代码 7-27：user-provider 服务提供者的 Pom 文件

```
<? xml version="1.0" encoding="UTF-8"? >
<project  xmlns="http: //maven.apache.org/POM/4.0.0"  xmlns: xsi="http: //www.w3.org/2001/XMLSchema-instance"
    xsi: schemaLocation="http: //maven.apache.org/POM/4.0.0 https: //maven.apache.org/xsd/maven-4.0.0.xsd">
    <modelVersion>4.0.0</modelVersion>
    <parent>
        <groupId>org.springframework.boot</groupId>
        <artifactId>spring-boot-starter-parent</artifactId>
        <version>2.2.8.RELEASE</version>
        <relativePath/> <! -- lookup parent from repository -->
    </parent>
    <groupId>com.xtgj</groupId>
    <artifactId>user-provider</artifactId>
    <version>0.0.1-SNAPSHOT</version>
    <name>user-provider</name>
    <description>Demo project for Spring Boot</description>

    <properties>
        <java.version>1.8</java.version>
        <spring-cloud.version>Hoxton.SR6</spring-cloud.version>
    </properties>

    <dependencies>
        <dependency>
            <groupId>org.springframework.boot</groupId>
            <artifactId>spring-boot-starter-web</artifactId>
        </dependency>
        <! --mybatis 依赖 -->
        <dependency>
            <groupId>org.mybatis.spring.boot</groupId>
        <artifactId>mybatis-spring-boot-starter</artifactId>
            <version>2.1.0</version>
```

```
        </dependency>
        <dependency>
            <groupId>org.springframework.cloud</groupId>
            <artifactId>spring-cloud-starter-netflix-eureka-client</artifactId>
        </dependency>
        <dependency>
            <groupId>mysql</groupId>
            <artifactId>mysql-connector-java</artifactId>
            <scope>runtime</scope>
        </dependency>
        <dependency>
            <groupId>org.springframework.boot</groupId>
            <artifactId>spring-boot-starter-test</artifactId>
            <scope>test</scope>
        </dependency>
        <dependency>
            <groupId>org.springframework.cloud</groupId>
            <artifactId>spring-cloud-starter-config</artifactId>
        </dependency>
    </dependencies>
    <dependencyManagement>
        <dependencies>
            <dependency>
                <groupId>org.springframework.cloud</groupId>
                <artifactId>spring-cloud-dependencies</artifactId>
                <version>${spring-cloud.version}</version>
                <type>pom</type>
                <scope>import</scope>
            </dependency>
        </dependencies>
    </dependencyManagement>
    <build>
        <plugins>
            <plugin>
                <groupId>org.springframework.boot</groupId>
```

```
                <artifactId>spring-boot-maven-plugin</artifactId>
            </plugin>
        </plugins>
    </build>
    <repositories>
        <repository>
            <id>spring-milestones</id>
            <name>Spring Milestones</name>
            <url>https://repo.spring.io/milestone</url>
        </repository>
    </repositories>
</project>
```

第二步，创建 model 文件夹，创建 User 实体类，实体类的字段同对应的数据表，如示例代码 7-28 所示。

示例代码 7-28：User 实体类

```
public class User {
  private String name;
  private String password;
  private String count;
  private String iphone;
// 省略 getter 和 setter 方法
// 省略 toString 方法
}
```

第三步，创建 dao 文件夹，创建 UserDao 接口，作为用户模块的接口，用于调用对应的方法与数据库进行交互，如示例代码 7-29 所示。

示例代码 7-29：UserDao 接口

```
@Mapper
public interface UserDao {
  // 用户注册接口
  int register(@Param(value = "password")String password,
          @Param(value = "name")String name,
          @Param(value = "iphone") String iphone);

  User login(@Param(value ="name") String name);// 用户登录接口;
}
```

第四步，创建 controller 文件夹，创建 UserController 控制器，实现 Dao 中注册的方法，如示例代码 7-30 所示。

示例代码 7-30：UserController 控制器

```
@RestController
public class UserController {
    @Autowired
    UserDao userDao;
    @RequestMapping(value = "/register", method = RequestMethod.GET)
    public int register(@RequestParam(value = "password") Stringpassword, @Request-
Param(value = "name") String name, @RequestParam(value = "iphone") String iphone) {
        return userDao.register(password, name, iphone);
    }
    @RequestMapping(value = "/login", method = RequestMethod.GET)
    public User login(@RequestParam("name") String name) {
        return userDao.login(name);
    }
}
```

第五步，在 resources 目录下创建 mapper 文件夹，创建 UserMapper.xml 用于 Mybatis 配置数据访问接口，如示例代码 7-31 所示。

示例代码 7-31：UserMapper.xml 文件

```
<? xml version="1.0" encoding="UTF-8"? >
<! DOCTYPE mapper PUBLIC "-//mybatis.org//DTD Mapper 3.0//EN"
    "http://mybatis.org/dtd/mybatis-3-mapper.dtd">、
<mapper namespace="com.xtgj.userprovider.dao.UserDao">
<! --id 为 dao 层对应的方法名 -->
    <insert id="register" parameterType="java.lang.String">
        INSERT INTO USER(name, password, iphone) VALUES
        (#{name}, #{password}, #{iphone})
    </insert>
    <select id="login" resultType="com.xtgj.userprovider.model.User"
            parameterType="java.lang.String">
        select * from user where name=#{name}
    </select>
</mapper>
```

第六步，编写 application 配置文件，连接数据库，如示例代码 7-32 所示。

示例代码 7-32：application.yaml 配置文件

```
server:
```

```
  port：7025
spring：
  datasource：
    url：jdbc：mysql：//localhost：3306/test？serverTimezone=UTC&useSSL=false
    username：root
    password：root
    driver-class-name：com.mysql.cj.jdbc.Driver
  application：
    name：user-provider-dev
eureka：
  client：
    service-url：
      defaultZone：http：//localhost：8761/eureka/
  instance：
    hostname：localhost
mybatis：
  mapper-locations：classpath：/mapper/*
```

第七步，编写启动类 UserProviderApplication，在类上编写 @EnableEurekaClient 注解，如示例代码 7-33 所示。

示例代码 7-33：UserProviderApplication 启动类

```
@SpringBootApplication
@EnableEurekaClient
public class UserProviderApplication {
    public static void main（String[] args）{
            SpringApplication.run（UserProviderApplication.class，args）；
    }
}
```

第八步，依次运行 eureka-server 和 user-provider，打开浏览器访问 login 接口，由于没有完成 user-consumer 服务消费者的编写，直接采用注册接口拼接参数的方式对方法进行访问。例如，访问“http：//localhost：7025/login？name=zhang”地址，即访问登录接口，获取用户信息。由于返回的是字符串类型，在页面上会显示所查询出来的结果，如图 7-27 所示。

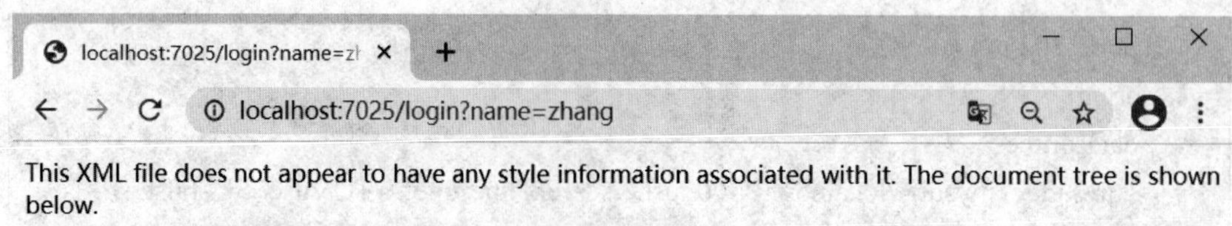


```
▼<User>
   <name>zhang</name>
   <password>123456</password>
   <iphone>15677543213</iphone>
 </User>
```

图 7-27 获取数据库数据

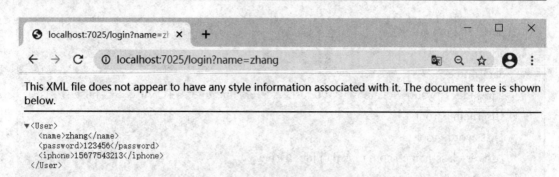

本任务讲解了 Spring Cloud 的基本构成，采用了服务注册、服务提供者和服务消费者，搭建了基本的微服务项目，实现了服务提供者（user-provider）的基本查询和新增功能，编写了用户接口。通过本任务，读者加深了对于 Spring Cloud 的理解，掌握了基本的 Spring Cloud 技术。

Provider	提供者	Release	发布
Dashboard	仪表盘	Eviction	逐出
Transport	运输		

一、选择题

1. 以下说法错误的是（　　）。

A. Eureka 是服务注册中心，主要用于服务管理

B. Ribbon 是客户端用于负载均衡的组件

C. Zuul 是分布式处理中心

D. Hystrix 是服务容错保护机制

2. 以下不属于 Spring Cloud 组件的是（　　）。

A. Eureka　　B. Ribbon　　C. Nginx　　D. Zuul

3. 关于负载平衡，下列说法错误的是（　　）。

A. Spring Cloud 构建的微服务架构中，Ribbon 有两种使用方法，分别与 Feign、RestTemplate 相结合，它们都默认集成了 Ribbon

B. 除了 Ribbon 之外，Nginx 和 Zookeeper 也可以作为负载均衡器进行使用

C. Ribbon 提供了很多负载均衡的算法，如轮询、随机算法等，同时也可自定义算法

D. 软件负载的解决方法是指在一台或者多台服务器上安装一个或者多个附加软件进行负载均衡

4. 对于 @FeignClient 注解的阐述，下列说法正确的是（　　）。

A. @Feign 注解应用在接口上，用于指定该接口所要调用的服务名称

B. @FeignClient 注解的 name 属性必须与 provider 服务提供者中配置文件指定的服务提供者名称保持一致

C. @FeignClient 注解的类上不能使用 @RequestMapping 注解

D. 以上说法均正确

5. 关于 @FeignClient 注解中 name 属性的说法，下列说法正确的是（　　）。

A. name 属性必须与 provider 服务提供者的项目名称保持一致

B. Name 属性值和服务提供者中全局配置文件（application）中指定的服务提供者名称保持一致

C. Name 属性值可以和服务提供者的项目名称一致，也可以和服务提供者全局配置文件中指定的服务提供者名称一致

D. Name 属性值必须和服务提供者项目名称、全局配置文件中指定的服务提供者名称保持一致

二、简答题

1. 微服务的优点有哪些？

2. Spring Cloud 的核心组件有哪些？

项目八 深入学习 Spring Cloud

通过学习 Spring Cloud 熔断器、Zuul 网关以及 Spring Cloud Config 配置中心的知识，熟悉相关组件的使用方法和应用场景。编写完善应用 Spring Cloud 的用户注册登录服务。

- 掌握 Hystrix 容错保护。
- 掌握 Zuul 网关服务。
- 掌握配置中心 Spring Cloud Config。

【情境导入】

在微服务项目运行时，其需要与之配合的组件。例如，检测和观察各微服务之间通信状态需要相关组件，访问不同服务也需要特殊的网关名称等。Hystrix 熔断器、Zuul 网关和 Spring Cloud Config 配置中心就为微服务项目提供了这些功能。

【功能描述】

● 搭建 user-consumer 服务。

技能点 1　Hystrix 容错保护

在 Spring Boot 项目中，虽然采用了服务发现和客户端负载均衡，对整体服务进行了配置，但对于一个复杂的项目，这依旧是不够的，当存在多个服务层之间的相互调用时，若基础的服务层发生故障，就会导致整体服务系统的崩溃，如图 8-1 所示。

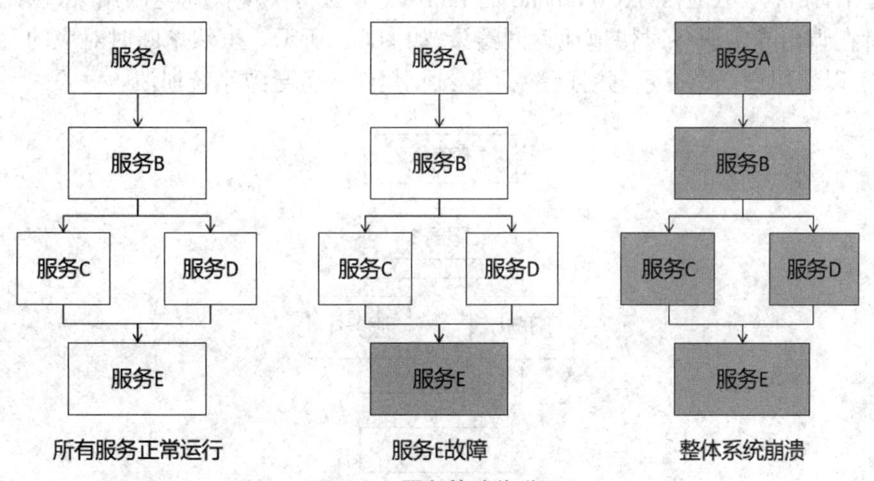

图 8-1　服务故障传递图

图 8-1 中，由于服务 E 出现故障，导致服务 C 和服务 D 的请求被中断，进而造成一系列的故障，最终导致系统的崩溃，造成不可估量的损失。造成这种情况的原因有很多，如硬件故障、流量激增、缓存穿透、同步等待、程序 BUG 等。

为了解决这种问题，可采用 Hystrix 中的熔断器（circuit breaker），引入熔断机制来对服务进行保护，这种保护机制类似于家中的电路保险丝对应的短路保护机制，保险丝可以在发生短路的瞬间切断电源以防止事故的发生。

1. Hystrix 熔断器简介

Hystrix 是一个容错管理工具，它通过熔断机制和第三方库的节点来控制服务，从而对延迟和故障提供更强大的容错能力，Spring Cloud 中默认整合了 Hystrix，只要开发人员利用 Hystrix 在项目中使用 Feign，系统就会使用熔断器处理所有的请求。

Hystrix 在运行过程中会向每个对应的熔断器报告失败、成功、超时和拒绝的状态，熔断器维护并统计这些数据，并根据这些统计信息来决策熔断开关是否打开。如果熔断器打开，则直接切断后续请求，快速返回 fallback 方法。经过一段时间之后，熔断器尝试进入半开状态，只允许一个请求通过，相当于对依赖服务的一次检测，如果请求成功，则熔断器关闭。若失败率超过一定值，该服务的断路器会打开。Hystrix 的工作原理如图 8-2 所示。

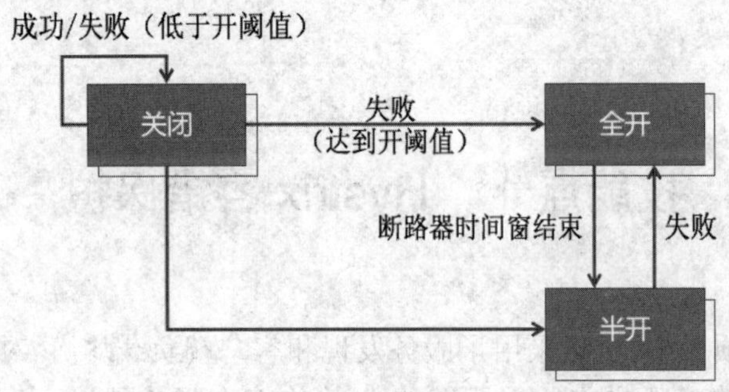

图 8-2 Hystrix 熔断器工作原理

通过 Hystrix 对应 HystrixCommand 的 fallback 方法可实现快速返回结果，以此保证服务调用者在调用异常服务时快速地返回结果，如图 8-3 所示。虽然熔断时对应的服务不可使用，但可以给用户一个提示，并且避免了继续调用该服务导致系统崩溃。

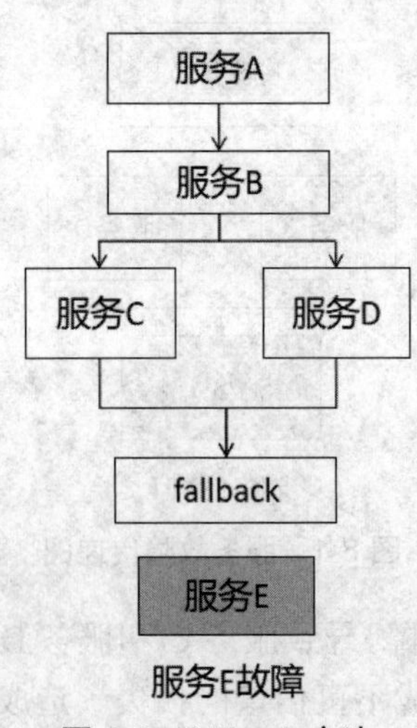

图 8-3 fallback 方法

2. Hystrix 熔断器使用方法

若要在服务中运用 Hystrix 熔断器，则需要在 Pom 文件中添加依赖，并在启动类上添加 @EnableHystrix 注解，代码如下。

```
<dependency>
    <groupId>org.springframework.cloud</groupId>
    <artifactId>spring-cloud-starter-netflix-hystrix</artifactId>
</dependency>
```

Hystrix 熔断器应用 @HystrixCommand 注解来完成 fallback。该注解所对应的常见参数，见表 8-1。

表 8-1　HystrixCommand 注解参数

参数名称	参数详情
groupKey	HystrixCommand 命令所属的组的名称：默认注解方法类的名称
commandKey	HystrixCommand 命令的 key 值，默认值为注解方法的名称
threadPoolKey	线程池名称，默认定义为 groupKey
fallbackMethod	定义回调方法的名称，此方法必须和 hystrix 的执行方法在相同类中
defaultFallback	定义回调方法：但是 defaultFallback 不能传入参数，返回参数和 hystrix 的命令兼容

【案例】应用 Hystrix 熔断器，在 eureka-provider 关闭之后，加载 fallback 对应的方法内容。

第一步，新建 my-hystrix 子模块，在其目录下创建 config 文件夹，加载 RestTemplate，如示例代码 8-1 所示。

示例代码 8-1：my-hystrix 子模块的 HystrixConfig 配置类

```
@Configuration
public class HystrixConfig {
    @Bean
    @LoadBalanced
    public RestTemplate restTemplate()
    {
        return new RestTemplate();
    }
}
```

第二步，在目录下创建 service 文件夹，并创建 HystrixService 类，应用 RestTemplate 对象，应用 restTemplate.getForObject()方法，调用 eureka-provider 的方法。在 GetStr()方法上使用 @HystrixCommand(fallbackMethod = "FallbackTest")注解，fallbackMethod 属性指向该类中的方法名称，即表示发生熔断时，调用该方法。注意，此方法的返回值和参数必须与原方法一致，如示例代码 8-2 所示。

示例代码 8-2：my-hystrix 子模块的

```
@Service
public class HystrixService {
    @Autowired
    private RestTemplate restTemplate;
    @HystrixCommand(fallbackMethod = "FallbackTest")// 指定方法名称
    public String GetStr(){
        return restTemplate.getForObject("http://EUREKA-PROVIDER/getprovider/test",
String.class);
    }
    // 覆写 fallbackMethod 中指定的方法，此方法的返回值、参数必须与原方法一致
    public String FallbackTest()
    {
        return " 该服务暂时故障 ";
    }
}
```

第三步，创建 controller 文件夹，在其下创建 HystrixController 类，负责接收请求，并调用 Service 中的方法，如示例代码 8-3 所示。

示例代码 8-3：HystrixController 控制器

```
@RestController
public class HystrixController {
    @Autowired
    private HystrixService hystrixService;
    @RequestMapping("/getStr")
    public String GetStr()
    {
        return hystrixService.GetStr();
    }
}
```

第四步，编写 my-hystrix 的配置文件，配置端口号、service-url 和项目名称，如示例代码 8-4 所示。

示例代码 8-4：my-hystrix 的配置文件

```
spring.application.name=my-hystrix
server.port=7011
eureka.client.service-url.defaultZone=http://localhost:8761/eureka/
```

第五步，编写 MyHystrixApplication 启动类，在类上编写 @EnableEurekaClient 和 @EnableCircuitBreaker 注解，@EnableHystrix 注解用来开启断路器功能，如示例代码 8-5 所示。

示例代码 8-5：MyHystrixApplication 启动类

```
@EnableEurekaClient
@EnableHystrix
@SpringBootApplication
public class MyHystrixApplication {
    public static void main(String[] args) {
        SpringApplication.run(MyHystrixApplication.class, args);
    }
}
```

第六步，依次运行 eureka-server、eureka-provider 和 my-hystrix，打开浏览器，访问"http://localhost:7011/getStr"地址，结果如图 8-4 所示。

图 8-4　访问 getStr 方法

可以看到，对应显示的信息以及端口号。之后，关闭 eureka-provider 服务，人为造成服务故障的情况，待服务关闭之后，再次访问"http://localhost:7011/getStr"地址，结果如图 8-5 所示。

图 8-5　故障时调用 fallback 方法

根据两次访问的结果可以看到，由于服务关闭，导致了系统错误，触发了断路器，显示了设置的 fallback。至此，成功地运用 Hystrix 对服务进行了部署。

3. Feign 整合 Hystrix 熔断器

之前学习了 Feign 的相关知识。当引入 Feign 依赖便整合了 Hystrix，因此可以在应用 Feign 的基础上，对 Hystrix 熔断器进行编写，只需在配置文件中设置 feign.hystrix.enabled=true 即可开启。Feign 整合 Hystrix 的依赖如图 8-6 所示，在图中可看到对应的 Hystrix 依赖。

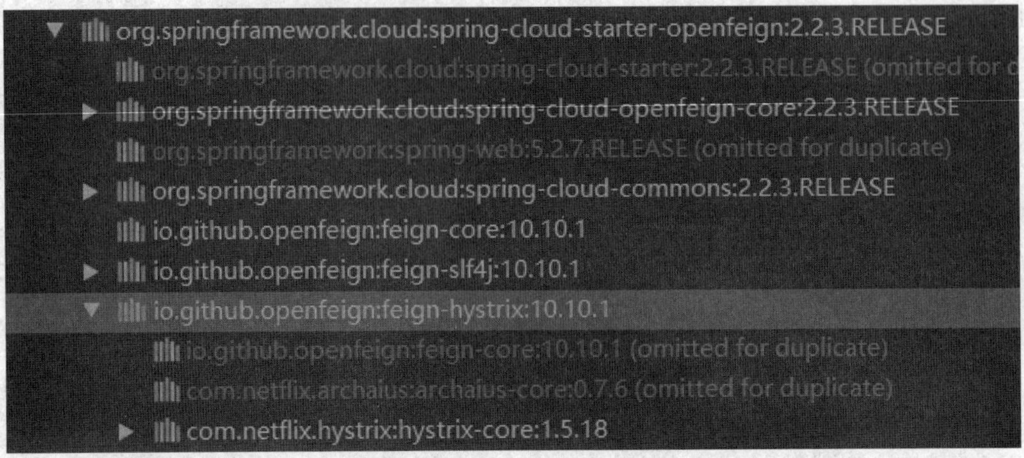

org.springframework.cloud:spring-cloud-starter-openfeign:2.2.3.RELEASE
org.springframework.cloud:spring-cloud-starter:2.2.3.RELEASE (omitted for d
org.springframework.cloud:spring-cloud-openfeign-core:2.2.3.RELEASE
org.springframework:spring-web:5.2.7.RELEASE (omitted for duplicate)
org.springframework.cloud:spring-cloud-commons:2.2.3.RELEASE
io.github.openfeign:feign-core:10.10.1
io.github.openfeign:feign-slf4j:10.10.1
io.github.openfeign:feign-hystrix:10.10.1
io.github.openfeign:feign-core:10.10.1 (omitted for duplicate)
com.netflix.archaius:archaius-core:0.7.6 (omitted for duplicate)
com.netflix.hystrix:hystrix-core:1.5.18

图 8-6 Feign 整合 Hystrix

【案例】在 Feign 中使用 Hystrix 熔断器。

在项目七的 Feign 技能点中，编写了 my-feign 子模块，下面继续沿用该模块完成任务。

第一步，在 service 包下，创建 Impl 文件夹，编写 HystrixFeignServiceImpl 类，作为熔断器类，实现接口 FeignService 中的方法，对应的返回值即为 fallback 返回的内容，如示例代码 8-6 所示。

示例代码 8-6：HystrixFeignServiceImpl 类

```
@Component
public class HystrixFeignServiceImpl implements FeignService {
    @Override
    public String getprovider() {
        return "服务出现错误，请稍后访问";
    }
}
```

第二步，修改 FeignService 接口，@FeignClient 注解添加 fallback 属性，指向 HystrixFeignServiceImpl 类，如示例代码 8-7 所示。

示例代码 8-7：FeignService 接口

```
@Service
@FeignClient(name = "eureka-provider", fallback = HystrixFeignServiceImpl.class)
public interface FeignService {
    @RequestMapping(value = "/getprovider/test", method = RequestMethod.GET)
    String getprovider();
}
```

第三步，完成 FeignController 类，在之前的案例中已经配置完成，无须额外修改，如示例代码 8-8 所示。

示例代码 8-8：FeignController 控制器

```
@RestController
public class FeignController {

    @Autowired
    private FeignService feignService;

    @RequestMapping("/getprovider/test")
    public String getpros(){
        return feignService.getprovider();
    }
}
```

第四步，修改 Feign 的配置文件，将 feign.hystrix.enabled 属性设为 true 开启 Hystrix。如示例代码 8-9 所示。

示例代码 8-9：Feign 配置文件

```
spring.application.name=my-feign
server.port=7010
eureka.client.service-url.defaultZone=http://localhost:8761/eureka/
feign.hystrix.enabled=true
```

第五步，MyFeignApplication 启动类，配置 @EnableEurekaClient 和 @EnableFeignClients 注解，之前已经配置完成，无须额外修改，如示例代码 8-10 所示。

示例代码 8-10：MyFeignApplication 启动类

```
@EnableEurekaClient
@EnableFeignClients
@SpringBootApplication
public class MyFeignApplication {

    public static void main(String[] args) {
        SpringApplication.run(MyFeignApplication.class, args);
    }
}
```

第六步，依次启动 eureka-server、eureka-provider 和 my-feign，打开浏览器，访问 http://localhost:7010/getprovider/test 地址，结果如图 8-7 所示。

现在访问的是provider,端口号为：7006

图 8-7　访问 getprovider/test 方法

页面显示了方法对应的信息，关闭 eureka-provider 服务，人为地制造服务故障，之后再次重新“访问 http://localhost:7010/getprovider/test”地址，结果如图 8-8 所示。

图 8-8　Feign 整合熔断器

发现系统调用了熔断器失败的 fallback 处理方法，表示熔断器已经启动，从而实现了在 Feign 中使用 Hystrix 熔断器。

技能点 2　Zuul 网关服务

在学习了微服务的基础组件之后，已经可以搭建简易的微服务架构系统，并实现各服务之间的调用。考虑到实际情况，各服务通常不会部署在一台服务器上，势必会有不同的网络地址，而用户在使用时，单独的服务不能满足用户的全部需求，需要调用多个架构的接口，才能完成一次请求。对于这种情况，现有的技术不能满足这个需求，Zuul 网关服务由此产生。

1. Zuul 简介

Zuul 是 Spring Cloud 中的微服务 API 网关，其是通过 Servlet 实现的，与 Eureka 整合。所有从设备或网站来的请求都会经过 Zuul 到达后端的应用程序，同时从 Eureka Server 中获取其他微服务架构的实例信息，Ribbon 和 Hystrix 等组件的结合能够提供动态路由、监控、负载均衡和安全支持等功能。

当用户请求多个服务时，没有应用服务网关的微服务架构，如图 8-9 所示。

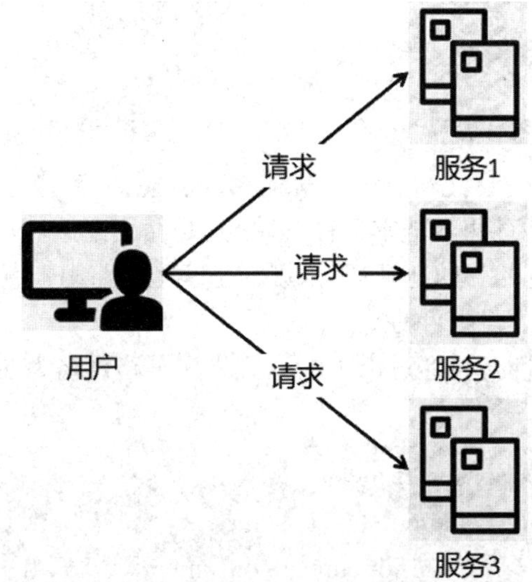

图 8-9　没有应用服务网关时用户请求多个服务

在这种情况下，会存在诸多问题，例如，用户请求多次不同的服务，客户端会变得更加复杂。在存在跨域请求条件下，当前端获取后台信息时，由于服务名称、端口号不同，导致在开发时会更加复杂。随着服务的更新迭代扩展，一些服务需要再次拆分，由于重构项目，服务信息将被修改，用户所访问的请求端口都会被修改，使得重构变得困难重重。

使用服务网关可以很好地解决上述问题。服务网关相当于介于客户端和服务端的中间层，所有外部请求都会经过服务网关进行调度和拦截过滤，如图 8-10 所示。

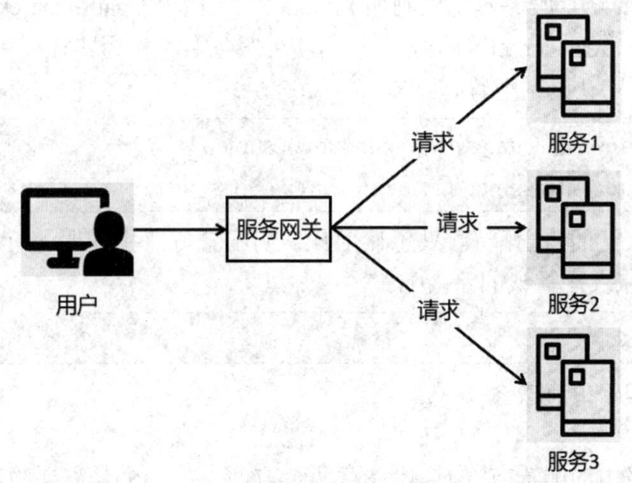

图 8-10　应用服务网关时用户请求多个服务

2. Zuul 应用方法

在服务中，应用网关服务 Zuul，需要在 Pom 文件中添加依赖，并在启动类上添加 @EnableZuulProxy 注解，完成网关服务的配置，代码如下。

```
<dependency>
    <groupId>org.springframework.cloud</groupId>
    <artifactId>spring-cloud-starter-netflix-zuul</artifactId>
</dependency>
```

对于 Zuul 配置路由，有两个基本的概念，分别是路由配置和路由规则。路由配置是配置某请求路径路由到指定的目的地址；路由规则是匹配到路由配置之后，再进行自定义的规则判断，规则判断可以更改路由目的地址。

在使用服务路由配置时，可通过以下的方式来配置路由名为 eureka-consumer 的服务实例。

```
// 转发至名为 eureka-consumer 的服务实例上
zuul.routes.< 路由名 >.path=/eureka-consumer/**
zuul.routes.< 路由名 >.service-id=eureka-consumer
```

上述方法可以通过更简便的方式进行整合，即 zuul.routes.<service-id>=<path>。其中，<service-id> 为服务名称，<path> 为配置匹配的请求映射地址，代码如下。

```
// 运用 zuul.routes.<service-id>=<path> 方式整合
zuul.routes.eureka-server=/eureka-server/**
```

在之前的项目中，可以使用"主机地址 + 端口号 / 服务名称 / 方法名称"的方式调用方法，虽然所创建的服务没有进行服务路由的配置，但却能够成功返回结果，原因是 Zuul 服务网关在默认情况下会以服务名作为 ContextPath 的方式创建路由映射，根据 ContextPath 查找对应的服务，需要结合服务发现机制如 Eureka 等。比如将 path：/eureka-server/** 的请求转发到 service-id=eureka-server 的服务上。当使用服务名称作为路径前缀时，实际上会配置如下代码所对应的默认路由配置。

```
zuul.routes.eureka-consumer.path=/eureka-consumer/**
zuul.routes.eureka-consumer.service-id=eureka-consumer
```

若不想使用默认的路由配置，只想让用户访问配置好路由的服务，可以在 Zuul 的服务中使用如下代码，来关闭默认的路由配置。

```
zuul.ignored-services="*"
```

3. 路由匹配和路由前缀

（1）路由匹配

在使用服务路由的配置方式时，要注意两个方面。第一个是需要为每一个路由规则定义路由表达式，即 path 参数；第二个是在参数配置时需要设置通配符。常用通配符及其说明和使用实例见表 8-2。

表 8-2　通配符

通配符	说明	实例
?	匹配单个字符	/user/?
*	匹配任意数量字符，不支持多级目录	/user/*
**	匹配任意数量字符，支持多级目录	/user/**

如果有一个可以同时满足多个 path 的匹配，此时匹配结果取决于路由规则的定义顺序。

（2）路由前缀

通过配置文件属性 zuul.prefix 来进行设置，将属性值设置为“/getStr”，当访问对应的“/getStr/eureka-consumer/a”路径时，其请求将被转发至“/eureka-consumer/a”。设置 prefix 后，Zuul 会把代理前缀从默认的路径中移除。如果要避免这种情况的出现，可以使用 zuul.routes.< 路由名 >.strip-prefix=false 属性来关闭代理前缀的动作，这时所编写的 prefix 前缀就不会被移除，而是直接进行转发，代码如下。

```
zuul.routes.eureka-consumer.path=/eureka-consumer/**
// 配置前缀
zuul.prefix=/getStr
// 设置关闭代理前缀
zuul.routes.eureka-consumer.strip-prefix=false
```

4. Zuul 应用实例

【案例】应用 Zuul 网关服务，完成通过网关访问 eureka-consumer 服务的方法。

本例沿用前文中所编写的 eureka-consumer、eureka-provider 和 eureka-server 服务，对于部分涉及修改的代码会进行说明。案例所要访问的方法的代码如下。

```
@RequestMapping("/getprovider/test")
  public String test()
  {
    System.out.println(" 调用 eureka-provider 方法 ");
    return restTemplate.getForObject("http://eureka-provider/getprovider/test", String.
class);
  }
```

第一步，创建 my-zuul 子模块，在 Pom 文件中，引入 Zuul 对应的依赖关系，如示例代码 8-11 所示。

示例代码 8-11：my-zuul 子模块 Pom 文件

```
<dependencies>
  <dependency>
    <groupId>org.springframework.boot</groupId>
```

```
        <artifactId>spring-boot-starter-web</artifactId>
    </dependency>
    <dependency>
        <groupId>org.springframework.cloud</groupId>
        <artifactId>spring-cloud-starter-netflix-eureka-client</artifactId>
    </dependency>
    <dependency>
        <groupId>org.springframework.cloud</groupId>
        <artifactId>spring-cloud-starter-netflix-zuul</artifactId>
    </dependency>
    <dependency>
        <groupId>org.springframework.boot</groupId>
        <artifactId>spring-boot-starter-test</artifactId>
        <scope>test</scope>
    </dependency>
</dependencies>
```

第二步，编写 my-zuul 子模块的 properties 配置文件，设置其端口号、服务名称、service-url 和路由等信息。需要注意的是，服务响应超时的问题，采用 Zuul 作为网关根据不同的访问路径进行微服务的访问，在这个过程中接口响应时间的不同可能会导致 Zuul 执行被熔断，出现访问超时的故障报错。对于这种情况，由于在 my-zuul 子模块中对 Zuul 组件进行了整合，只需设置 zuul.host.connect-timeout-millis 和 zuul.host.socket-timeout-millis 让 Zuul 的超时时间大于 Ribbon 的超时时间，就不会出现超时异常了，如示例代码 8-12 所示。

示例代码 8-12：my-zuul 子模块配置文件

```
server.port=7015
spring.application.name=my-zuul
eureka.client.service-url.defaultZone=http://localhost:8761/eureka/
// 应用 zuul.routes.< 路由名 >.path= 的方式配置
zuul.routes.eureka-consumer.path=/eureka-consumer/**
// 连接超时大于熔断的时间
zuul.host.connect-timeout-millis=15000
// socket 超时时间设置
zuul.host.socket-timeout-millis=60000
```

第三步，编写启动类 MyZuulApplication，在类上标注 @EnableZuulProxy 和 @EnableEurekaClient 开启 Zuul 网关服务，如示例代码 8-13 所示。

示例代码 8-13：MyZuulApplication 启动类

```
@EnableZuulProxy
```

```
@EnableEurekaClient
@SpringBootApplication
public class MyZuulApplication {
    public static void main(String[] args) {
        SpringApplication.run(MyZuulApplication.class, args);
    }
}
```

第四步，依次运行 eureka-server、eureka-provider、eureka-consumer 和 my-zuul 服务，并访问“http://localhost:7015/eureka-consumer/getprovider/test”地址，结果如图 8-11 所示。

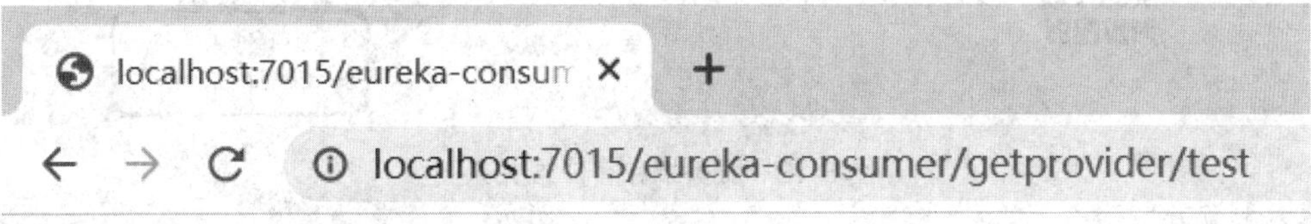

现在访问的是provider,端口号为：7006

图 8-11　应用网关访问 getprovider/test 方法

从图中地址栏可以看到，访问的是 eureka-consumer 中的方法，对应的访问路径是 ResquestMapping(“/getprovider/test”)，使用了 Zuul 网关服务，将访问的请求统一经过 Zuul 网关进行跳转，成功完成了这个案例。

技能点 3　Spring Cloud Config 配置中心

在本书之前的案例中，所搭建的配置文件都是通过全局配置文件进行配置的。但是在实际的开发中，对应的集群都是由独立的微服务框架组成，每一个服务都有相应的配置，当需要更改某个服务的配置时，会因此修改其他服务的配置，在修改完成后，需要进行重启才能使配置生效，这显然是很麻烦的，给后期维护、合并拆分项目带来了很多困难。为了便于集中配置，统一管理，Spring Cloud 集成了相关的组件，Spring Cloud Config 应运而生。

1. Spring Cloud Config 简介

Spring Cloud Config 适用于 Spring 应用程序，可以与其他语言编写的应用程序一同配合使用。最大的优势是和 Spring 的无缝集成，已有的 Spring 应用程序的迁移成本很低，对于 Spring Boot 依赖版本和约束规范能有更加统一的标准，避免了因为集成不同开发软件造成的版本依赖冲突等问题。主要是用来为分布式系统中的外部配置提供服务端(config server)和客户端(config client)支持。

服务端(config server)是一个可扩展、集中式的服务器，用于集中管理每个微服务架构环境下的配置，可以使用 Git 存储、SVN(Subversion)存储或是本地文件存储。

客户端（config client）微服务架构中的各个微服务应用通过指定的服务端配置中心来管理应用资源以及业务相关的配置内容，动态地从服务端获取并加载配置信息，达到无须重启即可完成配置信息的修改。这两者之间的关系如图 8-12 所示（使用 Git 存储）。

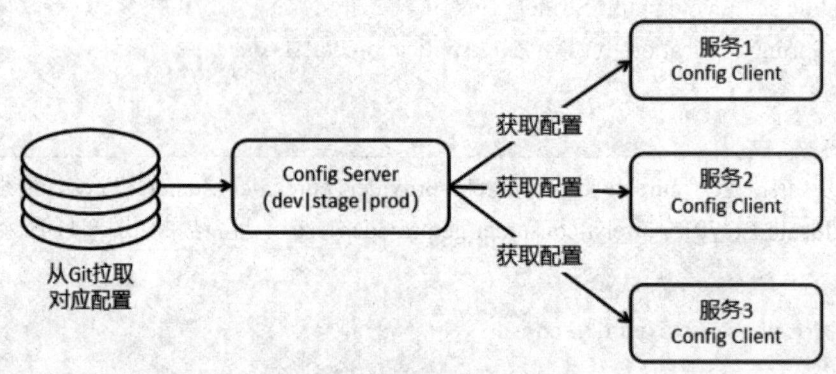

图 8-12　服务端和客户端关系图

从图 8-12 中可以看出，在实际开发中，首先将配置信息存放入 Git 中，在服务启动时，会请求 Config Server，对应获取不同的环境和配置信息，如 dev、stage 和 prod 等，服务端收到请求之后就会从 Git 上拉取对应配置好的信息，Config Client 只需从服务端获取需要的配置信息，缓存这些信息就可以达到提高性能的效果。

2. Spring Cloud Config 应用方法

应用 Spring Cloud Config 有两种方式来完成对于配置文件的统一管理。第一种是加载本地配置文件。这种方式将所有配置文件均放置在 Config Server 项目的 resource 目录下，Config Server 通过设置的 API 来读取配置文件信息。第二种是从远程的 Git 仓库读取配置文件。在这种方式中，每次修改配置文件，需要人工进行刷新，加载速度依靠网络，若网络不可使用时，存放在 Git 中的配置都将不可用。

服务端（Config Server）可以在 Pom 文件中引入相关的依赖，并在启动类上添加注解 @EnableConfigServer 来使用 Config Server 的功能，代码如下。

```
<dependency>
   <groupId>org.springframework.cloud</groupId>
   <artifactId>spring-cloud-starter-netflix-eureka-client</artifactId>
</dependency>
<dependency>
       <groupId>org.springframework.cloud</groupId>
       <artifactId>spring-cloud-config-server</artifactId>
</dependency>
```

在编写 Config 模块的配置文件时，多数情况下需要创建名称为 Bootstrap 的配置文件。由于在 Spring Cloud 启动时，会首先创建 Bootstrap Context，其优先级要高于 Application Context，即 Bootstrap 配置会被优先加载，有不会被本地配置所覆盖的特点。适用于在编写 Config Server 时，相关的配置项 spring.application.name 和 spring.cloud.config.server.git.uri 以及一些加密信息（encryption/decryption）。在 Spring Cloud 中，Bootstrap 属于引导配置，Application 属于应用配置。

以 properties 文件类型为例，对相关的核心配置信息进行说明，如下所示。

```
// 使用本地存储
spring.profiles.active=native
// 配置文件的存放的位置，默认为 resource 目录
spring.cloud.config.server.native.search-locations=classpath:/xxx
```

客户端（Config Client）可以在 Pom 文件中引入相关的依赖，来使用 Config Client 的功能，代码如下。

```
<dependency>
  <groupId>org.springframework.cloud</groupId>
  <artifactId>spring-cloud-starter-config</artifactId>
</dependency>
```

同样，需要编写 Bootstrap 文件，编写相关的核心配置信息，如表 8-3 所示。

表 8-3　核心配置项

名称	说明
spring.cloud.config.name	配置中心的服务端配置文件的名称，获取多个则以逗号隔开
spring.cloud.config.profile	配置文件的应用环境
spring.cloud.config.discovery.service-id	配置中心在注册中心上的名称，从该服务上抓取配置信息
spring.cloud.config.discovery.enabled	是否启动服务发现
spring.cloud.config.uri	不适用 Eureka 时，直接通过 URL 指定配置中心的地址获取配置信息

根据不同规则，对不同环境所创建的配置文件名也不同，如对于用于开发版本、预发布版本和测试版本的配置文件，一般可以按照“{application}-{profile}.properties（yml）”的形式，即“应用名称 - 环境名 . 格式”的规范来命名，如应用名称为 user 的配置文件命名。

- 开发版本：user-dev.yml。
- 预发布版本：user-prod.yml。
- 测试版本：user-test.yml。

除了按照“应用名称 - 环境名 . 格式”的方式，还有其他的常见命名方式。例如，{application} 表示应用名称，{profile} 表示环境名，{label} 是可选的代码如下。在使用 Git 配置时，需要设置 Git 的分支，默认为 master。

```
    /{application}/{profile}[/{label}]
    /{application}-{profile}.yml
    /{label}/{application}-{profile}.yml
  /{application}-{profile}.properties
  /{label}/{application}-{profile}.properties
```

3. Spring Cloud Config 应用实例

【案例】编写 Spring Cloud Config，采用本地获取配置的方法，通过客户端（Config Client）调用的方法，获取配置中心所设置的配置信息。

（1）服务端（Config Server）配置

第一步，编写 my-configserver 子模块，作为服务端配置中心。引入 Config、Web 和 Test 依赖关系，如示例代码 8-14 所示。

示例代码 8-14：my-configserver 子模块的配置文件

```
<dependencies>
        <dependency>
                <groupId>org.springframework.boot</groupId>
                <artifactId>spring-boot-starter-web</artifactId>
        </dependency>
        <dependency>
                <groupId>org.springframework.cloud</groupId>
                <artifactId>spring-cloud-config-server</artifactId>
        </dependency>
        <dependency>
                <groupId>org.springframework.cloud</groupId>
                <artifactId>spring-cloud-starter-netflix-eureka-server</artifactId>
        </dependency>

        <dependency>
                <groupId>org.springframework.boot</groupId>
                <artifactId>spring-boot-starter-test</artifactId>
                <scope>test</scope>
        </dependency>
</dependencies>
<dependencyManagement>
        <dependencies>
                <dependency>
                        <groupId>org.springframework.cloud</groupId>
                        <artifactId>spring-cloud-dependencies</artifactId>
```

```
                    <version>${spring-cloud.version}</version>
                    <type>pom</type>
                    <scope>import</scope>
                </dependency>
            </dependencies>
        </dependencyManagement>
```

第二步，在 resources 目录下创建 bootstrap.yml 文件（或 Properties 文件），编写相应的端口号、服务名称、使用本地存储信息，如示例代码 8-15 所示。

示例代码 8-15：bootstrap.yml 文件

```
// 设置端口号
server:
  port: 7016
// 设置服务名称
spring:
  application:
name: my-configserver
// 设置使用本地存储
  profiles:
active: native
// 编写注册中心
eureka:
  client:
    service-url:
      defaultZone: http://localhost:8761/eureka/
```

第三步，编写入口类 MyConfigserverApplication，在该类上编写 @EnableConfigServer 注解，开启服务配置功能，如示例代码 8-16 所示。

示例代码 8-16：MyConfigserverApplication 启动类

```
@EnableConfigServer
@SpringBootApplication
public class MyConfigserverApplication {
    public static void main(String[] args) {
        SpringApplication.run(MyConfigserverApplication.class, args);
    }
}
```

第四步，编写名为 config-dev 的客户端配置文件，即服务的配置文件。在 resources 目录下创建 config-dev.properties 文件（或 config-dev.yml 文件），编写其端口号、服务名称以及定义一个名为 name 的变量，如示例代码 8-17 所示。

示例代码 8-17：config-dev 配置文件

```
spring.application.name=config-dev
server.port=7021
name=zhang
```

完成以上步骤后，服务端 Config Server 就编写完成了，依次启动 eureka-server 和 my-configserver 服务，在浏览器打开如下格式的 URL“http：//localhost：7016/{applicaiton}/{profile}”发起请求，浏览器会以 JSON 字符串的形式显示应用名、环境名等信息。例如，访问“http：//localhost：7016/config-dev/dev”地址，查看对应的配置文件信息，结果如图 8-13 所示。

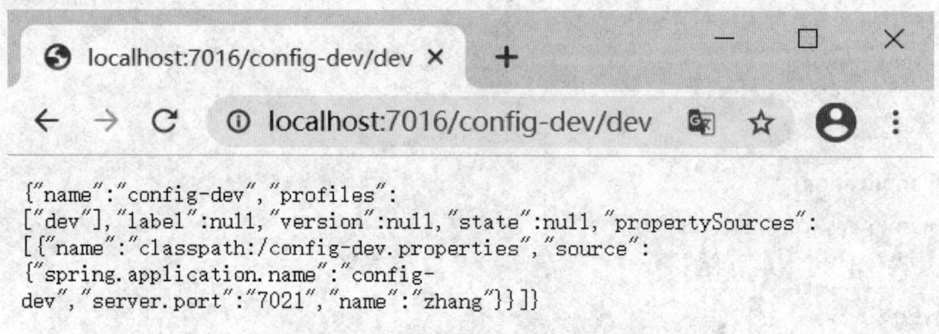

图 8-13 config-dev 配置文件信息

可以直接通过访问“http：//localhost：7016/config-dev.properties”地址的方式，查看对应的配置文件信息，结果如图 8-14 所示。

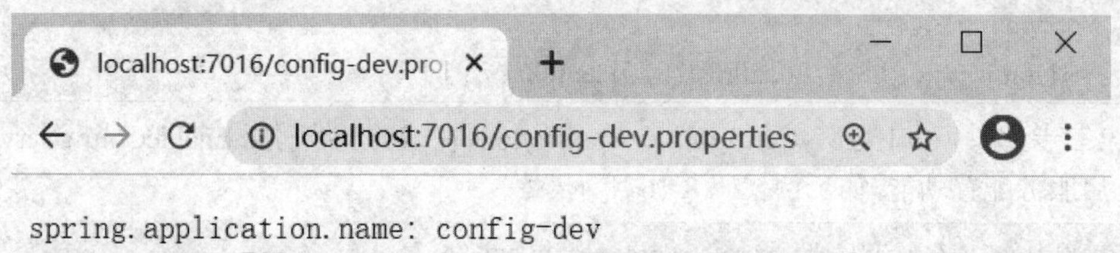

图 8-14 config-dev.properties 配置文件信息

在控制台中，访问对应的地址，同样会显示加载了对应的配置文件，结果如图 8-15 所示。

```
o.s.c.c.s.e.NativeEnvironmentRepository  : Adding property source: classpath:/config-dev.properties
```

图 8-15 加载配置信息效果

(2)客户端(Config Client)配置

第一步，编写 my-configclient 子模块，作为客户端的工程模块。引入 Config、Web 和

Test 依赖关系，其核心依赖如示例代码 8-18 所示。

示例代码 8-18：my-configclient 子模块配置文件

```
<dependencies>
  <dependency>
    <groupId>org.springframework.boot</groupId>
    <artifactId>spring-boot-starter-web</artifactId>
  </dependency>
  <dependency>
    <groupId>org.springframework.cloud</groupId>
    <artifactId>spring-cloud-starter-config</artifactId>
  </dependency>
  <dependency>
    <groupId>org.springframework.boot</groupId>
    <artifactId>spring-boot-starter-test</artifactId>
    <scope>test</scope>
  </dependency>
</dependencies>
<build>
  <plugins>
    <plugin>
      <groupId>org.springframework.boot</groupId>
      <artifactId>spring-boot-maven-plugin</artifactId>
    </plugin>
  </plugins>
</build>
</project>
```

第二步，创建 controller 文件夹，在其中创建 MyClientController 控制器，作为测试获取配置文件的方法，如示例代码 8-19 所示。

示例代码 8-19：MyClientController 控制器

```
@RestController
public class MyClientController {
  @Value("${name}")
  private String name;
  @RequestMapping("/hello")
  public String test()
  {
```

```
        return "Hello Config Client：name is "+name；
    }
}
```

第三步，在 resource 目录下创建 bootstrap.yml 配置文件，编写对应的配置信息。设置 spring.cloud.config.uri=http：//localhost：7016、spring.cloud.config.name=config-dev，即指向 my-configserver 配置中心的端口，用于接收 Config Server 中编写的名称为 config-dev 的配置文件信息。具体的配置项代码，如示例代码 8-20 所示。

示例代码 8-20：my-configclient 子模块 bootstrap.yml 文件

```
spring：
 application：
  name：my-configclient
 cloud：
  config：
   uri：http：//localhost：7016
   fail-fast：false
   discovery：
    enabled：true
   name：config-dev
   profile：dev
# 定义注册中心，指向 eureka-server
eureka：
 client：
  service-url：
   defaultZone：http：//localhost：8761/eureka/
```

第四步，入口类无须改动，依次运行 eureka-server、my-configserver 和 my-configclient。打开浏览器，访问"localhost：7021/hello"地址，结果如图 8-16 所示。

图 8-16 访问 hello 方法

从图中可以看到，在 Config Server 配置中心所定义的 name 属性、端口号成功被 my-configclient 加载，完成了通过配置中心获取客户端配置信息的案例。

运用 Spring Cloud 微服务知识，设计 user-consmer 服务消费者，实现用户注册功能。注册页面效果如图 8-17 所示。

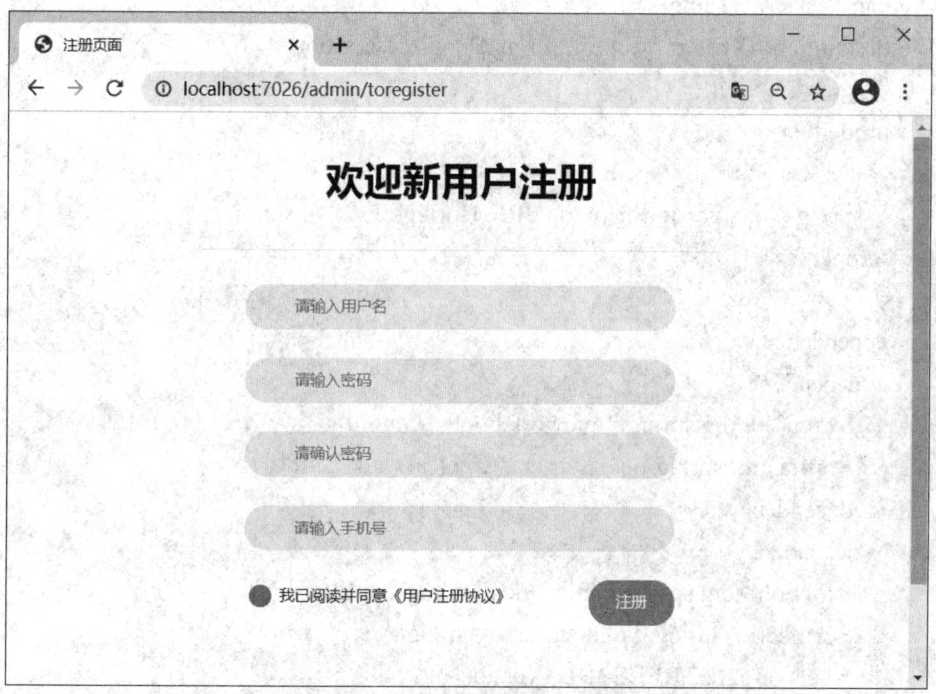

图 8-17　用户注册功能效果图

第一步，创建 user-consumer 项目，引入 Web、Thymeleaf、Feign、Eureka Client 和 Hystrix 依赖。在 resources 目录下加入静态资源。Pom 文件如示例代码 8-21 所示。

示例代码 8-21：user-consumer 工程的 Pom 文件

```
<? xml version="1.0" encoding="UTF-8"? >
<project  xmlns="http: //maven.apache.org/POM/4.0.0"  xmlns: xsi="http: //www.w3.org/2001/XMLSchema-instance"
         xsi: schemaLocation="http: //maven.apache.org/POM/4.0.0 https: //maven.apache.org/xsd/maven-4.0.0.xsd">
    <modelVersion>4.0.0</modelVersion>
    <parent>
        <groupId>org.springframework.boot</groupId>
```

```
    <artifactId>spring-boot-starter-parent</artifactId>
    <version>2.2.8.RELEASE</version>
    <relativePath/> <!  -- lookup parent from repository -->
  </parent>
  <groupId>com.xtgj</groupId>
  <artifactId>user-consumer</artifactId>
  <version>0.0.1-SNAPSHOT</version>
  <name>user-consumer</name>
  <description>Demo project for Spring Boot</description>

  <properties>
    <java.version>1.8</java.version>
    <spring-cloud.version>Hoxton.SR6</spring-cloud.version>
  </properties>

  <dependencies>
    <dependency>
      <groupId>org.springframework.boot</groupId>
      <artifactId>spring-boot-starter-thymeleaf</artifactId>
    </dependency>
    <dependency>
      <groupId>org.springframework.cloud</groupId>
      <artifactId>spring-cloud-starter-openfeign</artifactId>
      <version>2.1.2.RELEASE</version>
    </dependency>
    <dependency>
      <groupId>org.springframework.boot</groupId>
      <artifactId>spring-boot-starter-web</artifactId>
    </dependency>
    <dependency>
      <groupId>org.springframework.cloud</groupId>
      <artifactId>spring-cloud-starter-netflix-eureka-client</artifactId>
    </dependency>
    <dependency>
      <groupId>org.springframework.boot</groupId>
```

```
            <artifactId>spring-boot-starter-test</artifactId>
            <scope>test</scope>
        </dependency>
        <! --hystrix 熔断依赖 -->
        <dependency>
            <groupId>org.springframework.cloud</groupId>
            <artifactId>spring-cloud-starter-netflix-hystrix</artifactId>
            <version>2.1.2.RELEASE</version>
        </dependency>
    </dependencies>

    <dependencyManagement>
        <dependencies>
            <dependency>
                <groupId>org.springframework.cloud</groupId>
                <artifactId>spring-cloud-dependencies</artifactId>
                <version>${spring-cloud.version}</version>
                <type>pom</type>
                <scope>import</scope>
            </dependency>
        </dependencies>
    </dependencyManagement>

    <build>
        <plugins>
            <plugin>
                <groupId>org.springframework.boot</groupId>
                <artifactId>spring-boot-maven-plugin</artifactId>
            </plugin>
        </plugins>
    </build>
</project>
```

第二步，创建 model 文件夹，创建 User 实体类，实体类的字段同对应的数据表，与 user-provider 相同，如示例代码 8-22 所示。

示例代码 8-22：User 实体类

```
public class User {
    private String name;
```

```
    private String password;
    private String count;
    private String iphone;
  // 省略 getter()和 setter()方法
  // 省略 toString()方法
  }
```

第三步，创建 service 文件夹，在其下创建 UserService 接口，调用 user-provider 的接口，实现用户的注册。同时应用 @FeignClient 注解，指向 user-provider，并采用 Hystrix 熔断器，如示例代码 8-23 所示。

示例代码 8-23：UserService 接口

```
@Component
@FeignClient(value ="user-provider-dev", fallback = UserServiceHystrix.class)
public interface UserService {
   @RequestMapping(value = "/login", method = RequestMethod.GET)
   public User login(@RequestParam(value = "name") String name);

   @RequestMapping(value ="/register", method = RequestMethod.GET)
   public  int register(@RequestParam(value ="password")String password, @Request-
Param(value ="name")String name, @RequestParam(value ="iphone")String iphone);
}
```

第四步，创建 hystrix 文件夹，在其下创建 UserServiceHystrix 类，实现 Userservice 接口方法，如示例代码 8-24 所示。

示例代码 8-24：UserServiceHystrix 实现类

```
@Component
public class UserServiceHystrix implements UserService {

   @Override
   public User login(String name) {
      return null;
   }

   @Override
   public int register(String password, String name, String iphone) {
      return 0;
   }
}
```

第五步，创建 controller 文件夹，在其下创建 UserController 控制器，用于处理用户请求跳转页面以及接收参数等，如示例代码 8-25 所示。

示例代码 8-25：UserController 控制器

```
@Controller
@RequestMapping("/admin")
public class UserController {
    @Autowired
    UserService userService;
    // 跳转登录页面
    @GetMapping(value = "/tologin")
    public String tologin() {
        return "login";
    }
    // 跳转注册页面
    @GetMapping(value = "/toregister")
    public String toregister() {
        return "register";
    }
    // 用户登录
    @GetMapping(value = "/login")
    public ModelAndView login(String name, String password) {
        ModelAndView modelAndView = new ModelAndView();
        User users = userService.login(name);
       if(users! = null) {
            if(userService.login(name).getPassword().equals(password)) {
                modelAndView.addObject("user", name);
                System.out.println(" 登录成功 ");
                modelAndView.setViewName("success");
                return modelAndView;
            }
        }
        modelAndView.setViewName("login");
        System.out.print(" 登录失败 ");
        return modelAndView;
    }
    // 用户注册
    @GetMapping(value ="/register")
    public String register(String password, String name, String iphone) {
```

```
        if(userService.register(password, name,iphone)>0){
            System.out.print("注册成功");
            return "login";
        }
        System.out.print("注册失败");
        return "register";
    }
}
```

第六步，创建 config 文件夹，在其中创建 Config 类，作为配置类，用于控制项目中的静态资源，如示例代码 8-26 所示。

示例代码 8-26：Config 配置类

```
@Configuration
public class Config extends WebMvcConfigurationSupport {
    @Override
    public void addResourceHandlers(ResourceHandlerRegistry registry) {
     // 如下配置可以访问 src/main/resources/static 下的文件
            registry.addResourceHandler("/static/**").addResourceLocations(ResourceUtils.
CLASSPATH_URL_PREFIX + "/static/");
        super.addResourceHandlers(registry);
    }
```

第七步，编写 application.yml 配置文件，如示例代码 8-27 所示。

示例代码 8-27：user-consumer 项目的 application.yml 配置文件

```
// 服务端口
server:
  port: 7026
// 服务名
spring:
  application:
name: user-consumer
// 服务注册地址
eureka:
  client:
    service-url:
      defaultZone: http://localhost:8761/eureka/
  instance:
    hostname: localhost
```

```
thymeleaf：
 cache：false
 prefix：classpath：/template/
 suffix：.html
 encoding：UTF-8
// 开启 hystrix
feign：
  hystrix：
    enabled：true
```

第八步，编写启动类 UserConsumerApplication，在其类上添加 @EnableFeignClients、@EnableEurekaClient 和 @EnableHystrix 注解，如示例代码 8-28 所示。

示例代码 8-28：UserConsumerApplication 启动类

```
@SpringBootApplication
@EnableFeignClients
@EnableEurekaClient
@EnableHystrix
public class UserConsumerApplication {
    public static void main（String[] args）{
        SpringApplication.run（UserConsumerApplication.class，args）;
    }
}
```

第九步，编写前端页面 login.html、register.html 和 success.html。这三个文件的部分重点代码分别如示例代码 8-29、8-30 和 8-31 所示。

示例代码 8-29：login.html 页面

```
<form action="http://localhost:7026/admin/login" method="get">
            <div class="dly_right">
                <p class="tit"> 欢迎登录 </p>
                <div class="zc_username">
                    <span class="zc_tit"> 用户名 </span>
                    <input type="text" id="userName" name="name" class="username"/>
                </div>
                <div class="zc_password">
                    <span class="zc_tit"> 密    码 </span>
                    <input type="password" id="password" name="password" class="pass_word"/>
                    <label class="error" id="pass_word_error"></label>
                </div>
```

```
            <input id="btn-ty" class="btn-ty" type="submit" value=" 点击登录 ">
            <p class="cstyle"> 还没有账号？<a href='http：//localhost：7026/admin/toregis-
ter'> 立即注册 </a></p>
        </div>
    </form>
```

示例代码 8-30：register.html 页面

```
    <form action="/admin/register" class="form-horizontal">
            <span class="heading"> 欢迎新用户注册 </span>
            <div class="form-group">
                <input type="text" name="name" class="form-control"id="name" placehold-
er=" 请输入用户名 ">
                <i class="fa fa-user"></i>
            </div>
            <div class="form-group">
                <input type="password" name="password" class="form-control" id="pass_
word" placeholder=" 请输入密码 ">
                <i class="fa fa-lock"></i>
                <a href="#" class="fa fa-question-circle"></a>
            </div>
            <div class="form-group">
                <input type="password" name="pwdconfirm" class="form-control" id="pwd-
confirm" placeholder=" 请确认密码 ">
                <i class="fa fa-lock"></i>
                <a href="#" class="fa fa-question-circle"></a>
            </div>
            <div class="form-group">
                <input type="text" name="iphone" class="form-control" id="iphone" place-
holder=" 请输入手机号 ">
                <i class="fa fa-lock"></i>
                <a href="#" class="fa fa-question-circle"></a>
            </div>
            <div class="form-group">
                <div class="main-checkbox">
                    <input type="checkbox" value="None" id="checkbox1" name="check"/>
                    <label for="checkbox1"></label>
                </div>
```

```
            <span class="text"> 我已阅读并同意《用户注册协议》</span>

            <button type="submit" class="btn btn-default"> 注册 </button>
        </div>
    </form>
```

示例代码 8-31：success.html 页面

```
<body class="backb">
<h1 style="margin：12%；text-align：center" th：text="' 欢迎：'+${user}+' 登录本系统，正在跳转页面 '"></h1>
</body>
```

第十步，编写完成主体内容之后，应用 Spring Cloud Config 的知识，对该服务进行改进。原有 user-provider 服务是在本地进行配置，先将该配置文件放置在 my-configserver 服务中，统一进行管理。

在之前创建的 my-configserver 项目中的 resources 目录下创建名为 user-provider-config-dev.yml 的配置文件，内容与原有配置相同，如示例代码 8-32 所示。

示例代码 8-32：my-configserver 工程 user-provider-config-dev.yml 配置文件

```
server：
  port：7025
spring：
  datasource：
    url：jdbc：mysql：//localhost：3306/test？serverTimezone=UTC&useSSL=false
    username：root
    password：root
    driver-class-name：com.mysql.cj.jdbc.Driver
  application：
    name：user-provider-dev
eureka：
  client：
    service-url：
      defaultZone：http：//localhost：8761/eureka/
  instance：
    hostname：localhost
mybatis：
  mapper-locations：classpath：/mapper/*
```

第十一步，在 user-provider 项目的 resources 目录下创建 bootstrap.yml 配置文件，设置属性，获取 my-configserver 中 user-provider-config-dev 文件所编写的配置信息，将配置文件

进行统一管理（对 user-consumer 配置文件操作一致），如示例代码 8-33 所示。

示例代码 8-33：user-provider 项目 bootstrap.yml 配置文件

```
spring:
 cloud:
config:
// 指定 my-configserver 的地址端口
   uri: http://localhost: 7016
   fail-fast: false
   discovery:
    enabled: true
   // 指定文件名称
   name: user-provider-config-dev
   profile: dev
eureka:
 client:
  service-url:
   defaultZone: http://localhost: 8761/eureka/
```

第十二步，依次运行 eureka-server、my-configserver、user-provider 和 user-consumer 服务，打开浏览器访问“http: //localhost: 7026/admin/login”地址，可以看到相应的登录页面，效果如图 8-18 所示。

图 8-18 用户登录页面

点击“立即注册”跳转至注册页面，效果如图 8-19 所示。

图 8-19 用户注册页面

输入基本信息后，点击“注册”便完成注册并返回登录页面。填写用户名和密码后，点击“点击登录”，跳转至登录成功页面，效果如图 8-20 所示。

图 8-20 登录成功页面的跳转页面

采用 Spring Cloud 实现基本的用户注册功能就完成了。在此基础上可进行其他内容的扩展。

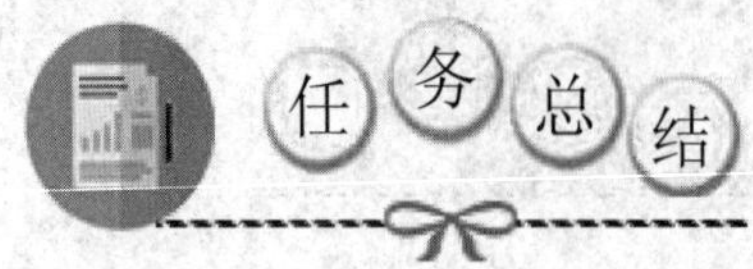

本任务讲解了 Spring Cloud 的注册登录功能，在实现了基础功能的基本上，搭建服务消费者，应用了熔断器以及配置中心的知识，实现了注册登录功能。通过本任务，使读者加深了对于 Spring Cloud 的理解，掌握了基本的 Spring Cloud 技术。

Fallback	应变计划	Prefix	前缀
Command	命令	Path	路径
Properties	属性		

一、填空题

1. ______________ 注解用于开启熔断功能。

2. ______________ 注解用于开启网关 Zuul 功能。

3. 配置文件中设置网关前缀应使用 ____________ 参数。

4. 对于 Config Server 的配置文件，使用 ___________ 属性来配置 Git 仓库的文件夹地址。

二、选择题

1. 以下说法正确的是（　　）。

A. @EnableHystrix 注解用于开启熔断功能

B. Feign 集成了 Hystrix，所以可以在 Feign 中直接使用 Hystrix，无须引入其余任何依赖，只需在配置文件中开启即可

C. Hystrix 中封装了很多依赖，当出现延迟时，会被限制在资源中，并且包含回退逻辑，该逻辑决定了出现故障时对应的处理响应方法

D. 以上说法均正确

2. 对于网关服务 Zuul 使用了 Aut 的定义风格，以下描述正确的是（　　）。

A.“？”表示匹配单个字符

B.“*”表示匹配任意数量字符，支持多级目录

C.“**”表示匹配任意数量字符，不支持多级目录

D.“*/?”表示匹配任意数量字符和单个字符

3. 关于 boootstrap.yml 和 application.yml 文件，下列说法正确的是（　　）。

A. boootstrap.yml 优先执行　　B. application.yml 优先执行

C. 两者无优先顺序　　D. 同时执行

三、简答题

1. 简要说明 Hystrix 熔断器的工作原理。

2. 简要说明 Config 配置中心完成对于配置文件的统一管理的两种方式。